Principles of Biogeography

McGRAW-HILL SERIES IN GEOGRAPHY

EDWARD J. TAAFFEE AND JOHN W. WEBB, *Consulting Editors*

BROEK AND WEBB *A Geography of Mankind*

CRESSEY *Asia's Land and Peoples*

CRESSEY *Land of the 500 Million: A Geography of China*

DEMKO, ROSE, AND SCHNELL *Population Geography: A Reader*

DETWYLER *Man's Impact on Environment*

FRYER *Emerging Southeast Asia: A Study in Growth and Stagnation*

FRYER *World Economic Development*

MURPHY *The American City: An Urban Geography*

POUNDS *Europe and the Soviet Union*

POUNDS *Political Geography*

RAISZ *General Cartography*

RAISZ *Principles of Cartography*

STARKEY AND ROBINSON *The Anglo-American Realm*

THOMAN, CONKLING, AND YEATES *The Geography of Economic Activity*

TREWARTHA *An Introduction to Climate*

TREWARTHA, ROBINSON, AND HAMMOND *Fundamentals of Physical Geography*

TREWARTHA, ROBINSON, AND HAMMOND *Elements of Geography: Physical and Cultural*

TREWARTHA, ROBINSON, AND HAMMOND *Physical Elements of Geography (A republication of Part 1 of the above)*

VAN RIPER *Man's Physical World*

WATTS *Principles of Biogeography: An Introduction to the Functional Mechanisms of Ecosystems*

Principles of Biogeography

DAVID WATTS
Lecturer in Geography, University of Hull

McGRAW-HILL BOOK COMPANY

New York   St. Louis   San Francisco   Dusseldorf   London
Mexico   Panama   Sydney   Toronto

To my Parents

Now let the song begin! Let us sing together
Of sun, stars, moon and mist, rain and cloudy weather.
Light on the budding leaf, dew on the feather,
Wind on the open hill, bells on the heather,
Reeds by the shady pool, lilies on the water. . . .*

*(From 'The Fellowship of the Ring' by J. R. R. Tolkien,
being the first volume of *The Lord of the Rings*, Allen & Unwin
London, 1954. Published in U.S.A. by Houghton Mifflin
Co., Boston. Reproduced by permission.)

Preface

It was about ten years ago that many of the aims of environmental science began to be reassessed, as new techniques of enquiry extended the range of related research. At that time, spheres of interest in disciplines such as geography, geology, and biology were still clearly delimited according to their traditional boundaries. That this is much less the case today is the result of a rapidly growing awareness that many links between the organic and inorganic world can be fully explained only by adopting a broader view of the biological and earth sciences, which takes little account of the rather artificial interfaces which had formerly been defined within them. Nowhere is this more true than in biogeography which, in seeking to interpret the differential patterns of distribution among organisms, and their changing relationships with each other and their environment both in time and space, must draw upon evidence derived from a wide range of sources. Much of the recent research in this complex field has been oriented specifically towards the elucidation of a better understanding of the workings of the biological world (the biosphere), particularly with respect to the way in which certain interactions between organisms and habitat might lead to a decline in its intrinsic stability and functional efficiency. This book is the first to present an overall view of these new trends. While it is planned as a basic, non-statistical text for geography and biological science students at university and college level, it may also be found useful by those social scientists who are interested in the overlap of ideas between their own subjects and those more appropriately concerned with the intricacies of plant and animal life. Moreover, many facets of it will appeal to intelligent laymen who wish to satisfy their curiosity about man's place in the natural world, and to examine the potential biological consequences of his frequently-demonstrated capacity for reducing the quality of the environment in which he lives.

The nature of the material included herein has determined to a considerable extent its arrangement under chapter headings. Following a brief introduction which emphasizes the holistic character of biogeography and discusses some fundamental concepts, chapter 2 deals with the receipt, transfer, and utilization of energy within the biosphere, and chapter 3 with the cyclical movement of chemical elements between atmosphere, earth and water bodies, and living organisms. Chapter 4 recognizes that there can be inherent environmental limitations to the growth and development of both individual life forms, and plant and animal assemblages; and chapter 5 analyses the demographic and competitive elements of restraint which might affect population size. Some aspects of ecological and evolutionary change are evaluated in chapter 6, and the final chapter is devoted to a consideration of the degree of man's dependence upon, and modification of, his physical and biological milieu. In each chapter, general principles are established which are applicable to all parts of the world, and these are supported and illustrated by many local and regional examples. It is perhaps unnecessary to say that the latter have not only been drawn from personal knowledge and experience, but also from the sifting of a large number of reports and papers presented by scientists of very different backgrounds. So as to detail the sources of this intelligence, to admit my debt to them, and to encourage the reader to follow up particular points of argument which may be of special concern to him, fairly comprehensive bibliographies have been placed at the end of every chapter. Most of the references included are available in the libraries of major universities, though some might be located only in the older centres of learning. However, the text will still be meaningful should the reader choose not to make use of such supplementary information.

It should also be made clear that this book could not have been written without the more direct assistance of many people. The ideas expressed within it have been crystallized over several years of indispensable formal and informal talks with mentors, colleagues, research students, and undergraduates, in and outside of a number of universities in Europe and North America. More specific recognition is due to those authors, research institutions and publishing companies who have granted permission for some of their copyright diagrams and tables to be redrawn or modified for my use, and I trust that an adequate acknowledgement to all who fall within this category has been given within the relevant captions. In particular, I acknowledge the use of figures 3.2 and 3.3 which have been reproduced from the third edition of *Principles of Modern Biology* by Douglas Marsland, copyright 1945, 1951 © 1957 by Holt, Rinehart and Winston Inc., reprinted by permission of Holt, Rinehart and Winston Inc., publishers, New York. I am also especially indebted to Professor H. R. Wilkinson and Dr D. J. Boatman, respectively of the Departments of Geography and Botany at the University of Hull, England, for their advice in the preliminary stages of planning and manuscript preparation; to Dr John R. Flenley, my biogeographical colleague at Hull, for his comments on much of the penultimate draft; to Professor Bruce G. Trigger and Dr Eric Waddell, respectively of the Departments of Anthropology and Geography, McGill University, Montreal, for their help with certain sections of chapter 7; to Dr Peter G. Holland of the Department of Geography at McGill, for his remarks on chapter 3; and to the McGraw-Hill Publishing Company and their advisers for their constructive criticisms. None of the above are, of course, responsible for any deficiencies in the text which might remain. Last but by no means least, thanks are also accorded to the task force of technicians and secretaries who were ultimately responsible for putting the manuscript together. Among these, Mr R. R. Dean, Mr K. R. Scurr, Miss W. A. Wilkinson, and other staff members of the Geography Department Drawing Office at the University of Hull, produced most of the diagrams; and Miss C. E. Hayward, Mrs P. E. Keene, Mrs J. M. Dealtry, Mrs B. Smith, and Mrs J. Liske cheerfully undertook between them the typing of the final draft copy.

David Watts

Notes

The metric system is used in the presentation of most measures of length, weight, volume, etc.

Most dates are recorded in years BP (before present), except where they refer to events recorded in historic time.

Contents

		Page
PREFACE		ix
CHAPTER 1	*INTRODUCTION*	1
1.1	Scope and origins of biogeography	1
1.2	Definitions: the nature of the biosphere	3
1.3	Definitions: ecosystems, other assemblages, and environment	5
References		6
CHAPTER 2	*ENERGY CONTROLS OF ECOSYSTEMS*	7
2.1	Introduction	7
Energy sources and transfer; some preliminary considerations		8
2.2	Energy: sources and availability	9
2.3	Laws of energy exchange	16
2.4	The measurement of energy exchange	18
Food chains and webs		19
2.5	Techniques for tracing food chains	19
2.6	Examples of simple food chains	22
2.7	Ecological pyramids: numbers, biomass, energy	24
2.8	The theory of energy exchange	27
2.9	Ecological efficiencies of energy transfer	28

Energy transfer within selected ecosystems   30

2.10 An example of energy flow   30
2.11 Grazing and detritus food chains   33
2.12 The importance of decomposers   35

Biological productivity in ecosystems   35

2.13 Primary and secondary productivity   36
2.14 The primary production of world ecosystems   40
2.15 The secondary production of world
       ecosystems   44
References   46

CHAPTER 3   BIOGEOCHEMICAL CYCLES WITHIN
              ECOSYSTEMS   52

3.1   Introduction   52

Chemical elements in living organisms   54

3.2   The relative abundance of elements   55
3.3   The measurement of relative abundance   58
3.4   Elements, cell growth, and nutrition   59

Cycles of element exchange   64

3.5   The cycling of phosphorus and sulphur   68
3.6   The cycling of carbon   72
3.7   The cycling of nitrogen   76
3.8   The importance of soil organisms   82

Man-induced modifications of element exchange patterns   85

3.9   The case of environmental deterioration   86
3.10 Pesticides: a worldwide problem   90
3.11 Radioactive substances   95
3.12 Air pollution   101

Patterns of element exchange in major world ecosystems   105

3.13 Causes of variability in element exchange
       patterns   106
References   110

CHAPTER 4   *ENVIRONMENTAL LIMITATIONS IN ECOSYSTEM DEVELOPMENT*   120

4.1   Introduction   120

The concept of tolerance   122

4.2   Techniques for measuring environmental conditions   124

Limits of tolerance in terrestrial ecosystems   126

4.3   The light factor   126
4.4   Heat and temperature   134
4.5   Humidity and moisture: the hydrological cycle   142
4.6   Energy/temperature—water relationships   157
4.7   Wind   171
4.8   Topography   173
4.9   Edaphic considerations   175
4.10   The biotic factor   184

References   188

CHAPTER 5   *POPULATION LIMITATIONS WITHIN ECOSYSTEMS*   197

5.1   Introduction   197

The demography of organisms   198

5.2   Methods of estimating population   199
5.3   Birth, death, and growth rates   201
5.4   Patterns of population growth   204
5.5   The importance of age   211
5.6   Genetic factors   213
5.7   The spatial arrangement of organisms   216
5.8   Migration   218
5.9   Density-dependent and density-independent controls   219

Competition between organisms   223

5.10   Intraspecific competition: territoriality   223
5.11   Interspecific competition: the ecological niche   227
5.12   Positive and negative competition   234
5.13   Competition and succession   235

References   237

Contents

CHAPTER 6   THE TIME FACTOR: DYNAMIC
             ASPECTS OF ECOSYSTEMS   242
      6.1   Introduction   242

      Ecological aspects of change   243
      6.2   World patterns of distribution among organisms 244
      6.3   Modes of dispersal of organisms   252
      6.4   Climax and polyclimax succession   257
      6.5   Climatic change and equilibrium within
             ecosystems   263

      Evolutionary aspects of change   275
      6.6   Evolution as a reaction to changing
             environments: general considerations   276
      6.7   Phyletic evolution and speciation   286
      6.8   The problem of extinction   294
      References   298

CHAPTER 7   MAN IN ECOSYSTEMS   306
      7.1   Introduction   306

      Some environmental restraints   307
      7.2   Human origins and diversity   308
      7.3   Environmental limitations   322
      7.4   The web of disease   327

      Man as an agent of ecosystem change   333
      7.5   The use of fire   334
      7.6   The domestication of plants and animals, and
             the ecological status of agricultural systems   339
      7.7   Human migration and its biogeographical
             consequences   351
      7.8   Environmental pollution   359

      The present status of world ecosystems   362
      7.9   The continued impoverishment of world
             ecosystems   363
      7.10   The need to maintain biological diversity   364
      7.11   The need for conservation   366
      References   368

      Author Index   379
      General Index   385

1

Introduction

1.1 Scope and origins of biogeography

Biogeography seeks to establish patterns of order from the apparent chaos of the multiplicity of life forms present upon the surface of the earth, and in its soil, atmosphere, and water bodies. In so doing, it is concerned with the mechanisms whereby both plants and animals originate, evolve, and organize themselves into assemblages which show particular distributions and affinities. It evaluates the challenge of environment, and the response to it by organisms of very different genetic structures, and takes into account the effects of environmental change, which can appreciably modify all organic relationships within a short space of time. Biogeographers wish to know why certain species may be found in some areas and not in others, and why associates can at times turn into competitors. They are also anxious to determine the exact means whereby all life forms are supported by the world's energy and chemical resources. Usually placed on the fringes of both disciplinary groups as they are currently conceived, biogeography is nevertheless located at an increasingly important meeting ground of the biological and earth sciences, and forms a recent growth area of scientific thought of no small significance to the future welfare of man.

Partly because of its very complex nature, there are a number of basic problems within this field of inquiry. All too frequently, the factual material which biogeographers consider is imperfect and incomplete. Thus, in many parts of the world, a full count of species and varieties of organisms has not yet been made, and the consequences of their interactions with each other are little understood.

1

Sophisticated techniques of measuring environmental components, such as those appertaining to the heat and moisture balances of the earth and atmosphere, have been introduced only recently, so our knowledge of the exchanges inherent in these is still somewhat imprecise; and the details of the pathways of energy and chemical element movement between organisms and their environment are far from being definitively interpreted. Under these circumstances, even the informed observer must to some extent use ideas for which positive proof is lacking, and so depend at times on personal judgements which may arouse controversy; however, the latter may perhaps be minimized if evidence is assorted within the framework of a flexible, though concise, system of related conceptual terminology, one scheme of which is presented in sections 1.2 and 1.3.

It is now generally agreed that biogeography differs from plant and animal ecology in that it does not confine its attention strictly to environmental-organism relationships *per se*, but places as much emphasis on the explanation of distributional inequalities among organisms. In this, it closely resembles early nineteenth century ideas of natural history (e.g., as presented by von Humboldt[4]), from which it is ultimately derived. The origins of these ideas can, in turn, be traced to several antecedent events of general scientific importance. One of the first of these took place in 1753 when the binomial system of organic classification was formulated by Linnaeus; then, in 1770, the discovery of oxygen led to the birth of modern chemistry. Both Benjamin Franklin and Thomas Malthus had applied their minds to theoretical models of population growth before 1800, by which time the first records of temperature and rainfall had also been collected. But perhaps the greatest stimulus of all to the study of natural history came from Charles Darwin's voyage of collection and observation in HMS *Beagle* between 1830 and 1835, for this led eventually to his presentation of general laws of natural selection, in *The origin of species* in 1859.[3] The evolutionary mechanisms which aided these were then clarified by Mendel's classic experiments into the nature of inheritance among pea populations.[5] Further details of these, and other contributions to the fundamental concepts of disciplines from which presentday biogeography has emerged, will be given later in the relevant chapters of this book.

1.2 Definitions: the nature of the biosphere

Since both the biological and earth sciences suffer at times from a surfeit of definitions, it is not the purpose of this, and the following section, to add to the unrestrained dissemination of new terminology, but rather to clarify that which already exists. The primary task is to determine a system of classification of organic entities which can serve as a foundation for subsequent discussion. But even though the most casual of travellers notices the presence of 'natural organic groupings', or 'natural regions' set apart from others by certain unique features found within their confines alone, the more precise delimitation of these is not an easy undertaking. Geographers have often attempted to present their essential characteristics by referring to components of the physical habitat, such as climate or relief, whereas biological scientists take more account of the distributional differentiation of floral or faunal elements, which may or may not coincide with the dissimilarities of their environment. Following (with some modifications) the precedent already set by Dansereau[2] and Cain,[1] the author has chosen to adopt a three-dimensional holistic view of these entities, in a manner which to some extent draws on the experience of both disciplinary groups, and yet has specific traits of its own. This involves the establishment of a hierarchy of plant and animal assemblages, in decreasing order of size and complexity, which has particular relevance to any explanation of those functional aspects of their existence which are emphasized throughout this text.

At the apex of the hierarchy stands the *biosphere*, or the complete, worldwide environment in which life of any sort is present. Initially introduced by the geologist Suess as an expression of the notion that the thin layer of living material which is supported by the earth's crust can considerably influence the patterns of geological erosion, the meaning of this term has had to be revised in order to take into consideration the recent discovery that life forms may be found not only in inner space, but also in oily liquids within rock at depths of several kilometres. However, it is still true that most of these reside within a zone of very narrow vertical extent close to the surface of the earth and oceans, where the supplies of energy and chemical elements through which they are maintained are most readily available to them. A schematic representation of the biosphere is shown in Fig. 1.1. Occasionally, it is subdivided into three *biocycles*, which include respectively all saltwater, freshwater, and terrestrial life forms; and in the last of these smaller *biochores* may be physiognomically delimited, such as grasslands, desert, and forest. Each biochore may then be split into *formations*, as in the case of the steppe, prairie, and savanna formations of the grassland biochore.

At times, too, the supplementary concept of a *noosphere* has been used to indicate the presence of a major subsidiary zone within the

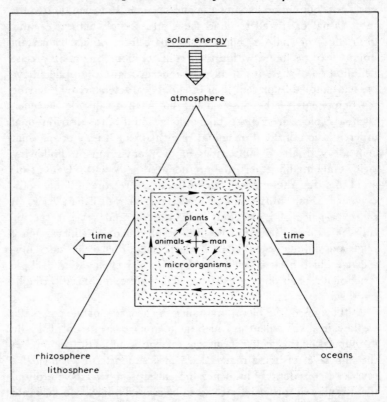

Fig. 1.1 A diagrammatic representation of the biosphere. The triangle indicates the environmental milieu of atmosphere, rhizosphere (plants, animals), lithosphere, and oceans, in which the organic constituents are placed. The latter are symbolized by a stippled square, whose boundary is meant to depict the demographic and environmental restraints which affect them all. The smaller square of arrowed lines suggests the continual cycling of chemical elements, and the interorganism reactions between plants, animals, man and microorganisms are also shown. The biosphere is controlled by the income of solar energy entering the atmosphere and its interactions change continually with time

biosphere, in which man's influence upon the environment is immediate and direct (Teilhard de Chardin[7]). Certainly, in view of his immense ecological significance, there are good grounds for regarding man as a special case among terrestrial organisms; however, the idea of separating noosphere and biosphere is, perhaps, somewhat abstract and unrealistic, since it is difficult today to find areas of the latter which are unaffected by man. Consequently, the term noosphere will not be employed in this book.

1.3 Definitions: ecosystems, other assemblages, and environment

Clearly, the massive scale of the biosphere in itself makes the study of environmental-interorganism reactions within it extremely difficult. Accordingly, practical considerations dictate that a large proportion of current biogeographical research is oriented towards the analysis of smaller component organic-environmental assemblages termed *ecosystems*.* These may range in size appreciably, but inevitably embrace several distinctive biotic (plant-animal) associations, comprising a number of organisms of very different life histories, living largely in harmony with each other, within a given milieu. Thus, in southern England, one may envisage a chalkland ecosystem, formed of woodlands and grasslands, with a variety of animal groups and the physical environment; or a smaller beechwood ecosystem within this. Minute pond ecosystems may also occur in many of the adjacent clay lowlands. The important point to remember is that all are, in effect, an intimate expression of the intricate relationships not only between organisms themselves, but also between organisms *and the environment*, and that within them a continual transfer of energy and chemicals is taking place between their organic and inorganic component parts. Since organisms may move into and out from particular ecosystems rather freely, their boundaries may at times be difficult to discern; indeed, it is often more common to find broad transitional zones or *ecotones* between them, rather than sharp dividing lines. But, despite this, ecosystems can be recognized as real entities within the biosphere and, as such, have been widely used in the presentation of examples and arguments within this book. They should not be confused with *associations* or *biocoenoses* of plants or animals, for, although there is some measure of similarity in the use of all three terms, the latter two are

* The term *ecosystem* was first defined by Tansley in 1935.[6]

Principles of Biogeography

more correctly applied in the analysis of biotic assemblages *per se*, *without* taking into consideration the further details of the environment in which they live. At lower levels of the hierarchy, it is customary to distinguish the plant and animal components of ecosystems separately as major *communities*, which usually (though not always) show a degree of stability in space and time. Within these, smaller groups, termed *societies* or *colonies*, are present locally, incorporating one or more species *populations*; and at the lowest point of the hierarchy comes the individual *organism* itself.

Some clarification may also be necessary with respect to the terms *environment* and *habitat*. It is true that these are often utilized interchangeably in a perfectly correct manner, this being the practice adopted generally within this work. However, some subtle differences of scale have been implied at times between the two, especially where the idea of habitat has been used in a much more specific sense than that of environment, as when a single small species population, or a very restricted area is being discussed. *Microhabitats* or *biotopes* can also be identified as the very smallest space occupied by a single life form, as when fungi grow on biotopes found in the hollows of uneven tree trunks. In biotope, habitat, and environment, all organisms live in *niches*, which support them and encourage their biospheric activities. But the idea of a niche is a difficult one to define in a few words; it is consequently to be accorded a more detailed examination later, along with some additional concepts of a less general applicability to the theme of this book than those outlined above.

REFERENCES

1. CAIN, S. A., Biotope and habitat. In *Future environments of North America*, eds., F. FRASER DARLING and J. P. MILTON, pp. 38–54, Garden City, New York. 1966.
2. DANSEREAU, P., *Biogeography: An Ecological Perspective*, New York, 1957.
3. DARWIN, C., *On the Origin of Species by Means of Natural Selection, or the Preservation of Favoured Races in the Struggle for Life*, London, 1859.
4. HUMBOLDT, A. VON, *Ideen zu einer Geographie der Pflanzen nebst einem Naturgemälde der Tropenländer*, Tubingen, 1807.
5. MENDEL, G., Versuche über Pflanzen-Hybriden, *Verhandlungen des naturforschenden Vereines in Brünn*, Vol. 4, pp. 3–47, 1865.
6. TANSLEY, A. F., The use and abuse of vegetational concepts and terms, *Ecology*, Vol. 16, pp. 284–307, 1935.
7. TEILHARD DE CHARDIN, P., The antiquity and world expansion of human culture. In *Man's Role in Changing the Face of the Earth*, ed. W. L. THOMAS JR., pp. 103–22, Chicago, 1956.

2

Energy Controls of Ecosystems

2.1 Introduction

Many of the most important relationships between living organisms and the environment are controlled ultimately by the amounts of available incoming energy received at the earth's surface, primarily from solar sources. It is this energy which helps to drive the complicated mechanisms of the biosphere and shape the growth of ecosystems. Even though the scientific study of energy and energy transfer (*energetics*) is, perhaps, more truly located within the realm of the physicist or physical meteorologist than in geography, the applied aspects of this field of inquiry are of such immense importance to the biogeographer than they should never be neglected by him. It is necessary, for example, to know all the potential sources of terrestrial energy which can influence ecosystems, and the way in which the earth's inclination of axis, the condition of its atmospheric milieu, and its surface features may give rise to patterns of areal differentiation in energy income and availability. It is even more important to be able to grasp the significance of variations in the quantity and direction of energy movement within ecosystems, and the relevance of this to their functional dynamics. All these facets of biogeographical inquiry are to be examined in this chapter.

Recently-acquired evidence strongly suggests that early interactions between the energy and chemical components of the earth must have been very delicately balanced in order to bring about the origins of life as we know it (see also pp. 54 to 55). Moreover, it has long been suspected that the later development of both the natural environment and organisms supported by it was affected to a considerable degree by the quantitative and qualitative conditions of energy inflow. In the light of this, it is perhaps surprising that the importance of energy availability as a controlling mechanism in ecosystem evolution was overlooked until very recently. Traditional viewpoints of natural history, particularly as expressed in the nine-teenth century, were always more concerned with the existence of apparently close relationships between ecosystems and certain parameters of meteorology and climate, more especially air and soil temperatures, atmospheric humidity and dew point, annual and seasonal precipitation, mean wind speed and the relative frequency of sunshine. But studies of such parameters in themselves always failed to account completely for the widely variable distributional patterns of plants and animals, and it was not until the theoretical implications of physical meteorology began to be examined, notably following Rossby's work in the 'twenties and later,[69] that the profound influence of the influx of energy on these matters first began to be fully appreciated. Since then, and particularly within the last two decades as techniques of instrumentation have become increasingly sophisticated, both measurement and analysis of the effects of energy inflow on ecosystem development have been extended at a very rapid rate, although it should be emphasized that the full significance of many related interactions are still far from clear. Indeed, this is hardly surprising when it is considered that, while the laws of physical energy exchange are well known, the large numbers of potential variables which can affect the biosphere are in contrast so great that it is exceedingly difficult to formulate comparable biological laws, which might clarify the functional status of organic and inorganic constituents within ecosystems with respect to energy flow.

ENERGY SOURCES AND TRANSFER: SOME PRELIMINARY CONSIDERATIONS

The energy which is available for use in ecosystems is present in several different forms or *states*, four of which are particularly significant to biogeographical studies. First, *mechanical* energy (or the energy of motion) is usually subdivided into *kinetic* (free) energy which an organism possesses from its movement, this being most frequently measured by the amount of work needed to bring it to rest; and *potential* energy, which is stored within the organism, often in a biochemical matrix, until it is converted into the kinetic state by means of movement, when it is released to do work. Second, *chemical* energy may be liberated by the rearrangement of assemblies of atoms, as in the oxidation process, when glucose breaks down to carbon dioxide and water:

$$C_6H_{12}O_6 + 6O_2 \rightarrow 6H_2O + 6CO_2 + 673 \text{ energy units (kcal*)},$$

or when fuel is burnt, or food consumed. Third, *radiant* energy is derived predominantly from the nuclear fusion of hydrogen to helium in the sun, and reaches the earth in a broad spectrum of electromagnetic waves. Fourth, *heat* energy results from a conversion of non-random to random molecular movements, when work is done. In nature, all types of work, ranging from relatively simple muscular contractions to the highly complex growth of organisms may produce such heat. Indeed, transformations from one state of energy to another are taking place continuously in the biosphere (e.g., incoming radiant energy $\rightarrow$ photosynthesis by green plants $\rightarrow$ oxidation of food by animals $\rightarrow$ energy for work), and within all but the smallest ecosystems they are immense, still exceeding by far the equivalent areal and temporal output of energy produced by man and his machines.

2.2 Energy: sources and availability

The energy absorbed by ecosystems comes from a variety of sources. Some is derived from beyond the solar system in the form of *cosmic* radiation from outer space and, though the amounts of this are not large, it may be of considerable biological importance to certain ecosystems (especially those in high mountain areas: see pp. 96 and 326) in that it can create ionizing effects on chromosome structure if received in too great quantities. In places, the existence of *radioactive* rocks, or excessive fallout, may contribute to some extent to radiation income; and in unstable crustal zones, *geothermal* energy from within the earth may be added to this, particularly in the vicinity of volcanoes and hot springs. But all of these are insignificant on a worldwide scale when compared to the steady inflow of *radiant* energy from the sun, an incandescent globe 93 million miles away, whose surface temperature is approximately 6000°C. Since Newton-

---

* A calorie is a unit used to indicate the amount of heat required to raise 1 g of water 1°C. Larger totals of heat and energy exchange are more often measured by kilogram-calories (kcal). 1 kcal = 1000 cal.

9

ian times, it has been known that this type of energy arrives along a range of different wavelengths, measured in microns, or millionths of a metre (μm). That received at sea level falls mainly between 0·3 and 2·0 μm, although there is some which is in excess of 3 μm (Fig. 2.1). Within this spectrum, visible light lies between 0·36 and 0·76 μm, wavelengths of less than 0·36 μm are ultraviolet, and those of more than 0·76 μm are infrared in structure. At the earth's surface,

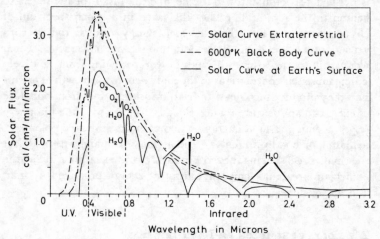

Fig. 2.1 The spectral distribution of incoming solar energy (from an original diagram in D. M. Gates[21]) (Reproduced by permission)

visible light rays comprise about 50 per cent of the total incoming radiation, with infrared amounting to an additional 40 per cent; accordingly, ultraviolet energy totals are always small.

At the outermost edges of the atmosphere, the mean intensity of incoming solar radiation is of the order of two calories per square centimetre per minute, a figure known as the *solar constant*. This has probably not shown any appreciable variation in historic time, although older, possibly minor, fluctuations have been inferred in the explanation of features such as climatic change, and in particular, ice-age incidence and decline.[76] Not all of this energy is destined to reach the earth's surface. Many of the very short waves are absorbed by ozone layers in the upper stratosphere and also, to a certain extent, by oxygen, while some infrared is lost to water vapour, carbon dioxide, and ozone. All of these atmospheric elements tend to absorb radiation along specific spectral bands, as seen in Fig. 2.1,

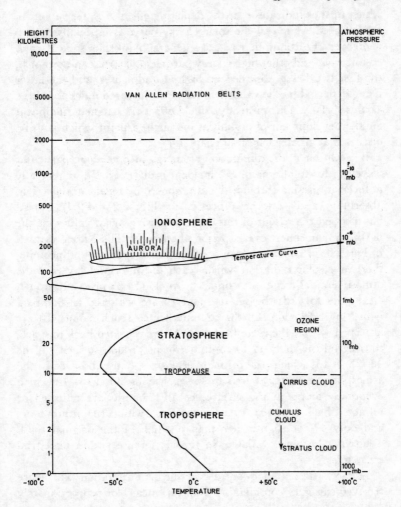

Fig. 2.2 The vertical stratification of the earth's atmosphere. Changes in mean temperature and pressure with height are shown, together with the main zones of aurora activity and ozone accumulation, and the van Allen radiation belts. Most of the earth's 'weather', and those processes leading to cloud formation, are confined to the *troposphere*, a vertical zone lying close to the earth's surface in which temperature decreases with an increase in height. A contrasting situation exists above the *tropopause* (an imaginary line marking the top of the troposphere), in which *stratospheric* temperatures increase with an increase in height. In the *ionosphere*, the electrical conductivity of air is very much greater than elsewhere. The *van Allen radiation belts* consist of charged particles emitted from the sun, or by cosmic radiation, which are trapped by the earth's magnetic field at these altitudes

11

which further indicates that the depleting effect of water vapour in the infrared section of the solar energy curve is especially severe. Moreover, the atmosphere acts as a general blanket for solar X-rays, cosmic rays, and other high energy particles (electrons and protons), the first two being absorbed by high-altitude gases, and the latter being deflected by the van Allen belt a few hundred miles above the earth's surface. This protective shield effect is extremely significant for the development of organisms on earth, a point which is to be discussed at greater length in chapter 4.

In addition to the diminution of energy income along particular wavelengths, total amounts of incident radiation are also generally reduced in passing through the atmosphere by means of reflection, absorption, and scattering processes (see Figs. 2.2 and 2.3). Gates[21] has assessed the extent of this depletion, using mean values for the northern hemisphere over a period of one year .These are presented in Fig. 2.3, in which one can see that out of 100 units of incoming shortwave solar radiation which reach the outermost layers of the atmosphere, 52 come into contact with clouds, 33 proceed through clear skies towards the surface, and 15 are scattered as 'indirect' light,* by refraction. Of the 52 units which reach clouds, 10 are absorbed within them, and another 25 are reflected back to space, thus permitting only 17 to continue to the ground surface. Of the 48 (33 + 15) which pass through clear skies, 9 are absorbed in the atmosphere and 9 scattered to space, leaving 30 (24 + 6) which eventually arrive at the surface. In all, 34 units are reflected or refracted back to space, 19 are assimilated within the atmosphere, and only 47, or slightly less than one-half of the original total, eventually become available, in theory, for use within terrestrial ecosystems.†

In fact, the actual quantities of incident radiation which are received can vary considerably from this mean, for proportionately

---

* An example of indirect light is that present at twilight, and before dawn, when the sun is not visible. It is, of course, also present throughout the daytime period.

† Figure 2.3 also emphasizes that, in order to preserve the incoming and outgoing radiation balance of the earth, 66 units of terrestrial (10 units) and atmospheric (56 units) longwave radiation must leave for space. Other units of longwave radiation circulate between the atmosphere and earth so as to maintain the heat balance in a way which leaves a net loss of 14 units from the earth's surface; this is induced either by reflection from plant or other surfaces, or following the emission of degraded heat energy from plants and animals (see p. 18). Additional radiation loss from the earth's surface is incurred in the processes of evaporation and condensation (23 units), and conduction (10 units).

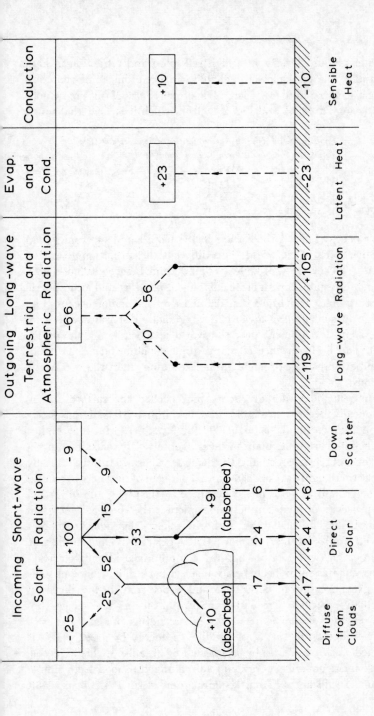

Fig. 2.3 The mean energy exchange of earth and atmosphere for the northern hemisphere (from an original diagram in D. M. Gates[21]) (Reproduced by permission)

greater amounts are lost through reflection and refraction in high latitudes, due to the lower angle of incidence of incoming radiation here as compared to elsewhere. The direct effects of this are demonstrated in Table 2.1 and Fig. 2.4, from which it is clear that much

Table 2.1   *Incoming energy received at three selected stations (data from J. Phillipson[57])*

Station	Latitude	Energy received (cal/m²/yr)
UK	ca. 53° N	2.5×10^8
Michigan	ca. 43° N	4.7×10^8
Georgia	ca. 33° N	6.0×10^8

more energy reaches the surface in the tropics and subtropics than in other parts of the world. This differential distribution gives rise, in turn, to several well-known climatic and biogeographic effects, aiding the establishment of worldwide temperature and atmospheric pressure belts, which are best defined over the major oceans (Fig. 4.6), and contributing towards the regionally disparate patterns of ecosystem development over the world as a whole. Compare, for example, the luxuriance of many tropical plant and animal communities with the extremely limited development of those in Arctic regions.

Once incident radiant energy has reached the surface and so technically become available for entry into terrestrial ecosystems through green plants, a further 95 to 99 per cent of it may be lost immediately from the plant surfaces themselves by reflection, and in the form of sensible heat and the heat of evaporation; thus, it is only a very small figure of between 1 and 5 per cent of the energy received at the ground which is finally absorbed by the plant, and so used in the processes of photosynthesis to replenish the chemical energy of plant tissues. Indeed, most authorities (e.g., Schröder,[73] Rabinowitch,[63] and Wassink[83]) estimate that 1 per cent is close to the true mean value within most natural communities, although certain crops, such as sugar beet,[20] may reach a peak of efficiency of solar energy conversion of between 5 and 6 per cent during the most productive period of growth (see also pp. 30 and 38). Fogg[19] has also recorded similar figures for the mass culture of algae.

The photosynthetic transformation of energy by green plants is termed *primary production* by *autotrophs*, i.e., organisms which are able to fix incident light energy and so manufacture food from simple chemical substances. Total accumulations of such energy per unit

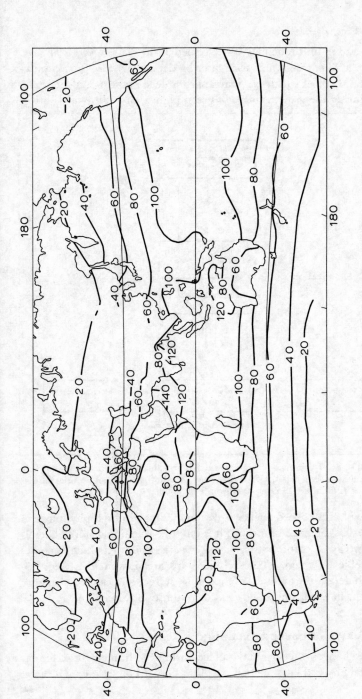

Fig. 2.4 Annual values of net incoming radiation at the earth's surface. All values in kcals/m²/year (from an original diagram in M. I. Budyko[13])

area and in unit time may be described as the *gross primary production*. But not all of this can be used later by dependent *heterotroph* populations (i.e., those which eat, rearrange, or decompose organic matter), for during the growth stages of green plants, an additional 10 to 20

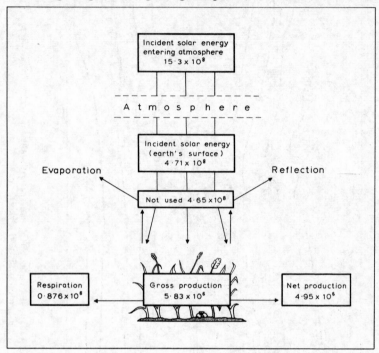

Fig. 2.5   Solar energy used in the perennial grass-herb vegetation of an old-field community in Michigan, USA. All values in cal/m²/year (from an original diagram in J. Phillipson,[57] using data from F. B. Golley[24])

per cent of absorbed energy is commonly lost through physiological processes, such as respiration; it is only the remaining stored chemical energy of plant tissues, or *net primary production*, which becomes available for them. These relationships are illustrated in Fig. 2.5, which details the amounts of incoming, stored, and expended energy present in a grass-herb old-field community of Michigan.

2.3 Laws of energy exchange

All exchanges of energy take place within the framework of fundamental physical laws, namely, the laws of thermodynamics. Of these, the first two are directly applicable to the theme of this book. The *first law* summarizes the physical changes which take place as energy is supplied to or taken from a closed system.* It is also called the *law of the conservation of energy* and states that, within such systems, energy may be altered from one form into another, but may neither be created nor destroyed. Accordingly, changes in the internal energy structure of such a system must be compensated for by means of equivalent modifications in the amount of available heat created or absorbed, and the work done by the system. This may be expressed as below, in an equation which holds true for any essentially closed

$$\Delta E \quad = \quad Q \quad + \quad W \qquad (2.1)$$

ΔE	Q	W
Decrease in the internal energy of the system	Heat given off by the system	Work done by the system

system, whether very large (as the world *in toto*) or small (as a minute pond). Thus, the amount of radiant energy entering the earth's atmosphere (see Fig. 2.3) is identical to that which leaves it, and we shall look later at very much more restricted ecosystems in which the conservation of energy may be seen to be just as severely maintained. Indeed, the latter case may be exemplified by reconsidering Fig. 2.5, in which values of gross production, net production, and respiration can be restated in terms of the first law of thermodynamics as follows:

$$\Delta E \quad = \quad Q \quad + \quad W$$

ΔE	Q	W
Radiant energy assimilated by vegetation	Heat energy of respiration	Growth of vegetation
$5\cdot83 \times 10^6$	$0\cdot876 \times 10^6$	$4\cdot95 \times 10^6$

While establishing the basic controls of energy movement, the first law does not, however, suggest how complete the transfer of energy in any given ecosystem might be. Indeed, in view of the fact that disorder among the elementary particles of matter is the normal condition,[62] and since heat is the only state of energy which results directly from this, direct transformations of solar energy to heat are the only ones which are entire in nature, and all others give rise to some form of energy loss.

The *second law* of thermodynamics deals indirectly with these considerations in stating that, in a spontaneous transfer of energy, there is always a degradation from a concentrated (non-random) to a dispersed (random) form, leading to the inevitable dissipation of

* A closed system is one in which energy is able to enter or leave, although the total amount of component matter remains at the same level.

some into heat, no matter what the states of energy involved in the transfer may be. In other words, the *entropy* (or randomness) of matter will always tend to increase at the expense of the efficiency of the exchange. This second law is fundamental to any study of energy flow through ecosystems, since it helps to explain why some radiant energy can never be used by plants or animals, in view of the fact that in its conversion to chemical compounds in plant leaves some must be degraded into heat, at which point it passes out of the biosphere. The full implications of both these laws to the study of energy exchange within ecosystems may be seen later in this chapter, particularly with respect to the development of food chains and webs, trophic levels, efficiencies of energy transfer, and biological productivity.

2.4 The measurement of energy exchange

It will have been observed that the basic units of energy exchange measurement used so far have been the calorie and kilogram-calorie (kcal). Inasmuch as all forms of energy may be converted to heat equivalents, these are valuable not only in themselves as measurements of heat exchange, but also in comparative studies involving other types of energy transfer. It is known, for example, that each chemical reaction involves a transfer of heat which is quantitatively linked to the number of reacting molecules. All organic compounds have a particular heat of combustion, which is the amount given up to their surroundings as one gram-molecule of the substance is burned, at the expense of molecular oxygen. Some indications of this have already been presented in the case of the equation illustrating the oxidation of glucose to carbon dioxide and water:

$$C_6H_{12}O_6 + 6O_2 \rightarrow 6CO_2 + 6H_2O + 673 \text{ kcal.}$$

In other words, the *molar enthalpy* of this reaction (or the heat given off from the system) is equivalent to *minus* 673 kcal.* This is a low figure, indicating that the basic units of CO_2 and H_2O are simple and stable, with a relatively small energy potential. Hydrolysis usually involves an even more minute energy loss, e.g.:

$$\text{sucrose} + H_2O \rightarrow \text{glucose} + \text{fructose} + 4.8 \text{ kcal,}$$

as does also the neutralization of H^+ and OH^- ions, e.g.,

$$NaOH + HCl \rightarrow NaCl + H_2O + 13.8 \text{ kcal.}$$

* This is a minus quality since it involves a heat loss.

Moreover, it may even be possible in certain ionizations for heat to be absorbed from the environment. In contrast, many other chemical exchanges involve much greater amounts of heat loss.

It is not proposed here to discuss the several techniques by which energy exchange or income may be measured. For these, the reader is directed to several standard manuals, or recent summaries such as that by Gates.[21]

FOOD CHAINS AND WEBS

Once energy has been fixed within plants, or within similar organisms in an aquatic environment, it passes through the ecosystem by means of a *food chain*, a term used to describe the successive consumption of one type of organism by another. Many types of food chain exist, one of the simplest being in oceanic waters, where energy-rich plankton are eaten by fish, which, in turn, are caught and eaten by man. Others are more complex, so that there may be a *food web* in which a series of different organisms are partially dependent on each other. At each stage of food transfer in the chain or web, potential energy is lost through dissipation to heat, resulting in a continued diminution of available energy. Steps in the transfer of food within the chain or web are termed *trophic levels*: thus, green plants will (on land) usually form the first trophic level of *producers*, a second level will be comprised of *herbivores*, a third of *carnivores*, and there may be fourth and fifth levels consisting of further carnivores. There are rarely more than five trophic levels, for by this stage the amounts of available energy are usually so greatly reduced that little is left to support additional organisms; but exceptions do occur, since the *efficiency of energy transfer* from one trophic level to another varies enough for this to become technically possible.

2.5 Techniques for tracing food chains

The passage of energy through an ecosystem is neither obvious nor easily determined, and so must usually be examined by indirect means. This is not to say that direct observations of food chains are not feasible, but, where these do take place, they tend to result in an understanding only in outline of the complex relationships involved. Thus, in oceanic waters it is known that plaice are dependent for food on a bivalve mollusc which, in turn, eats debris; and whales feed on euphasid crustacea, which consume a variety of smaller plankton.

19

On land, Elton[18] has pointed out that a very simple type of food chain may be seen in groves of birch saplings within the New Forest of southern England, in which energy passes from birch twigs to flocks of aphids, which are then milked by ants for their sweet excretions. However, the full intricacies of these interactions only rarely emerge from superficial inspections such as these: for example, in the case of the aphids, one might not easily discover whether other insects, which might normally have formed part of a food web, had been driven from the twigs by the ants; or whether aphids were eaten by organisms other than ants from time to time; or whether the ants themselves by-passed the aphids whenever sweet sap was available directly to them from broken twigs.

It is therefore preferable to use indirect means for estimating the transfer of energy between organisms at different trophic levels. One method is to analyse gut contents of species, assuming that, if a sufficiently large sample is taken, one can establish the nature of the diet and, therefore, the energy income of each. This technique is not entirely satisfactory, in that even the hard parts of certain organisms may not be recognizable in the digestive tracts of other animals, and it may be completely impossible to trace other types of food, such as plant juices, body fluids, and invertebrates. In addition, some types of food digest more quickly than others and so may not be detectable in the gut after a certain period of time. Some of these drawbacks have been surmounted by means of chemical tests which are centred around the matching of liquified smears of gut, or of whole crushed organisms, with liquified antiserums* of one particular species, whose role as a food source it is desired to check. If the latter is present in the gut content of the other organisms, antigens and antibodies will form between the two matching liquids in laboratory analysis: thus, it is relatively easy to determine the presence or absence of such a species. These ideas have been employed with success by Dempster,[15] working with predators of the broom beetle *Phytodecta olivacea*, and more recently by Reynoldson and Young[65] and Young *et al*[92] in gut analyses of certain British lake-dwelling triclads.

One may also track food chains by means of chemical or radio-active tracers. Indeed, some recent inadvertent examples of the

---

* An antiserum, antigen, or antibody, is a substance produced biochemically by an organism as a reaction to the introduction of other substances which are normally alien to it.

transfer of elements along natural food chains in this way have become *causes célèbres* among those who have rightly pointed out the dangers of interfering too greatly with the natural environment through adding chemicals to the soil. In Scotland, toxic dieldrin has passed from grassland through the bodies of sheep and smaller field animals into golden eagles (*Aquila chrysaetos*), many of which have died in consequence. In this, and similar cases, the toxic additives usually become more concentrated as they move through the food chain, and thus mortality rates among carnivores may be potentially much greater than among herbivores. Birds which feed on small mammals or fish appear to be particularly vulnerable, and the case of the golden eagle is certainly no isolated example. Thus in the Clear Lake area of the Coast Ranges of California, the chemical DDD (a close relative of DDT) was applied in 1957, at a rate of 1/50 parts per million to the water of several lakes, as a means of controlling swarms of gnats. The ensuing food chain reaction was as shown in Table 2.2.

Table 2.2   Relationships between food chains and DDD concentration in Clear Lake, California. (Taken from statistics in R. Carson.[14])

Organism	DDD concentration, ppm
(Source)	1/50
Small plankton	5
Fish	40 to 300
Grebes	1600

At this point, the birds died, due to the severe accumulation of DDD in their bodies, and the food chain stopped. There are many other similar examples.

More controlled experiments usually necessitate the use of radio-active isotope tracers of which the first significant example was that described by Odum and Kuenzler in 1963.[46] For this, plant foliage was sprayed with phosphorus$_{32}$, following which the P_{32} intake of selected feeder insects and animals was measured at regular intervals. Results suggested that the earliest consumers of plant food were relatively small herbivores, especially a cricket (*Oecanthus* sp.) and an ant (*Dorymyrmex* sp.), to be followed later by larger herbivores, such as grasshoppers (*Melanoplus* sp), and finally by their predators, including several spiders. Additional experiments, such as those undertaken by Ball and Hooper,[1] Marples,[38] and Wiegert, Odum,

21

and Schnell,[90] have confirmed that the patterns of P_{32} can be indicative of the trophic position of a population of consumers in the field (see Fig. 2.6). Thus, populations known to be strictly herbivorous

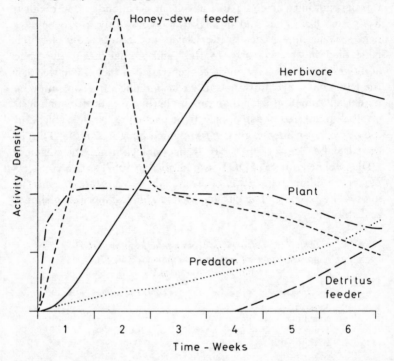

Fig. 2.6 Uptake of phosphorus in selected feeders, using activity-density measurements (from an original diagram by R. C. Wiegert, E. P. Odum, and J. H. Schnell[90])

reach a peak of radioactivity very quickly, while known predators, such as spiders,[56] show delayed uptake curves. Intermediate patterns could then be interpreted as indicating uptake from more than one trophic level.

2.6 Examples of simple food chains

In nature, many simple and obvious food chains may be demonstrated. On the chalk grasslands of Britain, the following is common:

Grass (e.g., *Festuca ovina*) → rabbit → man;

and, in many freshwater streams of the same country, diatoms form the basis of a similar arrangement:

Diatoms (*Navicula viridula*) → Mayfly (*Baetis rhodani*)
→ Caddis fly (*Rhyacophila* sp.).

In both cases, three steps in the transfer of energy may be delimited as follows:

plant → herbivore → carnivore.

Of course, relationships in food chains are rarely as obvious as those presented above, and often more than three trophic levels are present:

plant → herbivore → carnivore$_1$ → carnivore$_2$ → carnivore$_3$.

But it is rare to find more than five, and Elton[17] has noted that there is a definite upper limit to the number of steps within any one chain. Moreover, most organisms consume a wide variety of food, and, in turn, may be eaten themselves by several other species, so that simple food chains usually comprise part of a larger food web and a much more complex arrangement of energy movement. Figure 2.7 shows part of such a food web in a small stream community in South Wales which was analysed by Jones in 1949.[30] Here, in addition to the

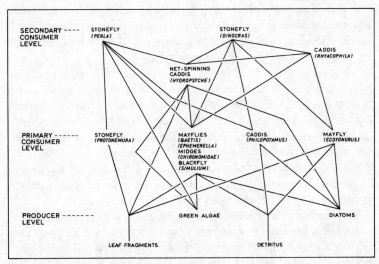

Fig. 2.7 A food web in a freshwater stream community of South Wales (modified from an original diagram in J. R. E. Jones[30])

23

previously noted uncomplicated patterns of transfer from diatom to mayfly to caddis fly, many further exchanges between these and other species are described. Moreover, while most of the organisms group themselves conspicuously into particular trophic levels, one (the net-spinning caddis fly which feeds on both animal and plant material) does not, thus increasing the intricacy of the web and illustrating the difficulty of attempting to categorize too strictly many such animals within a food chain system. Another point of importance, to be expanded later, is that in this and many other instances, some organic matter enters from outside the immediate area under consideration: in other words, it is only rarely in nature that one deals with completely closed systems of energy transfer.

2.7 Ecological pyramids: numbers, biomass, energy

As a means of examining patterns of energy exchange, the delimitation of food chains and food webs has its drawbacks. The most significant of these are the large number of food chains in any one ecosystem and the complexity of most food webs of any size. Some generalization is therefore necessary as an aid to analysis, and, since it has already been established that organisms within food chains and webs show a tendency to group themselves into trophic levels, this is best accomplished by creating cartographical or statistical representations of the numbers, biomass, or energy exchange totals of all organisms within each trophic level of an ecosystem. Such representations are termed *ecological pyramids*.

The precise form taken by *pyramids of numbers* depends on the type of ecosystem under consideration, but usually one of three situations is encountered (see Fig. 2.8 a–c).[43] The normal pyramid first noted by Elton in 1927[17] is one in which many small primary producers support a relatively large number of herbivores and carnivores. On occasions, the number of primary producers is itself small (Fig. 2.8b), as when one or two trees maintain a large insect population of primary consumers. Where parasitic food chains exist, inverted pyramids of numbers may also develop (Fig. 2.8c).

Pyramids of numbers have one major disadvantage: no allowance is made for the size of organisms which are present. Thus, one may be dealing with, for example, primary producers which range in size from small plankton to the giant *Sequoia* redwoods of California. This problem is overcome to some extent by the creation of *pyramids*

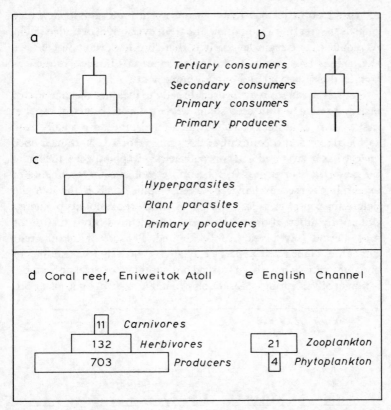

Fig. 2.8 Pyramids of numbers (a–c) and biomass (d–e). The following situations are indicated: 2.8a, small-sized primary producers; 2.8b, large primary producers; 2.8c, a plant-parasite food chain; 2.8d, a normal pyramid of biomass; 2.8e, an inverted pyramid of biomass. All values for 2.8d and 2.8e in g/m². (Redrawn from original diagrams in J. Phillipson[57] and E. P. Odum,[43] using data from H. T. Odum and E. P. Odum,[49] and H. W. Harvey[25])

of biomass, in which the weight of organisms in each trophic level is measured and plotted. Normally, one would expect that the weight of producers should always exceed the weight of consumers, as in figure 2.8d.[49] But very occasionally misleading inverted pyramids of biomass exist (see Fig. 2.8e).[25] Odum [43] has pointed out that this apparent paradox may only be found when most of the producer organisms are small (e.g., algae) or when one is measuring a very restricted sample area. In the latter case, especially in oceanic waters, organisms of different trophic levels may have very dissimilar rates

25

of reproduction, or may move into or out from a sample area very quickly, so creating a temporary apparent overweight or underweight of biomass for certain levels. It is therefore important to consider the *turnover time* (birth-life-death) of organisms in such studies, as well as their seasonal and annual movements.

Biomass pyramids may be used directly to measure the amounts of energy present within ecosystems when calorific values of organic material are uniform. For plants and animals, these are usually close to 4 kcal and 5 kcal per ash-free dry gram respectively, though once energy has been stored and concentrated, as in plant seeds, they may be raised to 7 or 8 kcal.[45] But a more sophisticated technique of examining energy exchange between trophic levels is through the determination of *pyramids of energy*, in which the amounts of energy utilized by different organisms over a sample area (say 1 metre) for a set period (say 1 year) are evaluated. Because of the inherent difficulties of measuring energy exchange, relatively few examples of energy pyramids have been worked out, although one classic example is that of Silver Springs, Florida, shown in Fig. 2.9. In some instances,

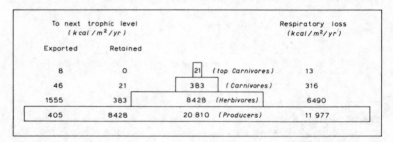

Fig. 2.9 Pyramid of energy for Silver Springs, Florida, USA. All values in kcal/m²/year (redrawn from an original diagram in J. Phillipson,[57] using data from H. T. Odum[47])

a comparison of the pyramids for biomass and energy reveals that the most interesting relationships may often be found within the decomposer group of organisms, where a small biomass is often responsible for a very large exchange of energy. It has been suggested by Russell and Russell[70] that this may be explained by their very rapid turnover time, which may even accelerate as their activity becomes greater; accordingly any increase in the rate of energy exchange may not necessarily lead to a complementary augmentation of their biomass.

26

2.8 The theory of energy exchange

Analyses of energy exchange by means of pyramids are useful in that they can be applied to every type of ecosystem, assuming that data are readily available. The knowledge so obtained may also be fitted into a general theory of energy exchange between trophic levels, as originally formulated by Lindemann in 1942.[36] In this, the total energy content of any one trophic level (i.e., *the standing crop*) is expressed by the symbol Λ (lambda), along with a subscript as follows:

$$\Lambda_1 \text{ producers}$$
$$\Lambda_2 \text{ herbivores}$$
$$\Lambda_3 \text{ carnivores}$$

Moreover, if Λ_n represents the herbivore level, Λ_{n-1} naturally refers to an assemblage of producers.

Since it is known that energy is moving continuously into and out from particular trophic levels, it is also essential to describe the quantity in motion at any one time: this is indicated by the use of a small lambda (λ). Thus, the energy being transferred from Λ_n to Λ_{n+1} is shown by λ_{n+1}. That which is dissipated as heat is defined as R, R_1, and so on. The loss of heat (R), together with the moving energy λ_{n+1} between Λ_n to Λ_{n+1}, is ultimately represented as $\Lambda_{n'}$, being that which at any one trophic level does not go into biomass production. These relationships are presented in diagrammatic form in Fig. 2.10. Following on from this, one may write a generalized equation for the rate of energy exchange from individual trophic layers as follows:

$$\frac{\delta \Lambda_n}{\delta_t} = \lambda_n + \lambda_{n'}. \qquad (2.2)$$

In other words, this is equivalent to the rate at which energy is taken up by that level (e.g., in a standing crop) *minus* the rate at which energy is lost from it, for it is clear that $\lambda_{n'}$ is always a negative quantity.

The proposal of this theory marked a turning point in the field of ecosystem energetics, not only because it was the first reasoned attempt to summarize energy exchange principles, but also since it encouraged the application of these to the very intricate patterns of transfer in land ecosystems; for before Lindemann, even though an understanding of the basic energy controls of individual land based organisms (man and animals) had been glimpsed by Lavoisier as

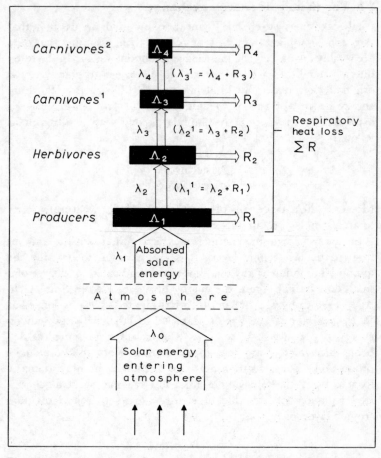

Fig. 2.10 A diagrammatic representation of the Lindemann theory of energy exchange (redrawn from an original diagram in J. Phillipson[57])

early as 1777,[34] energy studies in natural ecosystems, first stimulated by Birge and Juday's pioneer work in 1922,[4] had been restricted largely to the relatively simple, and easily defined confines of certain aquatic environments.

2.9 Ecological efficiencies of energy transfer

The influence of Lindemann's theory may be seen in the very few studies of laboratory ecosystems which have been undertaken so far.

Thus, Slobodkin in 1959,[74] in defining the interactions of a producer (*Chlamydomonos reinhardi*), a primary consumer (*Daphnia pulex*), and a secondary consumer (man) under controlled conditions, was especially interested in the efficiency of energy transfer from one trophic level to another, and also as to whether the ratios between net productivity of prey and consumer could remain constant. His evidence indicated the existence of a general linear relationship between the energy contents of food consumed and the total population of consumers (*Daphnia pulex*), but when the consumers themselves were subjected to differential rates of predation, some variation in the efficiencies of energy transfer began to appear, as indicated by changing yields in *Daphnia*. From this, Slobodkin inferred that there is an optimum rate of energy exchange from one trophic level to another which, at the predator level, is determined to some extent by the rate and amount of predation. If this should be too great, as in some cases of overhunting, overcropping, or overfishing, energy transfer efficiencies may fall to very low values, and possibly to less than 25 per cent of the potential.

These interactions may be formalized in two basic equations. When food from one trophic level is exploited by predators from another, the efficiency of energy transfer (or the *food chain efficiency*) is determined as follows:

$$\frac{C_p}{C_s} \times 100\%, \qquad (2.3)$$

in which C_p refers to the calories of prey eaten by the predator, and C_s to the calories of food supplied to the prey. This only rarely approaches 100 per cent, but when it does so a new situation develops in which predators are sufficiently numerous to eat all the available food, and it is accordingly possible to measure the maximum, or *gross ecological efficiency*, by means of a slightly different equation:

$$\frac{C_p}{C_e} \times 100\%, \qquad (2.4)$$

in which C_e refers to the calories of food eaten by the prey. In Slobodkin's experiments, the latter proved to be only of the order of 13 per cent, the remainder of the energy having been dissipated during the process of transfer. Not enough is known yet to establish whether this low figure is representative of energy movement between trophic levels in all natural or man-made ecosystems, but, if this

29

should prove to be so, it would have important implications for the exploitation of energy in ecosystems everywhere.

ENERGY TRANSFER WITHIN SELECTED ECOSYSTEMS

It has already been determined that green plants, in the process of photosynthesis, fix approximately one per cent of the total incident solar radiation reaching the earth's surface. Of course, this is an average value, and certain extremely productive communities, such as some types of agricultural crops, forests, or algae in shallow water basins, may have local seasonal or annual efficiencies of fixation which are much greater than this (see pp. 39 to 41), while, in contrast, radiation which reaches the surface of deep oceans may be scattered so greatly as to reduce these values at times to less than 0·10 per cent.[66]

However, once energy has been fixed within primary producers, the natural efficiency of energy transfer to other trophic levels is usually reasonably good. Studies completed to date (see Table 2.3) show this well and, even though the detailed figures for different locations do not always completely coincide, one may often take these to reflect dissimilarities in the relatively unsophisticated techniques of analysis rather than an indication of the existence of real major variations on an areal basis.

Table 2.3 *Efficiencies of energy transfer in three aquatic ecosystems (taken from E. P. Odum, 1959[43])*

Trophic level	Energy intake efficiency (%)		
	Cedar Bog Lake Minnesota (from [35])	*Lake Mendota, Wisconsin (from [31])*	*Silver Springs, Florida (from [47])*
Photosynthetic plants (Producers)	0·10	0·40	1·2
Herbivores (Primary consumers)	13·3	8·7	16·0
Small carnivores (Secondary consumers)	22·3	5·5	11·4
Large carnivores (Tertiary consumers)	absent	13·0	5·5

2.10 An example of energy flow

Some of the factors affecting the efficiency of energy exchange within natural ecosystems may be clarified by looking at a detailed and relatively uncomplicated example in Massachusetts, analysed by Teal in 1957.[79] This consisted of a small spring, about 2 m in diameter and between 10 to 20 cm deep. Temperature conditions were almost constant for the period of study. Both floral and faunal components were few, the former being comprised mainly of diatoms, different forms of algae, and *Lemna minor*, a duckweed, and the latter (40 in number) being predominantly herbivorous. There were, however, two predators. The small size of the ecosystem, the temperature

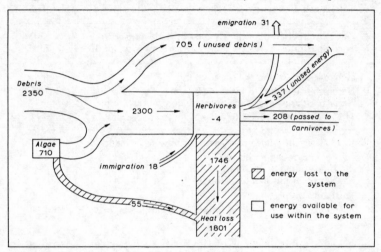

Fig. 2.11 Energy flow at Root Spring, Concord, USA. All values in kcal/m²/yr; numbers in boxes represent changes in the standing crop (redrawn and modified from an original diagram in J. M. Teal[79])

patterns, and the poverty of flora and fauna all closely resembled conditions in an artificial laboratory ecosystem, and thus enabled real comparisons to be made with previous experimental studies.

The results of Teal's examination are presented (partially) in Table 2.4, and Fig. 2.11. Most of the incoming energy (2300 kcal/m²/yr) was derived from sources outside the spring, especially in the form of leaves, fruits, and branches from green plants which drifted or were blown into it as debris. The energy was then transmitted through the ecosystem in a normal way, and from the patterns of its movement one may test the laboratory theory that gross ecological efficiencies of transfer have maximum values approximating 13 per cent. Since, here, λ_2 equals 2300 kcal/m²/yr, and λ_3 equals 208 kcal/m²/yr, the natural gross ecological efficiency of transfer between

31

Table 2.4   Stages in the transfer of energy at Root Spring, Massachusetts
(taken from data in J. M. Teal.[79]*)*

Stages of energy transfer	Totals of energy kcal/m²/yr
Initial stages	
1. Gross production from photosynthesis by autotrophs	710
Autotroph respiration (estimated)	55
Net autotroph production	655
2. Energy entering the system from plant debris	2350
3. Total amount of energy entering system $(1 + 2)$	3005
4. Plant material not immediately used by herbivores	705
5. Total amount of energy consumed by herbivores, λ_2 (or $3 - 4$)	2300
How was this energy used?	
6. Herbivores: heat lost through respiration	1746
7. Herbivores: amount not used immediately by carnivores	337
8. Therefore, remaining energy (or $5 - (6 + 7)$)	217
9. In addition, the following adjustments need to be made:	
(a) For emigrating herbivores $\qquad -31$	
(b) Immigrating herbivores $\qquad +18$	
(c) Net change in standing crop $\qquad +4$	
Total adjustments	-9
10. Therefore, amount of energy passed on to carnivores, or λ_3	208

these two trophic levels is of the order of 9 per cent. As noted in Table 2.3, a similar study by Odum[47] for a much more complex, but still limited, ecosystem at Silver Springs, Florida, recorded comparable efficiencies as follows:

$$\text{Herbivores to carnivores}_1 \left(\frac{\lambda_3}{\lambda_2} \times \frac{100}{1} \right) = 11 \cdot 4 \text{ per cent}$$

$$\text{Carnivores}_1 \text{ to carnivores}_2 \left(\frac{\lambda_4}{\lambda_3} \times \frac{100}{1} \right) = 5 \cdot 5 \text{ per cent}$$

Even though one may expect greater measurement errors to be inherent in the more complex Florida situation, it seems probable that there is a measure of agreement between the two natural ecosystems and the artificial experiments which have been noted previously. But much more research is needed before definite conclusions on this point may be reached.

2.11 Grazing and detritus food chains

While within Teal's study at Root Spring most of the energy available for use by primary consumers entered from outside the system in the form of debris, in contrast, at Silver Springs, Florida,[47] most energy was derived from living plants through the fixation of radiant energy by photosynthesis. These contrasting situations emphasize the fact that within natural ecosystems two major types of food chain may exist either separately or in conjunction with each other: the *grazing food chain* and the *detritus food chain.**

The grazing food chain involves a fairly direct and rapid transfer of energy from living plants through grazing herbivores to carnivores. In this case, the term *grazing herbivore* may be taken to mean any primary consumer, from large animals through to minute zooplankton. In contrast, the detritus food chain transmits energy much more slowly through the decomposing elements of dead plant and animal material to other organisms which feed on them. Differences between these two food chains are presented in Fig. 2.12,[44] in which a marine grazing food chain, and a forest detritus food chain are analysed. However, it should be emphasized that it is not always easy to separate the effects of either in nature since, to a certain extent, both are dependent on each other. To take only one example, large living animals within the grazing food chain often help to break down particles of decaying humus or other organic matter through burrowing, scratching, or other means, so making this available to larger numbers of smaller organisms dependent on the detritus food chain. Both chains are important to energy flow, and neither should be underestimated, for high values of energy turnover for both have been recorded. Thus, Odum[45] has suggested that over 50 per cent of energy flow in grasslands may pass through the grazing food chain, while, in contrast, on *Spartina* salt marshes, over 90 per cent of energy is transferred through the detritus food chain. Similarly high figures for the latter have been obtained by Minshall[40] for a woodland spring-brook community, in which between 50 to 100 per cent of the ingested food of *Ephemeroptera* and the herbivorous *Diptera* was in detritus form; and between 30 to 100 per

---

* Food chains may also be differentiated into *predator chains*, in which energy moves from small organisms to larger animals; *parasite chains*, in which smaller organisms prey on larger ones; and *saprophyte chains*, where energy is removed from dead material by a series of microorganisms.

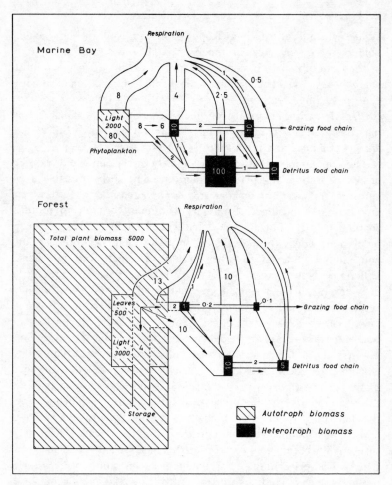

Fig. 2.12 Energy flow through the grazing and detritus food chains. All values of energy flow in kcal/m²/day; and of standing crop biomass in kcal/m² (modified from an original diagram by J. Phillipson,[57] using data in E. P. Odum[44])

cent of gut material in the crayfish *Orconectes rusticus* and *Cambarus tenebrosus* (the two largest invertebrates in the community) was terrestrial leaf material. Of the two, the detritus chain is much the more complex and least understood, but, paradoxically, appears to be the most significant to energy transfer is many natural ecosystems.

34

2.12 The importance of decomposers

Taking into account the potentially large amounts of energy which move through the detritus food chain, it is essential to be able to determine the major agents of transfer in this situation. Altogether, the numbers of heterotroph species are much greater here than in the grazing food chain, and the rapid turnover of some of these has already been noted. Taken as a whole, they may be divided broadly into two overriding groups, the first consisting of small detritus-feeding animals (soil mites, millipedes, worms, etc.), and the second including many of the bacteria and fungi of decay. The relative importance of each of these has not yet been determined, although it is known that on occasions microorganisms may be responsible for transmitting as much as 90 per cent of the energy flow through such a system. At other times, larger organisms, such as earthworms, seem to play an almost equally important role,[16] and it is also thought that the presence of larger animals in itself may accelerate the activity of microorganisms, so stimulating the turnover of energy, although the reasons for this are unknown. Certainly, more research is needed to clarify all these interrelationships.

BIOLOGICAL PRODUCTIVITY IN ECOSYSTEMS

Thus far, the processes of energy transfer within ecosystems have been considered largely from theoretical viewpoints, or within very limited natural or laboratory situations. Although measurements of energy transfer efficiency on a larger scale are much more difficult to make, it seems probable that those previously determined for small areas may also be applicable to *all* ecosystems of the earth, since each, taken as a whole, will react as an individual whose behaviour is ultimately controlled by the first and second laws of thermodynamics. Accordingly, the amount of energy entering any such system will be identical to the amount leaving it (the first law), and, at the same time, in passing through the ecosystem much energy will be degraded to heat, therefore becoming unavailable for growth processes within it. The practical implications of this are far reaching, particularly as they affect the possibility of estimating the rate of production of organic matter, or in other words, the *biological productivity*.*

* It is to be noted that biological productivity always refers to a *rate* of growth of organic material per unit time, and should not be confused with simple estimates of biomass at any one time.

2.13 **Primary and secondary productivity**

Most studies of energy transfer in large-scale ecosystems are concerned with evaluating the *gross primary productivity* (i.e., the *photosynthetic efficiency*) of green plants—in other words, they are seeking to establish the relative proportion of incoming energy directly involved in the processes of photosynthesis. It should be pointed out immediately that the techniques used to achieve this aim may vary widely and give rise to a certain amount of confusion, for although some research workers include totals of reflected incident radiation in their calculations, others do not; and while some utilize mean figures for periods of a year or more, others take values which are representative only for the length of time in which a plant is in leaf and, accordingly, in which photosynthesis is actively taking place. Indeed, in some instances, mean values can be extremely misleading, for the rates of new organic matter production can alter greatly even within the growth period. Thus in the case of *Sericia lespedeza*, a summer legume two to three feet tall, relatively little of the assimilated energy goes to produce fresh organic material after the months of rapid growth (April to July inclusive, see Table 2.5), most being dissipated through respiration. Rather similar declines in productivity, with an increase in crop age, have been noted for forest ecosystems,[52] and negative values can even be recorded during the fall of the year, as Penfound[55] and Porter[61] have indicated in discussions on tall-grass prairie in the USA.

Table 2.5   Primary production of lespedeza (kcal/m²)
(taken from E. F. Menhinick.[39])

Month	Respiration per month	Cumulative	Production per month	Cumulative
April	3	3	9	12
May	61	64	105	169
June	479	543	731	1274
July	1062	1605	1573	3178
August	1332	2937	1803	4740
September	1212	4149	2075	6224
October	460	4609	2346	6955

Moreover, the full patterns of productivity, whether seasonal or annual, may not be easily determined, since much can occur in the form of root growth underground. Only a few attempts so far have been made to assess the importance of root production in total

36

productivity figures, the earliest classic example being Weaver and Zink's approach of 1946.[84] Since then, Bray[9] has proposed that the importance of root growth to total productivity increases substantially as conditions become more xeric, and since in tall-grass prairie areas in Missouri root growth has been estimated to form over 50 per cent of total productivity,[33] the equivalent value for even drier environments may well be much higher. A predominance of underground productivity does not appear to reduce the annual photosynthetic efficiency, which in the case of the Missouri data remains at 1·1 per cent of the incoming visible light energy.

Further complications arise from the fact that differences in the rate of photosynthesis may also occur for various complex reasons. It used to be assumed (e.g., Gessner[23]) that chlorophyll content and productivity potential in different communities were largely the same, but recent work, summarized by Ovington and Lawrence[54] and Strickland,[75] has made it clear that large differences in both can occur. Some of the governing factors which control the rate of photosynthesis and the productivity potential of vegetation have recently been reviewed by Jahnke and Lawrence,[29] who maintain that they are affected by four factors: *first*, their relationships with the patterns of respiration for the whole plant body; *second*, by the amount of chlorophyll exposed to incoming radiation and the proportion of the year during which it is exposed; *third*, by the geometric forms of species and ecosystems, particularly with regard to the height of the crown and its reflectivity and transmissivity as determined by leaf shape, surface texture, thickness, orientation, and distribution throughout the canopy in relation to angles of incidence of incoming light; and *fourth*, by the density of species spacing which, in turn, reflects the geometric form of the individual and the amount of chlorophyll that can be displayed per unit area of land or water. Saeki and Kuriowa,[72] Saeki,[71] and Whittaker and Garfine[88] have paid special attention to the structure of plant communities, notably in relation to the effects of the amount and distribution of leaf tissue and the shape of leaves on photosynthesis, but, again, much more work needs to be completed before the full relationships between these can be determined. More recently, Hunt and Cooper[27] have proposed a chlorophyll index (amount of chlorophyll per unit of ground area) which, they claim, provides a useful indication of productivity among seven forage grasses; in particular, the relatively rapid accumulation of dry matter of *Festuca arundinacea*, although based in part on a larger canopy, has been proved to be associated with features such as specific leaf weight, and/or chlorophyll content per unit area of leaf, which may spread the incoming radiant energy over a larger photosynthetic apparatus within the individual leaf. Chlorophyll indices have also been used with similar results by Brougham,[11] and Nishimura, Okubo, and Hoshino.[42] Rates of photosynthesis and productivity in individual plants may vary not only on an annual, seasonal, and ecological basis, but also diurnally, for as Hodges[26] has noted, marked differences in photosynthesis occur in many species towards midday during clear weather. These may be attributed to changes in the leaf-water potential of plants which, in turn, are dependent on differences in the vapour-pressure gradient from leaf to atmosphere. Changes in the carbohydrate content of leaves may also affect rates of photosynthesis.

Whatever the complications, almost all measurements in natural ecosystems have to date revealed low rates of gross primary productivity. By including totals of reflected light, Bray[8] recorded mean annual values of 7·9 per cent for the conifer *Picea omorika*, while Odum[47] indicated a figure of 5 per cent at Silver Springs, Florida. But Phillipson[57] suggests that these two figures are, in fact, relatively high, and that for most producers in nature, rates of gross primary production may be much less than this. Even though this may be so, such rates are still sufficiently high to ensure that life in all terrestrial ecosystems receives an adequate supply of energy at the producer level. Of course, this is not to say that this will be fully utilized, for the actual amounts of organic material which result may be limited by many restricting factors, not only those noted previously in connection with the patterns of photosynthesis, but also related to the availability of water, carbon dioxide, nitrogen, phosphorus, various trace elements, and other environmental components, all of which are to be discussed in more detail in chapter 4.

Following the second law of thermodynamics, it is to be expected that much of the energy entering into the producer levels of ecosystems will be dissipated through respiration in the form of heat. The remainder will be converted into organic material, to be added to that already existing in plant tissues; the rate of growth of this increment is termed the *net primary productivity*. Equivalent amounts of energy storage in new organic matter at the consumer and decomposer levels of the grazing and detritus food chains are termed *net secondary productivity*, or *net secondary assimilation* respectively. In all cases, the last two rates will be much lower than that of the first, due to the further heat loss which occurs as energy is transferred to these secondary stages in the food chain. In nature, wide variations in the rates of net primary and secondary productivity are known, depending on the environmental circumstances and the type of organisms living therein. When all plants are developing quickly under ideal habitat conditions, as much as 90 per cent of the gross incoming energy may go to form new organic material. But since this situation is rarely found, the maximum growth rates of any organism usually last only for very short periods and, moreover, there may be highly dissimilar growth rates among different competing species within the same ecosystem. Indeed, frequently only one-half or less of all gross incoming energy is used to add to the accumulation of organic matter. Odum[43] has compared gross and net primary productivities for a few selected ecosystems, as shown in Table 2.6, which illustrate these features.

Table 2.6   Comparison of gross and net primary productivity (taken from E. P. Odum[43])

Ecosystem	Rate of production, g/m²/day Gross	Rate of production, g/m²/day Net	% of gross, lost in respiration $\left(\frac{gross\text{-}net}{gross} \times 100\right)$
Silver Springs, Florida yearly average (from [47])	17·5	7·4	57·5
Alfalfa, experimental plot, Maximum growth (from [81])	56·0	49·0	12·5
six months' average	30·1	18·7	38·0
Sargasso Sea, yearly average (from [67])	0·55	0·26	53·0

Primary productivity measurements are determined generally in one of two ways. The first involves the direct measurement of biomass produced per unit time, usually by means of harvesting vegetation and/or other organisms from sample plots at frequent intervals. Such measurements are usually positive, but may occasionally be negative when mortality rates exceed those of new growth (see Wiegart and Evans[89]). This method is particularly valuable in areas protected from grazing and where it is possible to harvest roots and rhizomes as well as the surface parts of plants, but where grazing severely hampers plant growth, a second method may prove to be more practicable. This is centred round the instrumentation of either unit areas or individual plants, in order to determine indirectly their rate of photosynthesis through the measurement of oxygen production or carbon dioxide consumption in a unit period of time, usually one year. While this is undoubtedly the best length of time for studies which are related to sustained rates of productivity over long periods, it should always be remembered that because many plants lie dormant for some months of the year, the annual rates so obtained may be very different from purely seasonal rates. Thus, while the net productivity of the visible parts of Arctic-tundra vegetation has been estimated to range from between 0·5 to 4 g/m²/day in a growing season of between 60 to 70 days, over a year these values would fall to between 0·06 to 0·6 g/m²/day.[6]

Further details as to the methods employed in estimating primary productivity may be obtained from several authors, notably Pomeroy[60] or (for forest areas) Newbould.[41]

2.14 The primary production of world ecosystems

Estimates taken from a wide range of locales suggest that substantial variations in the rates of primary productivity do occur and are dependent partly on differences in energy income on a latitudinal basis, partly on environmental limitations, and partly on the type of organisms which may be found there. However, it is likely that the annual primary productivity in most areas is quite high: to take but one example, the amount of harvested timber has recently reached 1700 million m³/yr,[51] and most of this is replaced in a short period of time. Impressive quantities of new biomass yield have been recorded generally for temperate forest areas (see, for example, Rennie[64]), but even so primary productivity and energy turnover here may be relatively small when compared to other ecosystems. Olson[50] has recently suggested that one must turn to the aquatic environment for some of the maximum rates of production, even where energy storage is limited, and these high marine rates have been confirmed in detail by Thienemann,[80] Ivlev,[28] and Talling.[77]

In 1959, Odum[43] attempted to produce a generalized summary of gross primary production in selected world ecosystems, as delimited in Table 2.7 and Fig. 2.13. In so doing, he determined that some are usually relatively unproductive, such as open oceans and land deserts, which have values of less than 1 g/m²/day; some are moderately productive (1 to 5 g/m²/day), including continental shelf seas, shallow lakes, moist grasslands, mesophytic and dry forests, and 'ordinary'

Table 2.7 Primary gross productivity in selected ecosystems
(taken from E. P. Odum[43])

Ecosystem	Production, g/m²/day
Deserts and semiarid grasslands	<0·5
Open oceans (probably deep lakes, too)	<1·0
Continental shelf open waters, shallow lakes and ponds, average forests, moist grasslands, ordinary agriculture	0·5 to 5·0
Coral reefs, estuaries, alluvial areas, intensive agriculture, evergreen forests	5·0 to 20·0
Maximum rates	<20·0

agricultural lands; and others, such as estuaries, coral reefs, moist forests, alluvial plains, and intensive agricultural areas, have productivities in excess of 5 g/m²/day. Productivity rates of greater than

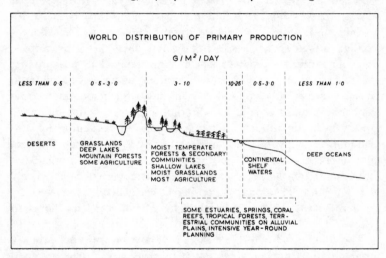

Fig. 2.13 The world distribution of primary production (redrawn from an original diagram in E. P. Odum[44])

20·0 g/m²/day were rare, but occasionally were found for limited periods only in very restricted natural communities, polluted waters, and experimental crops (Table 2.8).

Table 2.8 Maximum recorded rates of gross primary productivity
(taken from E. P. Odum.[43])

Ecosystem	Production, g/m²/day
Pond with untreated waste, So. Dakota, summer (from [2])	27
Silver Springs, Florida, May (from [47])	35
Marine turtle-grass flats, Florida, August (from [48])	34
Estuaries, Texas (from [43])	23
Mass algal culture, extra CO_2 added (from [78])	43

41

In addition, several summaries of the annual net rates of primary productivity in a broad variety of ecosystems are now available. Gerloff[22] has concluded that, for woodland environments, these range from 1·4 g/m²/day for fir forests of the northern taiga to 9·24 g/m²/day in tropical rain forests. Non-wooded ecosystems show a corresponding diversity from 0·28 (in arctic tundras) to 3·36 (meadows) g/m²/day, with one exceptionally high value of 21·56 g/m²/day for a grass swamp in a river valley of the USSR. Rather similar figures were presented by Westlake in an extensive analysis of 1963.[85] He proposed that the most productive communities, such as certain cultivated tropical plants (e.g., sugarcane) and tropical rain forests, created new growth material at an average rate in excess of 20 g/m²/day. Maximum values such as these occurred invariably in tropical areas, which have the greatest incidence of incoming radiant energy (see Fig. 2.4), and once outside the tropics, rates of net primary productivity fell appreciably. The most productive of temperate communities (reed swamps) added organic material at a net rate of only 7·5 to 11·25 g/m²/day, while maize and certain conifers produced over 8·5 g. Marine algal production in many areas may be almost identical to maximum values of temperate land communities. In contrast, some very low rates of productivity were also recorded, such as that of the blue-green alga of phytoplankton in Denmark (*Oscillatoria agardhii*), whose mean growth rate was only 0·55 g/m²/day. These suppositions have been generally confirmed by a more recent review undertaken by Menhinick[39], as shown in Table 2.9.

The comparability of these three estimates of world net primary productivity is further accentuated by considering their extremes (Table 2.10). In general, they also provide a general correlation with the known values of total litter production in different ecosystems, as recently determined by Bray and Gorham,[10] who suggest that these may range from between 2·3 and 3·55 g/m²/day in humid tropical forests, 0·475 to 1·125 g/m²/day in European woodlands, and 0·175 to 0·25 g/m²/day in arctic-alpine communities.

The figures of annual net primary productivity are also interesting from a practical viewpoint, especially in terms of world food production, for they show that certain agricultural systems appear to be less efficient users of solar energy than some natural communities. Those which do utilize energy most fully tend to have year-round

Table 2.9   Annual net productivity in selected vegetation communities (compiled from data in E. F. Menhinick.[39])

Community	Production, $g/m^2/day$
Forest areas	
Fir forests (from [85])	
Palm plantations (from [85])	8·6 to 10·1
Tropical forests (from [68])	
Subtropical broad-leaf forests (from [68])	6·7
Pine and alder (from [85])	
Beech (from [68])	2·9 to 4·4
Pine-oak mesophytic forest (from [87])	
Oak (from [68])	
Birch (from [85])	2·7
Central taiga spruce (from [68])	2·6
Forest heaths (from [87])	1·2 to 1·8
Cultivated crops	
Sugarcane (world average: from [43])	4·7
Sugarcane (exceptional: from [85])	22·7 to 25·7
Grain crops, potatoes, hay (from [43])	0·9 to 1·4
Others	
Spartina spp. (from [43])	8·7
Tall grass prairie (from [43])	
Steppes and savannas (from [68])	1·2 to 3·3
Sphagnum bogs (from [68])	0·9
Tundra (from [68])	0·3 to 0·7
Harsh desert (from [68])	0·2 to 0·3

Table 2.10   Comparisons of estimated maximum and minimum annual net primary productivity in natural ecosystems

Rates, $g/m^2/day$		Gerloff[22]	Menhinick[39]	Westlake[85]
	Max.	21·56	25·7	20·0
	Min.	0·28	0·2	0·35

cultivation and a dense crop canopy, so that bare soil is not seen from above. Systems which have a mixed tree-herb canopy, such as the olive-wheat/barley agriculture of the European Mediterranean, are almost as efficient, but, in contrast, the majority of western European and North American cultivation techniques, which tend to leave large open spaces between individual plants, give rise to a less complete transfer per unit area. Although good rates of biological productivity in agricultural systems are also partially dependent upon sound management techniques, it is clear that, even if these are beyond reproach, the consumption of plant food resources other than those obtained from agriculture may sometimes represent a

43

very effective means of accumulating energy within the human body. Moreover, Pirie[58] has shown that the concentrated protein extracted from wild plants is of an equivalent quality to that derived from many cultivated species or, indeed, from fish meal, leading Phillipson[57] to speculate that, if sufficient quantities of it can be obtained, this may be an extremely significant future food source in selected areas.

2.15 The secondary production of world ecosystems

It has already been established that the mean gross efficiency of energy transfer between producers and primary consumers (green plants to herbivores) is of the maximum order of between 10 to 13 per cent under laboratory conditions, but the secondary productivity of individual organisms within limited ecosystems may range from between 6 and 37 per cent for gross values, and between 5 and 60 per cent for net values.[57] With such great potential variability, and bearing in mind our currently limited knowledge as to the reasons which lie behind it, any analyses of secondary production in large-scale ecosystems must at present be regarded as largely pioneer works and their conclusions as being extremely tentative. Basilevic and Rodin[3] have also pointed out that these patterns of production may be further complicated by the fact that primary and secondary producers are usually very intimately related and may affect each other in diverse ways. Thus, while the secondary producers depend ultimately on the primary producers for food, they can exert some influence on the rate and form of primary production, for instance, by defoliating plants so that their photosynthetic efficiency is reduced; and this, in turn, can then react upon the secondary producers themselves.

But in most cases it is the primary producers which dominate overwhelmingly and directly. Indeed, Bourlière,[7] in a recent study of ecosystems in West Africa, has shown that the basic control of primary production may have unexpectedly farreaching consequences on terrestrial animal communities, and may to a large extent help to decide both the total numbers and social habits of populations within them. To develop this theme, he examined quantitatively several animal communities existing in rain forest, seasonal savanna, and thorn bush. Within rain forest (including derived savanna), food production is maintained continuously throughout the year, so that an abundant supply is available for resident consumer populations; these are able to build up their numbers almost to the limit of the carrying capacity of their environment, so that there is little room for any intrusion by non-forest animals or migrants. It is thought that this in itself may be responsible for the complete lack of any animals with palaearctic and nearctic affinities in these areas. Very different conditions are found in seasonal savanna and thorn bush communities, where food production is limited primarily to wet months (8 to 10 out of 12 in savanna, but only 3 to 4 in thorn bush). Although the growth of new organic material is then abundant, the amount of available plant food is very restricted during the dry season, so that only small numbers of resident consumers are present at that time. This low consumer population is unable to utilize fully all the food produced during the rainy months, with the result that a seasonal surplus is created, and the area is able to support incoming seasonal and long-range migrants for part of the year. Boulière suggests that migration and animal nomadism may well have originated at the drier edges of tropical rain forests as a response to a situation such as this.

Not enough is yet known about terrestrial animal communities to determine whether these conclusions may be applied on a worldwide scale, and many more observations need to be made before the detailed patterns of secondary productivity in natural ecosystems begin to emerge. However, from our existing knowledge, certain other general important implications are already clear, particularly in respect of the utilization of consumer populations as a human food source. Despite some cultural specializations, man has always been a largely omnivorous animal, eating most things which he can catch or cultivate or which come his way. For several practical reasons, he also seems likely to retain this trait, since he requires approximately 20 g of protein per 1000 kcal of food consumed, and this is most easily obtained from animal foods. Referring to the general principles of energy transfer, one may assume that it would be advantageous for him to obtain these from primary consumer (herbivore) populations wherever possible, since carnivores will be relatively much more deficient in energy resources. Indeed, some herbivores, such as rabbits and natural range animals (e.g., zebras), are known to have particularly high rates of growth efficiency and energy storage and these are accordingly very suitable for his use. Of course, considerations of energy transfer are not the only ones to

45

be heeded, for practical problems of rearing and slaughtering make it preferable for most such animals to be domesticated, and this often induces a reduction in their potential energy reserve. For instance, mature cattle have net and gross growth efficiency values of only 10·9 and 4·1 per cent respectively. However, as with plants, all animals have variable rates of growth,whose maximum usually occurs in the early years of life. The gross growth efficiency of young cattle may be as high as 35 per cent, and the same is true for chickens, at least until three to four months after hatching. The case of chickens is particularly interesting, in that they do not begin to produce eggs before the age of six months, by which time their growth efficiencies have rapidly declined; on these grounds, it has been argued by Kleiber[32] that chicken production for eggs is a very wasteful method of utilizing the available energy, being about 25 times less effective than grain farming and 3700 times less than potential algal crops. Be that as it may, a concentration of young animals for food will obviously increase the efficiency of energy transfer from animal to human populations and, along with a wider use of certain other techniques (for instance, the farming of herbivorous fish), may contribute towards a general reduction in the sometimes wasteful exploitation of food-energy resources which has taken place in many parts of the world, particularly in historic time.

REFERENCES

1. BALL, R. C. and HOOPER, F. F., Translocation of phosphorus in a trout-stream ecosystem. In *Radioecology* ed. by V. SCHULTZ and A. W. KLEMENT, Jr., New York, 1963.

2. BARTSCH, A. F. and ALLUM, M. D., Biological factors in the treatment of raw sewage in artificial ponds, *Limnol. and Oceanogr.*, Vol. 2, pp.77–84, 1957.

3. Базилевич, Н. И. и Родин, Л. Е., Типьт биологического круговорога эольных эпеменгов/и/азога в основных природньтх эонах северного молушария, *Генезис, Классификауия и Картография, почв СССР*, pp. 134–46, МОСКВА, 1964. (BASILEVIC, N. I. and RODIN, L. E., The biological circulation of ash elements and nitrogen of the basic natural zones in the northern hemisphere, *Genesis, Classification and Cartography of the soils of the U.S.S.R. . . .*)

4. BIRGE, E. A. and JUDAY, C., The inland lakes of Wisconsin. The plankton. Part I. Its quantity and chemical composition, *Bull. Wisconsin Geol. Nat. Hist. Surv.*, Vol. 64, pp. 1–222, 1922.

5. BILLINGS, W. D., *Plants and the Ecosystem*, London, 1964.
6. BLISS, L. C., Net primary productivity of tundra ecosystems, in *Die Stoffproduktion der Pflanzendecke*, ed. H. LEITH, Stuttgart, 1962.
7. BOURLIÈRE, F., The temporal pattern of secondary productivity in the tropics, *J. Ecol.*, Vol. 55, pp. 51–72, 1967.
8. BRAY, J. R., An estimate of a minimum yield of photosynthesis based on ecological data, *Pl. Physiol.*, Vol. 36, pp. 371–3, 1961.
9. BRAY, J. R., Root production, and the estimate of net productivity, *Can. J. Bot.*, Vol. 41, pp. 65–72, 1963.
10. BRAY, J. R. and GORHAM, E., Litter production in forests of the world, *Adv. Ecol. Res.*, Vol. 2, pp. 101–57, 1964.
11. BROUGHAM, R. W., The relationship between the critical leaf area, total chlorophyll content, and maximum growth rate of some pasture and crop plants, *Ann. Bot. N.S.*, Vol. 24, pp. 463–74, 1960.
12. BUDYKO, M. I., *Atlas of the Heat Balance*, Leningrad, 1955.
13. BUDYKO, M. I., *The Heat Balance of the Earth's Surface*, Trans. N. A. STEPANOVA, US Weather Bureau, US Department of Commerce, Washington, D.C., 1958.
14. CARSON, R., *Silent Spring*, New York, 1963.
15. DEMPSTER, J. P., A quantitative study of the predators on the eggs and larvae of the broom beetle, *Phytodecta olivacea*, Forster, using the precipitin test, *J. Anim. Ecol.*, Vol. 29, pp. 149–67, 1960.
16. EDWARDS, C. A. and HEATH, G. W., The role of soil animals in the breakdown of leaf material, In *Soil Organisms*, ed. by J. DOEKSEN and J. VAN DER DRIFT, Amsterdam, 1963.
17. ELTON, C. S., *Animal Ecology*, London, 1927.
18. ELTON, C. S., *The Ecology of Invasions by Animals and Plants*, London, 1958.
19. FOGG, G. E., Actual and potential yields in photosynthesis, *Adv. Sci.*, Vol. 14, pp. 395–400, 1958.
20. GAASTRA, P., Light energy conversion in field crops in comparison with the photosynthetic efficiency under laboratory conditions, *Meded. Landb. Hoogesch.*, *Wageningen*, Vol. 58, pp. 1–12, 1958.
21. GATES, D. M., *Energy exchange in the Biosphere*, Harper and Row, New York, 1956.
22. GERLOFF, G. C., Comparative mineral nutrition of plants, *Ann. Rev. Pl. Physiol.*, Vol. 14, pp. 107–24, 1963.
23. GESSNER, F., Die Chlorophyllgehalt im See und seine photosynthetische Valenz als geophysikalisches Problem, *Schweiz. Zflf. Hydrol.*, Vol. 11, pp. 378–410, 1949.
24. GOLLEY, F. B., Energy dynamics of a food chain of an old field community, *Ecol. Mon.*, Vol. 30, pp. 187–206, 1960.
25. HARVEY, H. W., On the production of living matter in the sea off Plymouth, *J. Marine Biol. Ass.*, *UK*, Vol. 29, pp. 97–137, 1950.
26. HODGES, J. D., Patterns of photosynthesis under natural environmental conditions, *Ecology*, Vol. 48, pp. 234–42, 1967.
27. HUNT, L. A. and COOPER, J. P., Productivity and canopy structure in seven temperate forage grasses, *J. Appl. Biol.*, Vol. 4, pp. 437–58, 1967.

28. IVLEV, U. S., *Experimental Ecology of the Feeding of Fishes*, trans. by D. SCOTT, Newhaven, Connecticut, USA, 1934.

29. JAHNKE, L. S. and LAWRENCE, D. B., Influence of photosynthetic crown structure on potential productivity of vegetation, based primarily on mathematical models, *Ecology*, Vol. 46, pp. 319–26, 1965.

30. JONES, J. R. E., A further ecological study of a calcareous stream in the Black Mountain district of South Wales, *J. Anim. Ecol.*, Vol. 18, pp. 142–59, 1949.

31. JUDAY, C., The annual energy budget of an inland lake, *Ecology*, Vol. 21, pp. 438–50, 1940.

32. KLEIBER, M., *The Fire of Life*, New York, 1961.

33. KUCERA, C. L., DAHLMANN, R. C., and KOELLING, M. R., Total net productivity and turnover on an energy basis for tall-grass prairie, *Ecology*, Vol. 48, pp. 536–41, 1967.

34. LAVOISIER, A. L., Experiences sur la réspiration des animaux et sur les changements qui arrivent a l'air en passant par leur poumons, *Mém. Acad. Sci., Inst. Fr.*, Paris, p. 185, 1777.

35. LINDEMANN, R. L., Seasonal food-cycle dynamics in a senescent lake, *Am. Mid. Nat.*, Vol. 26, pp. 636–73, 1941.

36. LINDEMANN, R. L., The trophic-dynamic aspect of ecology, *Ecology*, Vol. 23, pp. 399–418, 1942.

37. MACFADYEN, A., Energy flow in ecosystems, and its exploitation by grazing. In *Grazing in Terrestrial and Marine Environments*, ed. by D. J. CRISP, Oxford and Edinburgh, 1964.

38. MARPLES, T. G., A radionuclide tracer study of arthropod food chains in a *Spartina* salt marsh ecosystem, *Ecology*, Vol. 47, pp. 270–7, 1966.

39. MENHINICK, E. F., Structure, stability and energy flow in plants and arthropods in a *Sericea lespedeza* stand, *Ecol. Mon.*, Vol. 37, pp. 255–72, 1967.

40. MINSHALL, G. W., The role of allochthonous detritus in the trophic structure of a woodland spring-brook community, *Ecology*, Vol. 48, pp. 139–49, 1967.

41. NEWBOULD, P. J., Methods for estimating the primary production of forests, *I.B.P. Handbook No. 2.*, Oxford and Edinburgh, 1967.

42. NISHIMURA, S., OKUBO, T., and HOSHINO, M., Light transmission and chlorophyll amount in a sward as a substitute for leaf area index. *Proceedings 10th International Grassland Conference*, pp. 117–20, 1966.

43. ODUM, E. P., *Fundamentals of Ecology*, Saunders, Philadelphia, 2nd. edn, 1959.

44. ODUM, E. P., Relationships between structure and function in the ecosystem, *Jap. J. Ecol.*, Vol. 12, pp. 108–18, 1962.

45. ODUM, E. P., *Ecology*, New York, 1966.

46. ODUM, E. P. and KUENZLER, E. J., Experimental isolation of food chains in an old field ecosystem with the use of phosphorus[32]. In *Radioecology*, eds. V. SCHULTZ and A. W. KLEMENT, Jr., New York., 1963.

47. ODUM, H. T., Trophic structure and productivity of Silver Springs, Florida, *Ecol. Mon.*, Vol. 27, pp. 55–112, 1957.

48. Odum, H. T., Primary production measurement in eleven Florida springs and a marine turtle-grass community, *Limnol. Oceanogr.*, Vol. 2, pp. 85–97, 1957.

49. Odum, H. T. and Odum, E. P., Trophic structure and productivity of a windward coral reef community on Eniweitok Atoll, *Ecol. Mon.*, Vol. 25, pp. 291–320, 1955.

50. Olson, J. S., Gross and net production of terrestrial vegetation, *J. Ecol.*, Vol. 52, Supplement, pp. 98–118, 1964.

51. Ovington, J. D., Organic production, turnover, and mineral cycling in woodlands, *Biol. Rev.*, Vol. 40, pp. 295–336, 1965.

52. Ovington, J. D. and Heitkamp, D., The accumulation of energy in forest plantations in Britain, *J. Ecol.*, Vol. 48, pp. 639–46, 1960.

53. Ovington, J. D., Heitkamp, D., and Lawrence, D. B., Plant biomass and productivity of prairie, savanna, oakwood, and maize-field ecosystems in Central Minnesota, *Ecology*, Vol. 44, pp. 52–63, 1963.

54. Ovington, J. D. and Lawrence, D. B., Comparative chlorophyll and energy studies of prairie, savanna, oakwood and maize-field ecosystems, *Ecology*, Vol. 48, pp. 515–24, 1967.

55. Penfound, W. T., Effects of denudation on the productivity of grassland, *Ecology*, Vol. 45, pp. 838–46, 1964.

56. Phillipson, J., Respirometry and study of energy turnover in natural systems, with particular reference to the harvest spider, *Oikos*, Vol. 13, pp. 311–22, 1962.

57. Phillipson, J., *Ecological Energetics*, Arnold, London, 1966.

58. Pirie, N. W., Future sources of food supply: scientific problems, In Food studies and Population Growth, *J. Roy. Stat. Soc.*, Series *A*, Vol. 125, pp. 34–52, 1962.

59. Platt, R. B. and Griffiths, J., *Environmental Measurement and Interpretation*, New York, 1964.

60. Pomeroy, L. R., Productivity and how to measure it. In *Algae and Metropolitan Wastes*, US Department of Health, Education and Welfare, Public Health Service, Robert A. Taft Sanitary Engineering Centre, Cincinatti, 1961.

61. Porter, C. L. Jr., Composition and productivity of a sub-tropical prairie, *Ecology*, Vol. 48, pp. 937–42, 1967.

62. Porter, G., *The Laws of Disorder*, British Broadcasting Corporation, London, 1965.

63. Rabinowitch, E. I., *Photosynthesis and Related Processes*, Vol. II, pp. 603–1208, New York, 1951.

64. Rennie, P. J., The uptake of nutrients by mature forest growth, *Pl. Soil*, Vol. 7, pp. 49–95, 1955.

65. Reynoldson, T. B. and Young, J. O., The food of four species of lake-dwelling triclads, *J. Anim. Ecol.*, Vol. 32, pp. 175–91, 1963.

66. Riley, G. A., The carbon metabolism and photosynthetic efficiency of the earth, *Am. Sci.*, Vol. 32, pp. 132–4, 1944.

67. Riley, G. A., Phytoplankton of the north-central Sargasso Sea, *Limnol. Oceanogr.*, Vol, 2, pp. 252–70, 1957.

49

68. RODIN, L. E. and BASILEVIC, N. I., The biological productivity of the main vegetation types, *For. Abstr.*, Vol. 27, pp. 369–72, 1966.

69. ROSSBY, C. G., Relations between variations in the intensity of the zonal circulation of the atmosphere, and the displacement of the semi-permanent centres of action, *J. Mar. Res.*, Vol. 2, pp. 38–55, 1939.

70. RUSSELL, E. J. and RUSSELL, E. W., *Soil Conditions and Plant Growth*, 8th edn., New York, 1950.

71. SAEKI, T., Interrelationships between leaf amount, light distribution and total photosynthesis in a plant community, *Bot. Mag. Tokyo*, Vol. 73, pp. 55–63, 1960.

72. SAEKI, T. and KURIOWA, S., On the establishment of the vertical distribution of photosynthetic system in a plant community, *Bot. Mag. Tokyo*, Vol. 72, pp. 27–35, 1959.

73. SCHRÖDER, H., Die jährliche gesamtproduktion des grünen Pflanzedecke der Erde, *Naturwissenschaft*, Vol. 7, pp. 8–29, 1919.

74. SLOBODKIN, L. B., Energetics in *Daphnia pulex* populations, *Ecology*, Vol. 40, pp. 232–43, 1959.

75. STRICKLAND, J. D. M., Solar radiation penetrating the ocean. A review of requirements, data, and methods of measurement, with particular reference to photosynthetic productivity, *J. Fish. Res. Bd, Can.*, Vol. 15, pp. 453–93, 1958.

76. SUTTON, O. G., *Understanding Weather*, Penguin, Harmondsworth, Middlesex, 1960.

77. TALLING, J. F., Photosynthesis under natural conditions, *Ann. Rev. Pl. Physiol.*, Vol. 12, pp. 133–54, 1961.

78. TAMIYA, H., Mass culture of algae, *Ann. Rev. Pl. Physiol.*, Vol. 8, pp. 309–34, 1957.

79. TEAL, J. M., Community metabolism in a temperate cold spring, *Ecol. Mon.*, Vol. 27, pp. 283–302, 1957.

80. THIENEMANN, A., Der Produktionsbegriff in der Biologie, *Arch. Hydrobiologie (Plankt.)*, Vol. 22, pp. 616–22, 1931.

81. THOMAS, M. D., and HILL, G. R., Photosynthesis under field conditions. in *Photosynthesis in Plants*, eds., J. FRANCK and W. E. LOOMIS, Ames, Iowa, pp. 19–52, 1949.

82. TRANSEAU, E. N., The accumulation of energy by plants, *Ohio J. Sci.*, Vol. 26, pp. 1–10, 1926.

83. WASSINK, E. C., The efficiency of light energy conversion in plant growth, *Pl. Physiol.*, Vol. 34, pp. 356–61, 1956.

84. WEAVER, J. E. and ZINK, E., The annual increase of underground materials in three range grasses, *Ecology*, Vol. 27, 115–27, 1946.

85. WESTLAKE, D. F., Comparisons of plant productivity, *Biol. Rev.*, Vol. 38, pp. 400–19, 1963.

86. WHITTAKER, R. H., Estimation of net primary production of forest and shrub communities, *Ecology*, Vol. 42, pp. 177–80, 1961.

87. WHITTAKER, R. H., Forest dimension and production in the Great Smoky Mountains, *Ecology*, Vol. 47, pp. 103–21, 1966.

88. WHITTAKER, R. H. and GARFINE, V., Heat characteristics and chlorphyll in relation to exposure and production of *Rhododendron maximum*, *Ecology*, Vol. 43, pp. 120–5, 1962.

89. WIEGERT, R. G. and EVANS, F. C., Primary production and the disappearance of dead vegetation on an old-field in south-eastern Michigan, *Ecology*, Vol. 45, pp. 49–63, 1964.

90. WEIGERT, R. G., ODUM, E. P., and SCHNELL, J. H., Soil-arthropod food chains in a one-year experimental field, *Ecology*, Vol. 48, pp. 75–83, 1967.

91. WOODWELL, G. M., The ecological effects of radiation, *Scient. Am.*, Vol. 208, pp. 40–9, 1963.

92. YOUNG, J. O., MORRIS, I. G., and REYNOLDSON, T. B , A seriological study of *Asellus* is the diet of lake-dwelling triclads, *Arch. Hydrobiol.*, Vol. 60, pp. 366–73, 1964.

3

Biogeochemical Cycles Within Ecosystems

3.1 Introduction

In order to explain the functional mechanisms of ecosystems, it is important to realize that the complicated patterns of energy exchange serve also to stimulate reactions among the chemical components of living cells, which then encourage, restrict or modify the growth and interrelationships of organisms. Unlike the use of energy, which is derived mainly from a single source (the sun) and which is relatively quickly dissipated in passing through the food chain system, the chemical elements essential to life may be obtained from any one or more of several pool areas to be found in the atmosphere, ocean, soil, and bedrock, and, furthermore, are continually recirculated among ecosystems, so being reused many times. Accordingly, it is necessary for the biogeographer to know something of the nature of these circulatory elemental movements, termed *biogeochemical cycles* by Vernadskii in 1934,[183] and also of the associated changes in cell chemistry which may accompany them. In this, he must draw heavily upon the work of the research biochemist and soil scientist. It is the aim of this chapter partly to summarize some of their more recent conclusions and partly to project these further where they aid our understanding of detailed biogeographical interrelationships.

Usually, biogeochemical exchanges require the presence of a wide variety of living organisms, whose patterns of birth, life, and death all encourage the movement of elements. Therefore one may expect them to occur at every level of the biosphere and, indeed, some very minor exchanges have been detected not only in the upper troposphere, but also in bedrock several thousand metres below ground surface, where microorganisms live off oily liquids.[135] However, it is to be expected that most will take place in the immediate contact zone between the lower atmosphere and the uppermost parts of land and ocean in which organic life is abundant and the annual turnover of materials is enormous.* Here, all organisms take up a selection of elements from pool areas, process them to form new biological compounds needed for food and growth, and then return them to the environment through respiration, litter fall, and decay. For land plants, 17 elements are considered to be essential to life, 4 of which (C, H, O, and N) are derived ultimately from atmospheric sources, and the remaining 13 (K, Ca, Mg, P, N, S, Fe, Cu, Mn, Zn, Mo, B, Cl) from the soil and bedrock. Of course, not all of these are readily absorbed because of environmental and physiological restrictions; indeed, only hydrogen and oxygen are freely available in large quantities, and all the others depend in some measure on complicated patterns of cycling for their supplies to be fully maintained. This means that under normal circumstances neither the soil nor the atmosphere can, in themselves, sustain and support advanced life forms; interactions between the two are almost always called for.

Detailed studies of the pathways of element exchange have only recently been forthcoming and, because of this, it is true to say that our knowledge of many aspects of this immense field is still extremely limited. Although Liebig in 1840[90] discovered the basic principles of mineral nutrition in agricultural crops, and several German scientists helped to determine the uptake rates of selected elements by forest species before 1900,[189, 201, 47, 139] there were few systematic attempts to elucidate the implications of biogeochemical exchanges to ecosystem growth and development until the 'thirties. Much of the subsequent related research has been summarized recently by Soviet authors, particularly Rodin and Basilevic[148] and Perel'man[132], to whom reference will be made in several sections of this chapter.

---

* Mason[101] has calculated that, given a relatively constant mass of organisms in this contact zone, the total turnover mass for the period of known organic life (say 500 million years) would be roughly equivalent to the total mass of the earth, assuming an average life cycle of one year for all organisms.

CHEMICAL ELEMENTS IN LIVING ORGANISMS

It has already been suggested in chapter 2 that the chemical and energy balance of the early terrestrial environment must have been very delicate in order to stimulate the origins of life as we know it. Current thought supposes that the first primitive forms of life emerged shortly after the earth's surface had cooled sufficiently for the chemical composition of land, sea, and atmosphere to be broadly comparable to that of the present day. Under these conditions, and in particular shortly after the permanent establishment of oceanic waters, primitive living cells began to evolve whose chemical components were able to react fairly easily with the carbides, nitrides, and phosphates found in incoming meteorites, so creating hydrocarbons and other biologically-interesting compounds. In time, it is thought that these gave rise to a souplike mixture of ocean water and organic materials, the latter being concentrated to quantities of one per cent or more.[180] At roughly the same time, electric storms in the proto-atmosphere helped to produce chemical compounds with free hydrogen from gaseous mixtures with oxidized carbon;[1] eventually, through synthesis, these led to the formation of the first relatively simple amino-acids, and possibly later to primitive proteins. At first, all life was purely exploitative, living off existing organic matter and chemical elements, both of which were consumed in a non-renewable way. Moreover, no life as we know it could have emerged until some mechanism for precise genetic reproduction had been introduced, and for this the subsequent development of chemically complex nucleic acids was essential (section 3.4). Only then were the higher forms of life able to evolve, among which were photosynthetic plants and animals capable of using solar energy within a chemically-balanced ecosystem in a non-exploitative way.

Since many of these events date from early Pre-cambrian times in which the fossil record is extremely poor, their precise means of inception and their exact chronology may perhaps never be known. Although a few possibly photosynthetic organisms have been reported from rocks 2700 million years old,[28] it was not until the beginning of the Palaeozoic era that large and diverse numbers of chlorophyll-producing plants, and other associated herbivores, carnivores, and microorganisms, were present. Since then, these, and their successors, have helped to induce immense changes in the geochemistry of earth and atmosphere, particularly with respect to the replacement of

virtually all the atmospheric carbon dioxide by free oxygen. They
have also ensured that chemical elements remain mobile near to
the earth's surface and, in so doing, they have stimulated further
chemical reactions in the biosphere, so that the net effect has always
been to create ever more complex chemical exchanges between
living organisms and the environment in which they live. A generalized portrayal of the pathways of chemical exchange for some of the
more common elements is shown in Fig. 3.1.

3.2 The relative abundance of elements

Although all of the 92 natural elements are capable of being
absorbed by living organisms, usually only 3—oxygen, carbon, and
hydrogen—are to be found in appreciable quantities within them.
These 3 often comprise about 90 per cent of the dry weight of
organic material and, in addition, a combination of hydrogen and
oxygen atoms in the form of water will account for much of the total
weight of all the life forms. Thus wood will contain over 50 per cent,
most vertebrates about 66 per cent, some mammals as much as
85 per cent, and certain marine invertebrates over 99 per cent water
in their body weight. In consequence, most other elements, including many of those essential to life, are to be found in relatively small
or even very minute quantities, as shown in Table 3.1.

*Table 3.1   Mean chemical composition of all living matter,
in percentage totals (after A. I. Perel'man[132] and A. P. Vinogradov[184])*

Macroelements

O	70	N	0·3	S	0·05
C	18	Si	0·2	Na	0·02
H	10·5	Mg	0·04	Cl	0·02
Ca	0·5	P	0·07	Fe	0·01
K	0·3				

Microelements

Al	0·005	Zn	0·0005	Pb	0·00005
Ba	0·003	Rb	0·0005	Sn	0·00005
Sr	0·002	Cu	0·0002	As	0·00003
Mn	0·001	V	0·0001	Co	0·00002
B	0·001	Cr	0·0001	Li	0·00001
Ti	0·0008	Br	0·0001	Mo	0·00001
F	0·0005	Ge	0·0001	Y	0·00001
		Ni	0·00005	Cs	0·00001

Ultramicroelements

Se	< 0·000001	(10^{-6})
U	< 0·000001	
Hg	0·0000001 (10^{-7})	
Ra	0·000000000001 (10^{-12})	

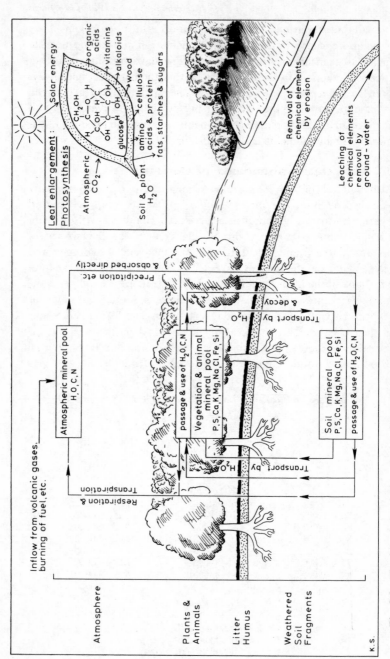

Fig. 3.1 The biogeochemical cycle (modified from original diagrams in A. I. Perel'man,[132] and in Arthur W. Galston, *The life of the green plant,* © 1961, used by permission of Prentice-Hall Inc.)

Most elements enter living organisms either in the gaseous state (e.g., O_2, N_2, CO_2) or as water-soluble salts (e.g., NaCl, KCl, etc.). Inevitably, those which have high or low ionic potentials with strong electric charges, and which respectively form suitable *cations* (positively-charged ions) and *anions* (negatively-charged ions) will be readily taken up, while those with intermediate ionic potential and no strong electrical charges will not. For a variety of complicated environmental and physiological reasons, there will also be inevitable differences in the intake of specific elements by different types of organisms, although the dominance of oxygen, carbon, and hydrogen always remains paramount. Table 3.2 illustrates some of

Table 3.2   *Mean chemical composition of selected organisms,*
in weight per cent (after B. Mason[101])*

Alfalfa		Calanus finmarchicus		Man	
O	77·9	O	79·99	O	62·81
C	11·34	H	10·21	C	19·37
H	8·72	N	1·52	N	5·14
		Cl	1·05	Ca	1·38
P	0·71	Na	0·54	S	0·64
Ca	0·58	K	0·29	P	0·63
K	0·17	S	0·14	Na	0·26
S	0·10	P	0·13	K	0·22
				Cl	0·18
Mg	0·08	Ca	0·04	Mg	0·04
Cl	0·07	Mg	0·03		
Si	0·009	Fe	0·007	Fe	0·005
Fe	0·003	Si	0·007	Si	0·004
Al	0·003			Zn	0·003
B	0·0007	Br	0·0009	Rb†	0·0009
Rb	0·0005	I	0·0002	Cu	0·0004
Mn	0·0004			Br	0·0002
Zn	0·0004			Sn	0·0002
Cu	0·0003			Mn	0·0001
F	0·0002			I	0·0001
Mo	0·0001				

these variations, observed in alfalfa, the marine invertebrate *Calanus finmarchicus*, and man. From this, it is clear that in all three instances the accumulation of certain elements exceeds the mean values of Table 3.1 by upwards of 10 times. This is the case for potassium in alfalfa; for sodium, chlorine, bromine, and iodine in the marine

* Data for *Calanus finmarchicus* from Vernadskii,[183] for alfalfa and man from Bertrand.[8]

† Mean figure for mammals.

invertebrate; and for sodium, sulphur, and chlorine in man. In addition, greater than normal quantities of nitrogen are also found. Where excessive, such accumulations of elements may be referred to as *selective chemical enrichment*.

Even more outstanding examples than these may be found, particularly among marine organisms, many of which are capable of extracting large amounts of calcium carbonate and silicon from sea water. Several other marine plants and animals, including sponges, corals, and certain seaweeds, may absorb large quantities of iodine, while oysters reach copper concentrations of over 200 times that of adjacent sea water. Instances such as these are less common on land, although classic cases may also occur here, as among plants such as locoweed (*Astralagus racemosus*) in the western USA[101] which can support up to 1·5 per cent of selenium in its body weight, a very high proportion. Practical demonstrations of the ability of certain plants to accumulate unusually large quantities of specific elements in this way have led in some instances to their successful use in the detection of otherwise unsuspected ore bodies beneath ground (see Hawkes[64]).

3.3 The measurement of relative abundance

No amount of speculation as to the chemical composition of living organisms or ecosystems can exceed in value the accurate measurement of specific elements within them. Recently, many research projects have been oriented towards this problem, the successful conclusion of which depends largely on detailed chemical laboratory analyses of the air, water, soils, and living organisms within ecosystems. To some extent, this has been achieved in the past by means of experimental cultures under controlled situations, in which known quantities of selected elements were added periodically to plants and animals in order to assess their effects on growth. But this method is such that it has never proved to be entirely satisfactory; indeed, it has now given way generally to the spectrometric handling of samples, an extremely rapid procedure which enables one person to process and analyse a large number of samples in a short space of time. For soils and living organisms, the necessary preliminaries to this technique involve the pulverization of samples after drying and the subsequent removal of coarse grains (e.g., quartz) in order to obtain homogeneity and also to increase the exposed surface area of the finer, metal-rich material. Organic fragments are then removed from the residue either by ignition, or through wet oxidation techniques, using one of the strong chemical oxidizing agents, such as HNO_3, $HClO_4$, or H_2O_2. Slightly different measures are used in the preliminary preparation of water samples. Following this, the inorganic fractions which remain may then be studied spectro-metrically through their exposure to the intense heat of electrical discharges, when almost all the elements give off specific radiations within the ultraviolet range. The determination of any element is decided by its representation on a particular wavelength, and its quantity by the intensity of the line which is produced. When measurements are complete, it is customary to present their relative abundance in terms of their weight per unit area, usually as kilo-grams per hectare (kg/ha).

Further details of these and other methods of analysis may be obtained from suitable textbooks (e.g., Piper,[133] and Hawkes and Webb[65]) and from a large number of research articles which deal with individual elements.

3.4 Elements, cell growth, and nutrition

When assimilated by plants and animals, chemical elements usually combine to form complex mixtures of substances within living cells. These cells are extremely diverse in appearance, ranging in size from less than one micron to well over three centimetres in diameter (the yolk of some eggs), and in shape from the simple spheres and ellip-soids of cells in primitive organisms to threadlike feelers several feet in length in the more advanced forms of life. Despite these dis-similarities, all cells within major groups of organisms have the same basic characteristics and, usually, (as in the plant cells shown in Fig. 3.2) four distinctive parts may be delimited—the cell wall,* the cytoplasm, the nucleus, and the vacuole. The *cell wall* is a rela-tively rigid and stable structure, comprising about one-third of the total dry weight of the cell material; it is formed mainly of cellulose, with an arrangement of C, H, and O atoms in the weight ratio of $7 \cdot 2 : 1 : 8$. Lying within this is the *cytoplasm*, containing a large number of subcellular structures, the purpose of which is to act as storage compartments for proteins and, to a lesser extent, for nucleic acids, fats, carbohydrates, and mineral salts. Both this and the *nucleus*, which is separated from the cytoplasm by a membrane, are

* The cell wall is absent in animal cells, where only a thin membrane exists.

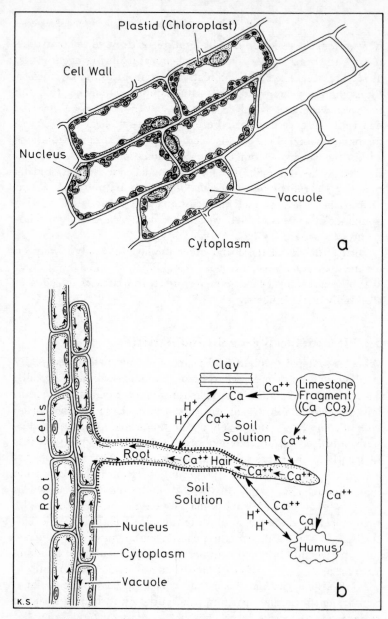

Fig. 3.2 The cell and root structure of plants: (a) the structure of living cells (from an original diagram in D. Marsland, *Principles of Modern Biology*, Holt, Rinehart & Winston, New York, 1965); and (b) the intake of nutrients by root hairs from the soil solution, and from exchangeable ions on a clay crystal and on humus (from an original diagram in R. L. Donahue,[43] *Soils, an introduction to soils and plant growth*, 2nd edn © 1965, using data of the Hawaiian Sugar Planters Association). In this instance, the nutrients are derived from limestone (Reproduced by permission)

vital to the life of any cell, for, while the cytoplasm may be thought of essentially as a food store, it is the nucleus which holds most of the living matter including the nucleic acids *DNA* and *RNA* (*deoxyribose nucleic acid* and *ribose nucleic acid*). The latter are essential to the two basic functions of the nucleus which are, first, to ensure the exact reproduction of the cell through replication of chromosomes and the genetic materials found in DNA structures, and, second, to control the activities of all other parts of the cell through the 'messenger' qualities of RNA which, in particular, help to determine the rates at which stored materials are used and replenished. Although the limits of nucleus and cytoplasm are usually clearly defined, the barriers between the two break down temporarily during the process of *mitosis* (cell-division, leading to cell multiplication and the growth of organisms), so allowing elements in both to mix freely. The fourth part of the cell consists of large internal vesicles termed *vacuoles*; these are filled with cell sap, an aqueous solution of mineral salts the presence of which plays a significant role in the preservation of the internal water balance.

Both the manufacture of cells and their component substances depend to a large extent on the adequate circulation of many chemical elements, the most abundant being oxygen, carbon, and hydrogen. These three are important not only in helping to form the basic cell structure, but also as major components in fats and carbohydrates; moreover, with nitrogen, they aid protein synthesis and, with nitrogen and phosphorus, they create many of the nucleic acids and much of the cytoplasmic material. In addition, the presence of several other elements may at times be critical for the healthy development of some cell substances. Sulphur, although needed in relatively small quantities, is required for the formation of amino-acids, without which proteins could not be synthesized. Calcium is important in the strengthening of cell walls and, were it not for magnesium, chlorophyll production would be seriously reduced. Phosphorus is essential to the transformation of energy through cells and also serves as a major structural component of the nucleic acids; and although their precise functions have not yet been fully determined, several other elements (especially K, Fe, Mn, Cu, Zn, Mo, B, Cl) appear to act as catalysts which speed up many of the complex chemical exchanges within cells.

Detailed investigations of the chemical arrangements and re-arrangements of relatively similar atoms and molecules inside cells

61

show that they may assume strikingly different forms. As in the production of cellulose, created from long, balanced chains of fibrous glucose residues, some may create substances which are immediately rather inert and stable, but many others need to react on more than one occasion so as to form ever more intricate molecular arrangements before some degree of stability is achieved. The latter case may be seen in the development of *proteins* from *peptides* and *amino-acids* held in the cytoplasm. In this instance, repeating groups of O, C, H, N, and often S, atoms combine organically to form the relatively simple amino-acids, of which 25 occur naturally, the least complicated being glycerine, or amino-acetic acid:

$$CH_2(NH_2) . COOH.$$

Since these 'acids' are both acidic ($-COOH$) and basic ($-NH_2$) in grouping, they may, and frequently do, join with each other in larger molecular arrangements termed *peptides*, in which hydrogen and oxygen atoms are removed to form water:

$$CH_2(NH_2) . COOH + CH_3CH(NH_2) . COOH$$
$$\text{(glycine)} \qquad\qquad \text{(alanine)}$$

$$\downarrow$$

$$CH_2(NH_2) . CO(NH) . CH(CH_3) . COOH + H_2O$$
$$\text{(Glycine-alanine)} \qquad\qquad\qquad \text{(water)}$$

In time, when the residues of several hundred amino-acid groups have been rearranged in specific sequences, the huge complex molecules of protein material may be derived directly from these. Chemical reactions such as this are extremely important to plant growth for, while proteins cannot pass through cell membranes, amino-acids can do so easily. Thus, the former need to be broken down by hydrolysis into their component amino-acids before they become capable of being moved from one part of a plant body to another.

Similar reactions encourage the formation of other organic materials in the cytoplasm. Thus, simple *fats* are usually compounds of an alcohol, often glycerol, commonly known as glycerine ($CH_2OH . CHOH . CH_2OH$), and several fatty acids, such as oleic acid ($C_{17}H_{33} . COOH$). In their simplest form, they contain only O, C, and H, but, since their more complex compounds also include nitrogen and phosphorus, they may be extremely important to plant growth. *Carbohydrate* compounds also result from combinations of O, C, and H, the H and O atoms being present usually in the same proportions as in water, while the number of carbon atoms varies. The least complicated carbohydrates are the sugars, the simplest of which is glyceraldehyde, related to glycerol with three carbon atoms:

$$CH_2OH . CHOH . CHO$$

Sugars may be joined together by virtue of their —OH groups, creating more complicated forms such as sucrose (cane sugar) and polysaccharides. The latter, in turn, have long chains of more stable sugar residues, from which both starch and cellulose are eventually obtained. Much more complex chemical transformations may occur in the nuclei of cells, in which combinations of O, C, H, N, and P with other elements may be found.

One of the most distinctive features of cell chemistry is the speed with which many of these reactions take place. Already, it has been confirmed by the use of radio-carbon tracing techniques that carbon taken in from the atmosphere may be built up by a living plant into very complicated substances, such as fats and sugars, within the space of 30 seconds,[52] while other relatively rapid reactions may develop only slowly outside the plant or—as in the case of glucose formation—not at all. Such impressive reactive speeds are triggered by organic catalysts known as *enzymes*, many of which require a specific chemical element in their structure: thus, molybdenum is involved in the functioning of the enzyme *nitrate reductose*, which reduces nitrate to ammonia. Crick[37] has noted that almost all biochemical reactions are catalysed by specific enzymes, and that all known enzymes are proteins.

In addition to these organic exchanges within cells, characterized by the presence of carbon, there may also be inorganic reactions between salts, acids, and bases which take place mainly in the cytoplasm; these are partially summarized in Fig. 3.3. Moreover, since animals differ fundamentally from green plants in their more varied nutritional requirements, it is to be expected that chemical reactions within their cells will also in some measure be dissimilar to those described above, despite the fact that identical elements are circulating in both plant and animal systems. This is emphasized further by the knowledge that animal cells in general have a more limited capacity to synthesize organic constituents than do those of plants; indeed, their metabolism often cannot be maintained unless they are provided with a wide selection of readily available organic foods,

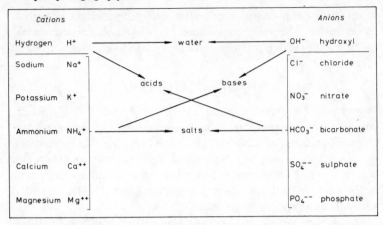

Fig. 3.3 The major inorganic ions in protoplasm, showing the origin of the four main types of inorganic compounds (from an original diagram in D. Marsland, *Principles of Modern Biology*, Holt, Rinehart & Winston, New York, 1965)

such as proteins (or amino-acids), vitamins, fats, and carbohydrates. These are usually obtained directly in their diet from chemical substances already synthesized in existing plant and animal materials.

Due to lack of space, this brief treatment obviously omits many details of the precise roles which particular elements play in the growth mechanisms and nutrition of cells; further details of these may be obtained in standard textbooks on cell biology and bio-chemistry.

CYCLES OF ELEMENT EXCHANGE

Natural cyclical movements of elements within present-day land ecosystems ensure that their uptake by plants and animals from pool areas in the soil, bedrock, and atmosphere does not take place irreversibly, but is largely balanced by the later degradation of living matter through the processes of decomposition and decay; these involve the breakdown of organic compounds into simple oxides (CO_2 and H_2O) and mineral salts, with a release of energy. Since plants usually create more organic material than they destroy, even though they are capable of oxidizing tissues when transpiring, this degradation is accomplished mainly through the actions of animals, and especially microorganisms, which in favourable soils may exist in densities of several million per gram. Once chemical breakdown has been achieved, many of the released elements will ultimately be recycled through the ecosystem. Thus, CO_2, H_2O, and NH_3 will yield C, H, O, and N, all of which may be taken up again by plants. On occasions when this does not occur, it is often due to the fact that further reactions following degradation have led to the production of compounds such as SiO_2, Fe_2O_3, and Al_2O_3 which may then interact to form secondary clay minerals, the basis of many new sediments.

Even though small quantities of chemical elements may be lost in this way, it is generally assumed that most biogeochemical cycles are essentially stable and in *steady state*. This is because such losses are usually counter-balanced by increments obtained indirectly from other sources, as in the circulation and ultimate deposition of dust and other particles [192] in and from the atmosphere; often these are released in rain water, as noted in Table 3.3. Of course, individual cycles of specific elements may vary considerably in their degree of completeness, particularly when examined on a seasonal or annual basis (note especially the seasonal variations in Fig. 3.4), or even over a number of years. Moreover, under ecologically changing conditions, such as the progressive infilling of marshes and bogs, the whole nature of the associated chemical cycles may be drastically modified in response to the altered environment. Like many other aspects of biogeography, the pathways of chemical cycling should not be considered as unalterable, but rather as subject to continuous evolution, though, essentially, always within a steady-state framework.

These well-balanced biogeochemical cycles on land are duplicated in the upper parts of most lakes and oceans, where the amount of incoming solar energy is still sufficient to stimulate the normal processes of photosynthesis and the creation of new organic material. However, in the deeper parts of oceans and lakes, and in the lowest soil layers, where due to poor light penetration little or no photosynthesis will occur, some modifications of cycling patterns may be seen. These are related particularly to the fact that although the breakdown of organic matter by microorganisms continues to take place at a normal rate, new organic substances (e.g., in microorganisms themselves, or deep sea fish) are formed only very slowly. In time, this will lead to an imbalanced situation in which many more chemical elements are created than can be used, and, in most cases, the surplus will either go indirectly to form new materials nearer to the surface, possibly carried by upwelling currents, or will be lost to ecosystems in sedimentation processes.

Table 3.3 Nutrients collected in rain water from selected stations (kg/ha/yr)
(taken from S. E. Allen, A. Carlisle, E. J. White, and C. C. Evans[a])

Station	Rainfall (mm)	Inorganic N	P	S	K	Ca	Mg	Year of publication	Source
Lerwick (Shetlands)	57·3	2·1	—	133·0	5·52	6·7	19·2	1959	(49)
North Lancashire, UK	161·69	6·28	0·43	35·34	2·96	7·3	4·63	1966	(23)
Kent, UK	84·0	—	<0·4	19·30	2·8	10·7	<4·2	1959	(98)
Denmark	60·7	6·9	—	16·1	3·1	6·5	3·0	1962	(75)
Anlarp, Sweden	64·1	6·14	—	7·45	1·37	7·31	1·65	1958–59	(49)
Sweden	57·3	—	0·1–0·2	2·8	1·2	4·6	—	1958	(170)
North Nigeria	106·7	54·7	2·58	60·30	36·76	1·01	2·91	1960	(77)
Yundum, Gambia	105·4	47·1	0·31	9·5	5·94	4·37	—	1965	(176)
Rotorua, New Zealand	160·0	—	0·33	31·9	6·09	3·34	—	1961	(109)
Katherine, North Territory, Australia	92·5	1·5	—	1·1	0·6	—	0·2	1963	(192)
Melbourne, Australia	98·22	—	<0·3	16·81	2·01	2·74	5·36	1966	(6)

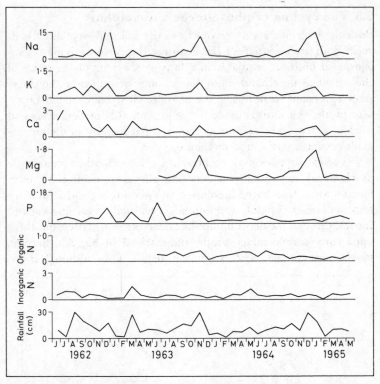

Fig. 3.4 The nutrient content of rain water (kg/ha/month) at Grizedale, England from June 1962 to May 1965 (from an original diagram in A. J. P. Gore[57])

Although it is now known that the annual rate of turnover of elements both on land and in water is extremely high, few details are available as to their precise routes through, or relative importance within, selected ecosystems except in a very small number of cases. Among the latter may be included Ovington's assessment of forests,[127] Tamm's[169] and Malmer's[99] studies of Scandinavian bogs, a scattered selection of reports from other areas (e.g., Park, Rawes, and Allen, 1962;[129] Newbould, 1960[121]), and the more comprehensive analyses of several Russian scientists to be examined later in this chapter. For the present, some of the major features of chemical cycling may be illustrated by an examination of the pathways of exchange of four major elements, two of which (potassium and sulphur) are released from the soil, and two (carbon and nitrogen) ultimately from the atmosphere.

67

3.5 The cycling of phosphorus and sulphur

Both these elements are derived from the soil and weathered bed-rock. Both are essential for the health and growth of living organisms, and both are available in relatively short supply in terms of their biological demand. Phosphorus, once described as the single most significant weak link in the chain of chemical elements necessary to life,[191] is indispensable in the formation of nucleic acids and nucleo-proteins, and sulphur is required particularly in the amino-acids cystine, cystein, and methionine.

Phosphorus supplies may be present in the soil either organically as chemical compounds, or inorganically as mineral salts. Both forms may be transferred in solution into plant roots, and from there into the bodies of plants and animals, eventually returning back to the soil through the decomposition of dead organic matter. But these apparently simple relationships, summarized in Fig. 3.5, become much more complex when studied in detail. Thus, although it has

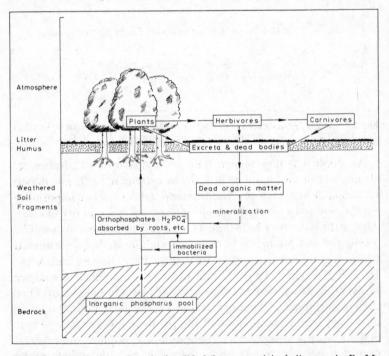

Fig. 3.5 The phosphorus cycle (modified from an original diagram in R. M. Jackson and F. Raw[74])

been known for many years that plants obtain most of their inorganic phosphorus requirements directly from orthophosphate ions $(H_2PO_4^-)$, it has more recently become clear that these are frequently present in quantities too small to meet demand, so that additional amounts of inorganic phosphorus need to be assimilated by other means, particularly through the activities of *mycorrhiza* existing on roots (as noted in Litav;[91] Rosedahl;[150] Melin and Nilsson;[107] Morrison;[116] Routien and Dawson [151]).Indeed, under certain circumstances, especially in the presence of excess $CaCO_3$, when phosphorus (and iron) tend to form highly insoluble salts, the amounts taken up by this method may be very considerable. The situation is further complicated by the fact that not all of the inorganic phosphorus is immediately available to plants, for some may be temporarily *immobilized* in the bodies of microrganisms which need it as food and so actively compete for it, often to the plant's detriment. However, the plant will still be able to absorb these supplies after the death of the competing organisms. In all cases, the inorganic intake is supplemented by large amounts of phosphorous released organically in the breakdown of excreta and dead material close to the earth's surface.

These general patterns of phosphorus movement become even more intricate when examined on a seasonal basis, or when interfered with by man's activities. It is now known that considerable seasonal imbalances of phosphorus may occur in many ecosystems, for the total quantities in circulation appear to be much greater in spring and early summer than at other times of the year.[93, 61,154] This may possibly result from its gradual storage over the winter months in the basal internodes of plant systems, from which it is released very quickly in relatively large amounts early in the growing season. The introduction of phosphatic fertilizers to the soil may also induce modifications, since the rates of chemical fixation may increase subsequently to such an extent that, ironically, the amounts of phosphorus available to plants are temporarily but significantly reduced. In time, however, these amounts will return to normal as the activities of fungi and bacteria help to convert sugars from root exudates into organic acids, which then aid the transformation of phosphorus back to available orthophosphates.[74] At the same time, other microorganisms have the ability to release inorganic phosphate from the organic phosphatic compounds, so ever more complex patterns of exchange are always being developed.

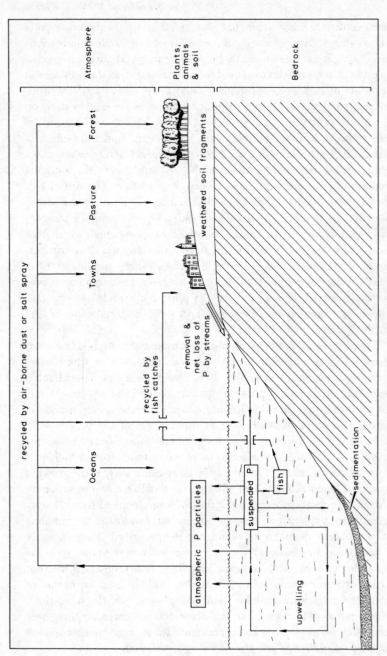

Fig. 3.6 Phosphorus exchange between land and ocean

Superimposed upon these general movements is a larger net loss of phosphorus from land to sea, some details of which are shown in Fig. 3.6. Each year, through erosion, large quantities of inorganic phosphorus are carried from continents by streams, eventually to become suspended in sea water. A small proportion of this is thrown back into the atmosphere from the sea surface, to be returned to continents by means of wind-transportation or in salt spray. More is concentrated in the bodies of fish and birds, especially in the vicinity of upwelling water, and this, too, may be given back to continents through the fish caught by man or through other activities, such as the dropping of guano by fish-eating birds (e.g., off the coast of Peru). But most of the suspended phosphorus will eventually be absorbed into oceanic sediments and so lost to terrestrial ecosystems. Odum[125,126] has concluded that this net loss of phosphorus from continents is rapidly becoming greater, being stimulated all the while by ever more intensive and efficient methods of mining reserves of phosphatic rock, and also by the very poor non-conservative and erosion-prone agricultural techniques employed in many areas. In illustrating this problem, Hutchinson[71] has suggested that the approximate average annual amount of phosphorus returned to continents in the form of fish catches and by other means is only in the order of three to six per cent of the average annual production of phosphatic rock, much of which is ultimately washed out to sea, so that, clearly, phosphorus deficiency must already be an extremely serious problem in many ecosystems.

Several similar features to these may also be seen in the *sulphur* cycle which is shown in Fig. 3.7. As in the case of phosphorus, most elemental soil sulphur is present in the form of inorganic compounds developed from weathered bedrock materials. Under anaerobic* soil conditions, these are converted to H_2S and sulphides, following which they are oxidized to sulphates (SO_4), and only then are they absorbed by plants in solution. The chemical reactions needed to create sulphides and sulphates from inorganic materials are stimulated by certain fungi, actinomycetes, bacteria, and other microorganisms, all of which tend to be widespread in anaerobic soils of low pH values. The most important of these is the autotrophic bacterium *Thiobacillus thiooxidans*, which obtains its energy from the chemical itself, and which has been shown by Waksman[186] and

---

* An *aerobic* soil condition is one in which free oxygen is present; in contrast, *anaerobic* soils, such as those which are waterlogged, have no free oxygen.

71

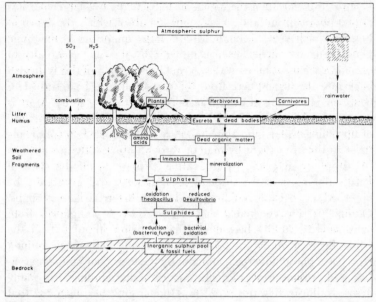

Fig. 3.7 The sulphur cycle (modified from an original diagram in R. M. Jackson and F. Raw[74])

Starkey[163] to be an efficient producer of sulphuric acid under natural conditions to a concentration of 10 per cent. Conversely, other bacteria, which are also particularly widely distributed in heavy, waterlogged soils, may restrict the amounts of available sulphate by reducing it to release H_2S and by using some of the oxygen to oxidize carbon materials needed for growth. The best known of these reducing bacteria are *Desulfovibrio desulfuricans* and *D. aesturii* which are mostly commonly found in river muds.

Like phosphorus exchanges, those of sulphur are frequently not entirely complete, since despite the presence of bacteria, much of the H_2S may never be oxidized, even under anaerobic conditions. In this case, the H_2S goes into a reservoir pool within the mud or soil, and remains unavailable for cycling for many years.

3.6 The cycling of carbon

The biogeochemical cycling of carbon, the most important single element of organic chemistry, represents a different type of exchange from those already described, since its main reservoir pool lies in the atmosphere and not the soil. Carbon is taken up fairly directly by plants, and converted into sugars and other substances by means of photosynthesis, after which some will be returned to the atmosphere through respiration, some absorbed into plant structures, and approximately one-half eventually added to the soil in the form of decomposing organic matter.[187] The total quantities of assimilated carbon are very large; indeed, one estimate suggests that an amount equivalent to one-thirty-fifth of the total carbon dioxide content of the atmosphere is taken up annually, often at a very rapid rate.[20]

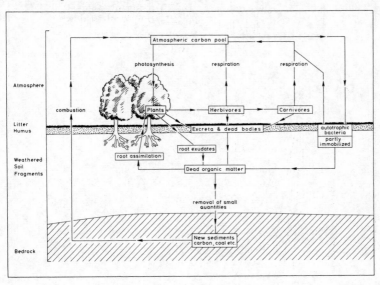

Fig. 3.8 The carbon cycle (modified from an original diagram in R. M. Jackson and F. Raw[74])

For the biogeographer interested in biogeochemical cycling, the relatively large movement of carbon between plants and a subsidiary organic pool in the soil needs further investigation. The pathways through which this is achieved may be several (see Fig. 3.8). Small amounts of carbon may be secreted from plant roots directly into the soil in the form of sugars, amino-acids, and other compounds. Much more is derived from the decay of plant roots, which in the case of tropical rain forest may result in 548 kg/ha of carbon being deposited annually.[74] It is usual, however, for most soil carbon to come from the decomposing surface litter of plants (911 kg/ha/yr in tropical rain forests), the rate and speed of breakdown of which may vary

73

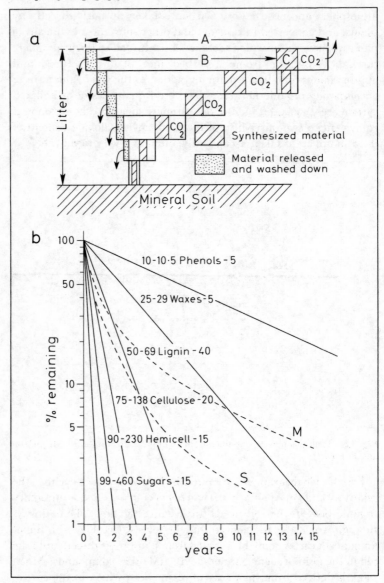

Fig. 3.9 Rates of decomposition of plant materials: (a) the progressive decomposition of fallen plant debris (from an original diagram in A. Burges[20]); (b) the decomposition curves of organic constituents in forest litter, if the decomposition is presented by a logarithmic function, i.e., the straight lines from the point 100 per cent (from an original diagram in G. Mindermann[111]). In (a), *A* represents the

substantially according to the nature of the organism: thus, as a general rule, trees will decay much less quickly than individuals in annual or perennial communities. It is, of course, to be expected that the cell constituents which most rapidly feed the pool of organic material in the soil are those which decompose most quickly, such as carbohydrates, fats, and proteins. The structural material of plant cells (cellulose) lasts much longer, but eventually succumbs to either fungi, bacteria, or soil animals which after digestion excrete a residue which serves as food for other soil organisms. If fungi attack the cellulose first, they may, in turn, be eaten by mycophagous animals, such as mites, nematodes, and springtails, which eventually form additional food for bacteria of decay, or more fungi. Similarly, should bacteria be the primary consumers of cellulose, these could be eaten by protozoa which, in turn, may be consumed by predators or eventually by bacteria of decay, and fungi. While these various subsidiary food cycles are under way in the general carbon cycle, all of the component organisms are respiring, thus releasing additional minor quantities of carbon dioxide to the soil and atmosphere. While the actual speed of decay appears to depend very largely upon the extent to which the fallen litter or dead roots are turned over initially by earth worms and other soil organisms, some of the material remains very resistant to decomposition and relatively unpalatable, and forms the humus fraction of the soil. This does, however, continue to decompose extremely slowly. Some of these reactions are presented in diagrammatic form in Fig. 3.9.

There are two important additional mechanisms which stimulate the exchange of carbon between the atmosphere, plants, and soil. First, certain autotrophic bacteria, which obtain energy from light or from oxidizing inorganic substances, have the ability to use atmospheric carbon from CO_2 directly in order to synthesize selected

initial amount of debris, and B the residual material after the first wave of decomposers has attacked it, converting or synthesizing some into CO_2 and body material. Some mineral material is released, and made available for plant roots. C refers to the dead bodies of decomposers, which, in turn, attract further decomposers. Subsequent layers in the diagram represent later periods of decomposition. In (b), the first number in front of the name of the constituent indicates the percentage loss after 1 year. The second number represents an exponential function of the decomposition rate of the type e^{-x}, in this case varying from 10·5 to 460. The number after the constituent refers to the percentage in weight of the original litter, in rough averages. The line S shows the summation curve obtained by annual summation of the residual values of the separate components. The line M gives an approximation of the probable course of decomposition of raw humus.

75

organic compounds: this tends to immobilize some of the carbon, which will therefore not reenter the general circulation until the organisms die. Some, however, is returned to the atmosphere through respiration of the same organisms. Second, the presence of predatory animals may expand the carbon cycle considerably, for this results in the passing of carbon materials from organism to organism down the food chain, some being returned to the atmosphere directly and some being deposited in the soil through excreta and the decomposition of dead bodies at every trophic level.

Not all of the pool of soil carbon will be recycled through plants, since small quantities can be removed at any time to assist in the formation of new sediments and rock groups. This loss has, however, been more than compensated recently by the increased rates of release of carbon to the atmosphere as more and more fossil fuels are utilized for home heating and industrial purposes. As a result, an imbalanced situation in the carbon cycle is being created, the ultimate consequences of which may be considerable (see section 3.12).

3.7 The cycling of nitrogen

Of all the cycles of chemical exchange in ecosystems, that of nitrogen is possibly the most complete, although not necessarily any less intricate than those of other elements. Since nitrogen comprises over 78 per cent of the gaseous weight of the atmosphere, it is perhaps to be expected that its main reservoir pool is to be found here, though, as in the case of carbon, much of its circulation actually takes place within the soil where the ratio of carbon to nitrogen is in the order of $10:1$. The important difference between the two lies in the fact that atmospheric nitrogen is unavailable to plants directly, so that extremely complicated mechanisms are required to convert it to a form which can be taken up by them. This involves the *mineralization* both of atmospheric nitrogen and of that contained in dead organic material, through the activities of bacteria and microorganisms. Following this, further bacterial and organic activity results in the production of nitrites (which may be toxic to plants when taken in excess) and eventually nitrates, which may be safely consumed. In time, additional bacterial activity may cause some of the nitrate to be denitrified, and so released to the atmosphere, from which it may again pass through the cycle. Only very minute traces of nitrogen are to be found in bedrock material. It should be emphasized that these relationships, which are summarized in Fig. 3.10, are still

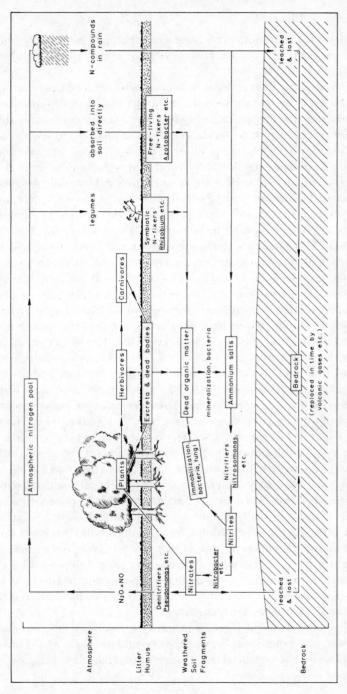

Fig. 3.10 The nitrogen cycle (modified from an original diagram in R. M. Jackson and F. Raw[74])

a subject of much controversy among ecologists and biochemists, and many of the precise mechanisms of exchange are still not fully understood.

It is perhaps most rational to begin a more detailed examination of this cycle by looking first at the transfer of nitrogen from atmosphere to soil. This has been of especial interest to scientists since the discovery of the basic mechanisms of transfer in the eighteen-eighties, not only because of its economic significance, but also due to the fact that, were it not for continued replenishment from the atmosphere, available supplies of nitrogen would rapidly become exhausted. Nitrogen may be added to the soil in two ways: first, by falling rain water and, second, through the process of *nitrogen fixation*. While the former involves the addition of very small totals (0·38 to 3·8 kg/ha/yr) as ammonium and nitrate ions, the second is of much greater importance to the healthy growth of organisms.

Nitrogen fixation is usually achieved through the activities of certain soil microorganisms, two main types of which may be distinguished: those which live *symbiotically* in association with higher plants, and those which have *non-symbiotic* relationships and so are not dependent directly on other forms of life. The former situation is perhaps best exemplified in the case of the bacterium *Rhizobium*, which becomes effective only when living in the root nodules of Leguminosae. Two main species of this have now been identified, *R. radicicola* and *R. leguminosarum*. There are, however, many strains, some of which may be located only on specific legumes: those which are found on lucerne may not thrive on clovers and may be ineffective on other legumes. Indeed, relatively few of the strains are generally capable of fixing nitrogen, except when they find a suitable host plant. Then, they often multiply very rapidly, after which freely-moving cells, attracted by root exudates, begin to penetrate root hairs so that eventually nodules are formed and stabilized. As nitrogen is fixed within the nodules, haemoglobin is also produced (an apparently unique occurrence in plants) thus giving the nodules a bright red to pink colour when they are cut open. It is interesting to note that, in contrast to the cycling of phosphorus, relatively little nitrogen is stored in the nodules,[93] and presumably most passes into the main plant body fairly quickly.

The importance of legumes in nitrogen fixation is by now well known, and experiments in widely varying parts of the world have emphasized the beneficial results which can follow their planting in an agricultural system. On clover pastures in New Zealand, where environmental conditions are close to the optimum for fixation, as much as 91 kg/ha of nitrogen may be fixed per annum. But while this is generally true, on occasions the use of legumes may produce unexpected consequences, especially when they stimulate the activities of microorganisms so quickly that much of the nitrogen is temporarily immobilized in their bodies, and plant growth is subsequently checked, a sequence of events which Witkamp and his collaborators have noted in kudzu (*Pueraria lobata*) pastures of the eastern USA.[199] However, this is probably a relatively unusual occurrence, and certainly one that has not been widely reported.

Through isotopic tracing and lysimeter experiments, it is now also known that several other higher plants have symbiotic relationships with so far unidentified bacteria which are capable of fixing nitrogen, though to a more limited degree than for legumes. Such is the case for alders (*Alnus* spp.),[13] sea buckthorn (*Hippophäe rhamnoides*)[14,112] and bog myrtle (*Myrica gale*)[74] in northern temperature latitudes, and for *Casuarina* spp. in the tropics. Many more examples may also exist elsewhere.

If environmental conditions are suitable, non-symbiotic fixation of nitrogen may be almost equally important in the general nitrogen cycle. Thus, experiments on tropical grasslands in Nigeria have indicated that between 18·2 to 23·7 kg/ha/yr of nitrogen may be fixed non-symbiotically, as compared to a symbiotic fixation of between 18·2 to 36·4 kg/ha/yr.[124] Several groups of organisms are capable of non-symbiotic fixation, the most widely distributed being *Azotobacter* bacteria, three species of which are known. These are most commonly found in well-aerated soils which have a pH* in excess of a value of 6, although some forms exist where soil pHs are just below 5. *Azotobacter* are particularly useful in desert or dune soils, where they provide a nitrogen supply which would otherwise be available in only very small quantities. In tropical areas, the more acid-tolerant genus *Beijerinckia* often replaces *Azotobacter*, and in the wet acid soils of temperate climates, the anaerobic bacterial genus *Clostridium* may be an equally effective non-symbiotic nitrogen fixer. Indeed, similar roles may be played by several other relatively common free-living organisms which use atmospheric nitrogen, notably *Aerobacter*

---

* The term *pH* refers to the relative concentration of hydrogen ions in a solution. The lower the value, the more acid a solution, and the more hydrogen ions it contains. *pH* 7 is neutral, less than this is acid; more, alkaline.

spp. and *Pseudomonas* spp. Nitrogen may also be fixed on surface areas or in water through the activities of free-living blue-green algae. Fogg and Wolfe[53] have listed 15 species of these organisms, which, unlike the other nitrogen fixers, can also photosynthesize. In the tropics where paddy rice is grown, such algae may multiply so quickly as to provide almost all the nitrogen requirements for the growing crop.

The nitrogenous compounds which result from fixation are added in time to the main pool of soil nitrogen formed from decaying organic material close to the soil surface. This needs further alteration through mineralization and nitrification before it becomes available to plants. Mineralization takes place when a wide variety of soil organisms (including *Bacillus* spp., *Proteus vulgaris*, and *Pseudomonas* spp., with other bacteria, fungi, and actinomycetes) help to change the nitrogenous compounds into organic ammonium salts by means of a process termed *proteolysis*, which first breaks down plant proteins into their amino-acids, from which mineral elements may ultimately be released. Occasionally, ammonia gases can be given off to the atmosphere at this stage, and some ammonia may even be absorbed by plants such as rice.[52] The timing of this breakdown varies considerably according to the nature of the organic compounds (see Fig. 3.9). It is probable, however, that little or no loss occurs during the early stages of decomposition, when the carbon to nitrogen ratio may be extremely high, ranging from $200:1$ in woody tissues to $30:1$ in leguminous and similar plants with a high nitrogen level,[20] and when the organisms attacking carbon need much more nitrogen than may be obtained from the debris, so that *all* the available nitrogen tends to be assimilated, and so immobilized. Eventually, with the continuous removal of carbon to the atmosphere in the form of CO_2, the carbon to nitrogen ratio is reduced until it reaches a factor of about $12:1$. From this point, the mineralization of nitrogen proceeds rapidly.

Once ammonium salts have been produced, nitrification processes stimulated by other soil microorganisms ensure that they are again modified fairly quickly to form nitrites and nitrates. Although a few heterotrophic organisms, such as bacteria, actinomycetes, and fungi, may be involved, the characteristic nitrifying organisms are usually true autotrophs, obtaining their energy from the oxidation of ammonia to nitrite and nitrate, while at the same time reducing atmospheric carbon dioxide to gain the organic compounds neces-

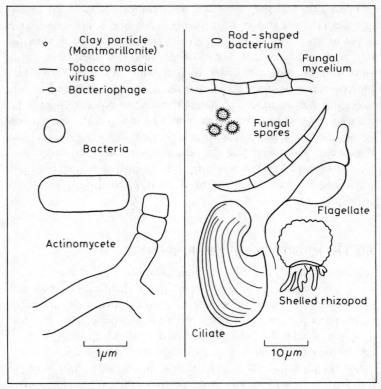

Fig. 3.11 The relative size of selected soil organisms (from an original diagram in R. M. Jackson and F. Raw[74])

sary for growth. It follows that most of the nitrifying organisms, among which are *Nitrosomas* spp. and *Nitrobacter* spp., live under aerobic conditions in which free oxygen is readily available, and Meiklejohn[103] has suggested that they also require certain amounts of calcium and copper in order to function satisfactorily. All are capable of oxidizing large amounts of ammonium or nitrite while making little growth. As noted previously, nitrite in more than very small quantities is toxic to plants, and occasionally harmful concentrations may occur either on a permanent basis in low-temperature alkaline soils in which large quantities of ammonium salts are present, or seasonally, as when soil aeration is markedly reduced in winter months. Usually, however, soil nitrite is oxidized fairly quickly so that this problem does not arise, and the resulting nitrate is then absorbed into the plant system.

These very complex chemical mechanisms are essential in the transformation of atmospheric nitrogen to a form which is available to plants and animals. But not all the nitrate so produced does, in fact, reach the plants. In heavy rains, much is washed downwards away from the root zone by leaching, to which it is very susceptible. Moreover, in soils which lack oxygen, especially when large amounts of organic matter have accumulated, nitrate may also quickly be reduced to gaseous form, or to oxides of nitrogen, in the process termed *denitrification*. Several microorganisms, including the bacteria *Pseudomonas denitrificans* and *Micrococcus denitrificans*, and fungi may stimulate this process by using nitrate as a source of oxygen, but their activities will cease once improved aeration conditions are again achieved.

3.8 The importance of soil organisms

It is clear from the preceding sections, and also from chapter 2, that many mechanisms essential to the cycling of chemical elements and the transfer of energy in ecosystems could not operate efficiently without the presence of a wide variety of soil organisms. The importance of these to the general health and development of ecosystems should therefore never be underestimated.

Soil organisms are extremely diverse in size, form, and function (see Fig. 3.11), and except in sterile soils may be present in very large numbers. Their major groupings may be summarized as follows:

1. *Macroorganisms.* (a) Mammals, such as mice, shrews, moles, rabbits, badgers, etc., all of which contribute to the large-scale turnover and aeration of the soil.

(b) Oligochaets and enchytraeids, of which earthworms form the most significant single group. Many species of earthworms are known (there are 25 in the UK, approximately 10 of which occur on cultivated land and the rest in specialized habitats), and usually more are present on permanent grassland than elsewhere. Very acid soils contain few earthworms, and only small numbers are found on arable land. As noted first by both Gilbert White,[194] and later Charles Darwin, earthworms are especially important in turning over soil particles and, at the same time, taking in fresh organic material and releasing nutrients in a form readily available to plant intake; these activities are enhanced on land where nitrogen and nutrient contents are reasonably high.[200,12] Enchytraeids are smaller organisms, often found in great numbers in moist acid soils where earthworms are rare, as on the heaths of Denmark and the UK, where densities of 200 000/m² have been recorded.[74] They also appear to encourage the decomposition of organic matter, and may play a significant role in the formation of humus.

(c) Nematodes are usually free-living, and often present in numbers of 10 to 20 million/m². The free-living forms may have a similar role to earthworms, but saprophytic and parasitic* forms also exist.

(d) Arthropods and molluscs. Arthropods are represented by many insects, spiders, centipedes and millipedes, the larvae of beetles, flies, ants and termites, all of which help to aerate the soil through burrowing, feed on decaying vegetation, and (in some cases) play important predator roles in soil food chains. Slugs and snails, the most numerous of soil molluscs, act as scavengers and are better able to digest the tougher parts of plant material (e.g., cellulose) than most other soil organisms.

2. *Microorganisms* (a) Protozoa usually prey on bacteria and some fungi, but may also be saprophytic. Three classes are present in soil: rhizopods or amoeba, and the free-swimming flagellates and ciliates, both of which move about in soil solutions. The relative functional importance of each group has not yet been satisfactorily determined.

(b) Bacteria, which are the most numerous organisms in soil, may occur in densities of up to 10^9 per g and can be autotrophic, heterotrophic, or symbiotic. They perform a wide variety of functions.

(c) Fungi, of similar significance to bacteria in their stimulation of organic cycling mechanisms within the soil, are if anything more important in acid soils than elsewhere. Many types may be present: these may be saprophytic, parasitic, or symbiotic.

(d) Actinomycetes. Closely related to both fungi and bacteria, these help in the decomposition of organic material, and may also produce antibiotics. They do not tolerate acid conditions.

(e) Algae may be present both on soil and in water, and contribute generally to the formation and processing of soil materials. They may be autotrophic or heterotrophic.

* A *saprophyte* is an organism whose food is derived directly from non-organic matter, whereas a *parasite* lives off other living organisms.

(f) Viruses, the smallest and simplest forms of life in the soil, have especial significance in that they can only multiply in the living cells of other organisms. Although they are therefore parasitic on a large number of life forms, their precise effects on soil populations are still not known.

While most of these are active throughout the weathered horizons of many soils (see section 4.9), their greatest concentrations almost always occur closest to plant roots, in an area which has been termed the *rhizosphere*. Within this zone, both saprophytes and parasites may become particularly well established, so that there is an intense competition for available nutrients not only among themselves and other free-living forms, but also, at times, with the parent plant; in particular, carbon-reducing organisms may actively compete with plants for available nitrogen, and phosphorus supplies may also be severely affected. In contrast, patterns of reciprocity between plants and soil fauna are known to exist to the mutual benefit of both; this is the case not only for symbiotically-living forms, but also where healthy plants emit many different types of root exudates, which then stimulate the activity of other soil organisms.[85] Similar relationships may also exist between macrofauna and microfauna within the soil, as suggested by van der Drift and Witkamp.[181] The situation is made more complex by the presence of soil enzymes, which are absorbed into clay particles, and may considerably modify specific relationships and chemical exchanges.[96, 82, 160] These and other features of the rhizosphere have recently been summarized in detail by Rovira and McDougall.[152]

Although it is often difficult to determine the exact relationships between the vast number of organisms which may be present in a soil or, indeed, their precise functions in the patterns of cyclical chemical exchange, one major attempt at clarification, portrayed in Fig. 3.12, has recently been presented by Birch and Clarke,[10] who have examined the food-web roles of soil organisms in a forest ecosystem. In this instance, three major groups of primary decomposers (bacteria, fungi, and animals) were considered to be arranged around a central block representing forest litter, and all were preyed upon, in turn, by several animal groups, and to a lesser extent by fungi. Usually, two tiers of predators were delimited, but relationships between them proved to be exceedingly complex, and often the detailed interactions between specific organisms still remained obscure.

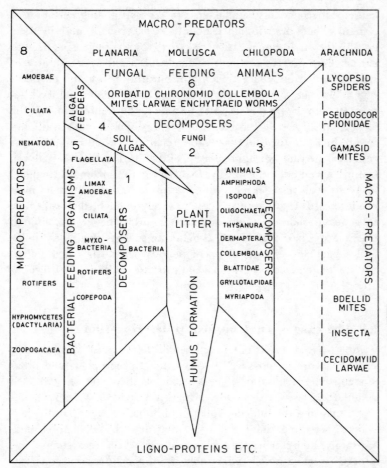

Fig. 3.12 The ecological function of soil organisms in forest soils, based on their feeding habits. For explanation, see text (modified from an original diagram in L. C. Birch and D. P. Clarke[10])

MAN-INDUCED MODIFICATIONS OF ELEMENT EXCHANGE PATTERNS

While most biogeochemical cycles are essentially stable under natural conditions, the pathways of energy flow among organisms, and so the potentialities for chemical cycling, will always tend to become more diverse (see also section 5.11). Usually, this is accomplished within a balanced framework, but, from time to time, and particularly within recent years, relatively rapid changes have caused biogeochemical imbalances to develop among certain ecosystems. These are stimulated mainly by excessive human interference in the environment, triggered partly by the increasing need to extract ever greater amounts of foodstuffs from agricultural ecosystems which have largely constant natural energy and chemical potentials, and partly through an impressive contemporary augmentation in the rates of industrialization and urbanization, and the associated large-scale problems of waste disposal. Both traits are by-products of the industrial and scientific revolutions and the rapidly increasing world population of the nineteenth and twentieth centuries, and both tend always to upset the well-balanced chemical and physical relationships which until recently existed in most ecosystems. If these disturbances are severe enough, the result may be general *environmental deterioration*, or even specific *pollution* through the excessive output of chemical materials such as pesticides, radioactive particles, and gasolene compounds, which the environment can no longer absorb in the normal processes of decay and recirculation.

3.9 The case of environmental deterioration

Symptoms and results of environmental deterioration are phenomena which are by now well known among geographers and other students in the natural sciences. Some of their effects have been summarized recently in a small paper by Wilkinson[196] who suggests in passing that the full consequences of human activities in the biosphere are so great that possibly they may never be fully appreciated. Certainly, many living organisms, and even entire ecosystems, have proved to be extremely vulnerable to human interference, as witnessed by the very restricted present-day distribution of natural vegetation, by the apparent dominance of domesticated plants and animals over many areas of the earth's surface, by the long list of wild mammals and birds which have become extinct in historic time, and so on.

The case for environmental deterioration is perhaps best exemplified in the wholesale destruction of forest over many parts of the world so as to prepare land for agricultural or pastoral activities. Man, the destroying biotype,[15] has used fire and axe over many centuries to achieve this aim (see also sections 7.5 and 7.6), often with the result that the circulation of nutrients is reduced or selectively impaired. In addition, as the vegetation cover is removed, rates of runoff, erosion, and elemental loss in stream flow may increase temporarily to devastating proportions; indeed, there is now ample evidence (see Douglas,[44, 45]) that these rates have also generally increased in historic time, presumably as a result of man's activities.

Such trends have been the subject of detailed study in only a few ecosystems, one of the most interesting cases of which may be found in northern England. Here, many areas above 1000 feet in elevation, which in Mesolithic times were covered by mixed oak forest and woodland,[42] are now, largely through subsequent clearance and grazing, bleak and desolate acidic and peaty moorlands whose vegetation is dominated by *Calluna vulgaris* (heather), *Erica* spp. (bell heather), and a variety of tough grasses and mosses. For the last 100 years, these moorlands have been used largely as sheepwalks or as game-bird reserves (particularly for red grouse), and since both sheep and birds find young shoots of *Calluna* extremely palatable, the moors have been burnt regularly during this period at intervals of from 8 to 15 years in order to clear away old and dead material and to stimulate young growth. The net effect of this has been to encourage *Calluna* growth generally at the expense of most other species, and to reduce further the chances of tree recolonization; consequently, trees are rarely seen at present.

Recent studies have now emphasized that the original clearing of timber from these areas has resulted in considerable environmental deterioration. Although we do not know what the precise rates of nutrient exchange were in Mesolithic forests, it has been established that the burning of *any* forest area immediately results in a large loss, both of volatile elements carried through smoke into the atmosphere, and also of ash elements removed by subsequent rainwater erosion from uncovered soil.[127] Moreover, Rennie [141] has shown that, since much greater quantities of most nutrient elements circulate in forest ecosystems than elsewhere (this being especially true for calcium), any forest clearance automatically reduces the totals of circulating elements. Indeed, in time, through the increased rates of leaching which follow the destruction of forest canopy, amounts of available elements may be so greatly reduced that soils may subsequently be unable to support trees beyond a certain stage of growth.

From this it may be inferred that at least in some areas, once environmental deterioration has begun, it tends to continue almost indefinitely unless strong measures are taken to prevent it. On moorlands today, a net loss of nutrients may still be observed, though the causes of this are a subject of some controversy. Allen[3] has suggested that the custom of moorland burning results in a continued removal of some carbon, nitrogen, sulphur, and other elements (though not phosphorus), of which that of nitrogen appears to be the most serious, since up to 45 kg/ha may be lost at one time. However, he also cautions that compensatory mechanisms may react against these effects, a point which has also been put forward by Fowells and Stephenson[54] and Tamm,[169] who consider that the increased activities of microorganisms which follow burning could mobilize additional nitrogen from the soil. In addition, it is known that young *Calluna* plants are very efficient in absorbing soil nitrogen[94,188,88] probably due to the presence of root mycorrhizas which are even capable of extracting this element from very deficient peats.[97] Since *Erica tetralix* and *E. cinera* also develop similar mycorrhizas, considerable quantities of nitrogen can be brought into moorland ecosystems by these methods alone. Additional compensatory mechanisms also reduce the apparent reduction in availability of other elements. Indeed, the consequences of present-day burning may be to produce only a temporary removal of volatile elements and a temporary increase in nutrient ions obtained from heather ash, following which a circulatory equilibrium may quickly be regained.

Heather moorlands are, of course, affected by human-oriented activities other than burning, the most important of which is the grazing of sheep. The results of this appear to differ widely from catchment to catchment, particularly with respect to its effects on the composition of vegetation cover, on erosion (through differential rates of treading), and on the distribution of nutrients. Of course, in all areas, sales of sheep and wool in themselves institute a net outward movement of elements, but this is usually insignificant. Much greater losses tend to follow from direct physical processes, such as leaching, the downstream flow of elements in solution, and the removal of peat by erosion—an accelerating feature at present.[130] Only when a detailed balance sheet is prepared, such as that in Table 3.4, will the relative importance of each of these factors become clear. Thus, in the Rough Sike catchment in the northern Pennines (an area which is not now directly affected by burning), Crisp[38] has suggested that the total net reduction in sodium, potassium, and calcium is not important since all these may be replaced partly by rainfall and, in greater quantities, from the underlying

Table 3.4   *Outline of balance sheet for Rough Sike catchment (83 ha),*
in terms of water and five elements of study year (after D. T. Crisp[38])

	Water '000 m³	Sodium kg/yr	K kg/yr	Calcium kg/yr	Phos. kg/yr	N kg/yr
1. Stream water output	1368	3755	744	4461	33	244
2. Evaporation	403	—	—	—	—	—
3. Peat erosion	—	23	171	401	37	1214
4. Drift of fauna in stream	—	Trace	0·01	Trace	0·01	0·2
5. Drift of fauna on stream	—	0·11	0·38	0·07	0·43	4·6
6. Sale of sheep and wool	—	0·16	0·44	1·58	0·98	4·4
7. Total output	1771	3778	916	4864	71	1467
8. Input in precipitation	1771	2120	255	745	38·57	681
Difference	—	1658	661	4119	14·33	786
Net loss/ha	—	20·01	7·97	49·68	0·17–0·40	9·48

rocks. The case of nitrogen is different in that the net estimated loss here is more than double the estimated input. Even though nitrogen is a notoriously difficult element for which to reconstruct balance sheets, since several of its compounds are gases at field temperatures, the evidence suggests that the fund of nitrogen in this catchment is being depleted rapidly, mainly through peat erosion which accounts for about 80 per cent of the total nitrogen output. The removal of phosphorus due to peat erosion is also substantial. It should be pointed out that, although some of these losses may not appear to be particularly significant in terms of the total amounts of nutrient elements held in the plants, litter layers, and soil, which in heather communities in Scotland[145] have been estimated to reach 200 to 600 kg/ha in the case of phosphorus, and about 2000 to 5000 kg/ha in the case of nitrogen, in time they may all reach considerable and effective proportions, thus contributing still further to the processes of deterioration which were initiated in Mesolithic times.

Despite the fact that environmental deterioration is an active process in many parts of the world, it must not be assumed that the activities of man have always operated to the detriment of natural ecosystems. One instance where the reverse is true may be found in the case of some tropical grasslands, where *environmental improvement* has occurred, particularly with respect to the nitrogen cycle. Although it had been known for many years that these areas were generally deficient in nitrifying bacteria under natural conditions,[124] the reasons for this remained obscure until Theron[174] put forward the notion that grass roots in the tropics secreted some substance which was toxic to such organisms. This theory was later supported in experimental work by Rice[143] and Munro[120] in Central Africa, who discovered that two very common species of *Hyparrhenia, H. filipendula* and *H. dissoluta*, as well as five other common African grass species, produced a water-soluble, heat-labile substance in their roots, which was toxic to both ammonium and nitrite oxidizers. Similar effects have been noted in Ghana and East Africa,[104,105] and on New Zealand grasslands.[147] At times, the effects of toxicity appear to vary seasonally, for Birch[9] has observed that there is often a flush in nitrification at the start of the rainy season.

Even though a few dissenting voices have been raised (e.g., Whyte[195]), there is now general agreement that the recent introduction of pasture management techniques in many tropical areas has resulted in increased rates of nitrogen production. Probably the most detailed evidence for this comes from Meiklejohn[106], who suggests that in Rhodesia, while the numbers of nitrifiers in a developing managed pasture are small at first, they build up very quickly over a period of two years. Under certain conditions, very rich organic soils, as in cattle pens or paddocks, could hold over 1000 times as many nitrifiers as soils under wild vegetation, the major reason for this being that nitrogenous organic matter provides plenty of ammonia, the essential substrate of nitrifying bacteria. In addition, organic wastes may also provide growth factors, such as biotin, which Krulwich and Funk[86] have shown to increase the growth of *Nitrobacter agilis* and possibly other nitrite-oxidizers, too. Of course, the evidence that pasture improvement techniques involving natural wastes will increase the number of nitrifiers is not restricted to the tropics, for Niklewski observed in the Soviet Union as early as 1910[123] that manured areas had many more nitrifiers than elsewhere, and Ulianova[178] has recently attempted to quantify this supposition by indicating that cattle manure may hold 100 million ammonia oxidizers per gram, as compared to 10 000 per gram in chernozems and only 100 per gram in sandy soils.

3.10 Pesticides: a worldwide problem

In many of the developed nations of the world, highly toxic chemicals have been used to control plant and animal pests for over half a century. As early as 1870, paris green was employed to combat the devastations of certain beetles on European potato crops. After 1900, arsenicals were widely dispensed, to such an extent that a few areas became sterile for crop growth after their use, while some organisms eventually developed strains which were resistant to them.[172] But it was only after the general release of DDT in 1945 that synthetic chemicals came to be used in agriculture on a worldwide scale, so that now no part of the continents, oceans, or atmosphere is entirely free of their effects. At the present time, over 54 000 pest control products, representing 500 synthetic chemical compounds, are registered with the US Department of Agriculture alone, and more than 1000 new products are being introduced to the environment every year. Many of these are partially toxic, and most are not specific, so that side-effects on a wide range of organisms are to be expected. Eventually, major modifications in the pathways of energy flow and chemical exchange within ecosystems may result from their continued use.

Today, synthetic chemical pesticides usually fall into one of three distinctive groups. The use of *organochlorine* products, or *chlorinated hydrocarbons*, dates from the Second World War when DDT (the first of these) began to be widely employed, particularly against mosquitoes and other insects in California. Once insect strains resistant to DDT appeared, stronger pesticides in this category were soon put on the market; at the present time, chlorinated hydrocarbons, such as aldrin and dieldrin, are several times more toxic to birds and mammals than the original compounds. All organochlorines are chemically stable, and so remain in the environment for relatively long periods of time.

Organophosphates, the second major group of pesticides, were introduced to world ecosystems later, as a response to the gradually increased resistance of organisms to organochlorines. In California, they were again first utilized between 1952 and 1955 for mosquito control; after 1964 they could be found almost to the complete exclusion of every other pesticide group. Although their chemical components break down and lose their potency more quickly than in the case of organochlorines, organophosphates are usually much more toxic to organisms while they are active. Taken as a whole, demetonmethyl and the related compounds demeton-S-methyl and oxydemeton-methyl are the most widely distributed of organophosphates, but it is interesting to note that despite their toxicity, some insect groups already appear to be immune to their effects.[18]

A third group of pesticides, known as *fungicides* and *fumigants*, are found in many parts of the world. They are composed mainly of ethyl and mercuric salts, and are usually less toxic than the organo-phosphates.

Although the actual quantities of pesticide employed in agriculture vary a great deal from area to area, it is probably true that either directly or indirectly organisms in every part of the world now have some accumulation of pesticide residues in their bodies. Traces of these have now been reported from widely different regions, even where they have not been directly applied, as in Antarctica,[55] although, in this case, they are very discriminately scattered, none being found in freshwater samples or in invertebrates, and most being located in predator birds, especially skuas, which travel widely. Elsewhere, residue totals are inevitably greater, though there are wide quantitative and qualitative differences depending on local circumstances. In some countries, such as Switzerland,[157] extremely toxic chemicals are banned, so that their residues are absent or, having drifted in from outside, present in only very small quantities. In others, such as the UK and many additional western European countries, where large-scale epidemics of foreign species are rare and agricultural systems have so far been ecologically well balanced, the demand for pesticides is relatively low[113], and about one-half of the quantity used is composed of non-toxic substances, such as sulphur.[168] This means that residue accumulation may generally be much less than, for example, in some areas of the USA, where certain monocultural crops, such as tobacco and cotton, are particularly vulnerable to pests, and where relatively few restrictions are applied to pesticide use; in the USA over 90 000 tons of pesticide may be employed per year,[179] with a greater additional amount of herbicides and other agricultural chemicals. In all countries, the degree of residue persistence in the ecosystem depends not only on the nature of the chemical compound, but also on climate and soil type, though usually relatively few substances last for more than one year.* There is no doubt as to the economic returns which may result from pesticide application, for in many instances, the increased value of the reaped crop may exceed the cost of pesticide by a factor of 10 or 12 times.

* The decay of pesticide chemicals is stimulated by the actions of actinomycetes fungi, and bacteria, catalysed by enzymes.[80]

Despite these benefits, the extensive use of pesticides may give rise to biological effects which represent a potentially severe environmental hazard. Even with low residue concentrations, negative imbalances of populations may rapidly develop,[198] although as Mulla[117] has pointed out, some organisms which become resistant may eventually return in even greater numbers than before. Occasionally, new resistant pests may take over from the old, especially if their predators are killed by the toxic elements.[187] Edwards[48] has noted that pesticides may generally depress growth patterns and cause wilting or cell and structural distortion within individuals. Moreover, as Moore[113] has suggested, these effects can also result in increased death rates and decreased birth rates among many populations. Soil microorganisms appear to be especially vulnerable[122] and, in particular, should sufficient numbers of nitrifying bacteria be killed off, ammonia gases may increase in the soil and temporarily encourage above normal rates of plant growth.[100]

Often, the consequences induced by the uptake of toxic elements are most clearly seen in predatory mammals, fish, or birds, for concentrations of tissue residues are increased as they pass down the food chain, as in the case of the grebes at Clear Lake, California, previously noted in section 2.5 (see also Moore and Walker[115]; Rudd[153]). Similarly, in Scotland, tissues of dead golden eagles (*Aquila chrysaetos*) have been found to contain large concentrations of dieldrin, originally introduced into the environment as a sheep-dip, and subsequently taken up by small mammals which, in turn, were consumed by the eagles themselves. Even greater tissue accumulations of organochlorine pesticides have been reported from populations of herons (*Ardea cinerea*) and the great crested grebe (*Podiceps cristatus*), which have the highest recorded increased death rates to be found in UK birds[136] within recent years. Although the precise mechanisms which give rise to such population losses are not yet known, it has been suggested that, as in the case of the osprey in southern New England, hatching failures following pesticide intake may be the prime cause;[5] moreover, the general reproductive capacities of organisms may also be reduced, as is known to be the case for other experimental bird populations whose diet included at least 90 ppm dry weight of DDT on a regular basis.[40, 81]

Detailed documentary evidence, to support the notion that many populations are rapidly reduced in numbers once pesticide accumulation in cell tissues has reached a danger level is now being received from many parts of the world. In Switzerland, experiments carried out in a forest area not previously subjected to pesticide spraying suggested that while DDT, which was directed against the larch-bud moth (*Eucosma griseana*), succeeded without the development of any side-effects, more toxic chemicals also killed 20 to 40 per cent of the original bird populations, involving 18 species in all —these, however, returned in their usual numbers after spraying had been discontinued.[157] Elsewhere, robins in the eastern USA have died as a result of eating earthworms sprinkled with chemicals used in an attempt to control Dutch elm disease.[7] In Holland, van Klingeren and his associates[182] have noted positive correlations between the aerial spraying of DDT and dieldrin to quantities of 0·5 kg/ha and the presence of reduced numbers of hares (*Lepus europaeus*). Many vegetables intended for human use have periodically become tainted and unusable through spraying,[24] and, often, milk samples within the USA have been found to contain residues of chlorinated hydrocarbons to concentrations of 1·5 ppm, not enough to affect the human body adversely, but sufficient to kill DDT-susceptible flies. Many other instances may be cited.

It is now clear that similar effects may also be felt in aquatic environments. Streams which have long been polluted by industrial and agricultural wastes (see chapter 7) may now, in addition, receive substantial quantities of pesticides washed down in runoff from adjacent cultivated fields, and the disposal of sheep-dips and related materials may add significantly to these stream residues.[72] Disastrous results can sometimes follow, leading to greatly increased death rates among aquatic organisms. In Alabama, after a series of intense rainstorms during 1950, enough pesticides were removed from nearby cotton fields to cause extensive fish kills in 14 adjacent tributaries of the Tennessee river;[203] since then, many similar kills have been recorded from streams throughout the USA (see Breidenbach and Lichtenberg[17]). The gills of dead salmon and trout in Scottish rivers have been found to contain concentrations of chlorinated hydrocarbons several thousand times greater than normal.[68] In lakes, some attempts have even been made to control the numbers of aquatic organisms through the direct application of pesticides to the water surface,[177] with unforeseen consequences, for while some of the soluble chemical elements may become widely distributed throughout the lake, others (e.g., copper carbonate) tend to precipitate and so become concentrated on the lake floor or on living organisms.[35] Usually, extensive side-effects may be expected in lakes, since in the relatively quiet waters enough CO_2 may be given off following plant deaths to remove most other organisms and micro-organisms nearby. Indeed, once pesticides have been introduced into freshwater ecosystems, multiple effects are as common as on land.

Either directly from rivers, or more indirectly from rain water,[193] some pesticide residues may also reach oceanic water where they are often found in oil to concentrations of 300 ppm. Usually, most are located in inshore waters, where they are absorbed particularly effectively by oysters, but they have also been recovered in plankton,[22] and from a wide variety of fish and molluscs whose distribution ranges from the Caribbean to Iceland and from Peru to Alaska.[67] The eggs of seabirds may also have residues within them, especially in the eastern Atlantic, and most of the evidence points to an increasing dispersal of pesticide elements throughout the world's oceans.

Of course, the full effects of pesticides on the structure of natural ecosystems are still not known. But it is already clear that major modifications to the pathways of chemical exchange may result from their use. So far, very little permanent damage appears to have been done, but man would perhaps be wise to restrict the application of both lethal and sublethal quantities of potentially toxic chemicals until their long-term effects on populations and element exchange patterns are better known. In the meantime, possible alternatives to their use need to be explored, particularly with respect to the ideas of biological pest control, or (as Stern *et al*[165] have suggested) integrated pest control, in which biological, cultural, and chemical measures are introduced concurrently.

3.11 Radioactive substances

The artificial release of large quantities of radioactive elements into the atmosphere, largely from atomic bomb tests, has also caused minor but important modifications in the chemical composition of organisms within ecosystems in the last 25 years. These may take the form of general irradiative effects or, more often, may follow from a relatively rapid uptake of three specific elements, strontium-90, iodine-131, and cesium-137, all of which may induce cell changes in body organs, and so affect the patterns of life in many populations.

The presence of radioactive elements on earth is not a new phenomenon, for background radiation has always been derived from the minute accumulations of radioactive substances in rocks, soils, and oceans. Cosmic radiation from outer space adds to this, but only in very small quantities, since most does not manage to penetrate the earth's atmosphere. Additional small totals of radioactivity may come from isotopes held within the carbon or potassium fraction of many organisms. Usually, the total incidence of natural radioactivity is slight, though it will vary, particularly with respect to altitude and local environmental conditions (Table 3.5). Thus, Alexander[2] has noted that a sailor who never comes to port will receive only one-fifth

Table 3.5   *Variations in income of natural radiation (after P. Alexander[2])*

Altitude	Income from cosmic radiation (mR/year)*	Normal background radiation income, including cosmic rays and natural radioactivity (mR/yr)*		
		Open ocean	Ordinary granite	Sedimentary rocks
Sea level	33	53	143	76
5000′	40	—	150	83
10 000′	80	—	190	123
15 000′	160	—	270	203
20 000′	300	—	414	347

of the natural radiation income of the inhabitants of high plateaux in Tibet; and those who drink tap water in some parts of Cornwall, England, may receive over three times as much irradiation through it as may be expected from water supplies elsewhere in Britain. In most cases, living organisms have adapted themselves to this background radiation and so are not adversely affected by it.

It is when man-induced radiation is added to that of the natural background that striking and rapid changes in the cell structure of living organisms may occur. Table 3.6 shows that, for man, a larger proportion of this originates from the use of X-rays in medical diagnoses than from any other single source; but, in the long term, and

* A rontgen (named after the discoverer of X-rays) is a measurement of energy, related to a dose of received radiation; it is equivalent to the uptake of approximately 100 erg/g of irradiated water or tissue. When it is considered that 42 million erg are needed to raise the temperature of 1 g of water 1°C, this is very small, but a few hundred rontgen may still be capable of killing a man. 1 mR (millirontgen) = 1/1000 rontgen.

Table 3.6 Comparative risks from principal sources of irradiation
(after UN Scientific Committee on the effects of radiation, and P. Alexander[2])

Source	Hereditary effects	Leukaemia	Bonecancer
Natural	1·0	1·0	1·0
Medical irradiation in highly developed countries	0·3	0·4 → 0·8	?
Fallout to December 1961	0·1	0·15	0·2
Continued fallout	0·2	0·3	0·4

in the environment as a whole, the fission materials released from atomic bomb tests may be much more significant, since all the radiation thus produced is capable of causing *ionization* of atomic particles (the release of an external electron from an atom). From this initial step, a well-defined pattern of biological effects may ensue within cell structures (see Fig. 3.13), leading to the potential malfunction of body organs, particularly those which are radiosensitive, such as the thyroid, spleen, appendix, and lymph nodes. At times, this may be followed by genetic modifications through chromosome breakage, or even death, if malignancies develop; indeed, it has now been shown that almost all human malignancies caused through irradiation (and those of many animals as well) are induced by means of the intake of artificially produced radioactive materials.[2]

Initially, fallout elements are often widely dispersed throughout the atmosphere, this being especially true of Sr-90 and Cs-137, both of which have gaseous origins following atomic explosions before they become absorbed on small dust particles (less than 40 μm in diameter), and so travel long distances at high altitudes, often many times round the world. The element I-131 is also produced in large quantities from any atomic reaction and distributed widely. All three frequently reach the earth's surface in rain water, after which they are taken in by plants and other organisms, when the Sr-90 begins to behave like calcium, and the Cs-137 like phosphorus. The initial method of intake varies according to the individual, and may be achieved through particulate accumulation on plant surfaces[149] or, less directly, through plant roots.[156] Once assimilated, all will pass down the food chain, although the precise reactions of each vary in detail: Sr-90 tends to accumulate steadily in bone structures, I-131 is common in children's thyroids, while Cs-137 remains only for a relatively short time before breaking down.

97

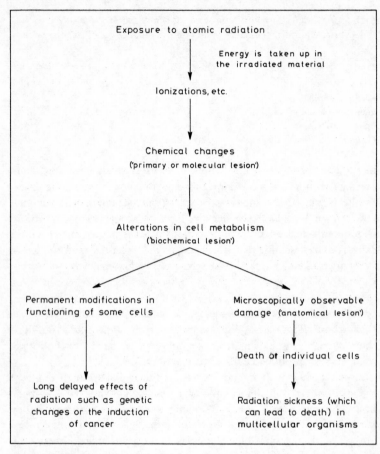

Fig. 3.13 The pattern of radiation injury to living cells (from an original diagram in P. Alexander[2])

These differential reactions themselves account, in part, for some of the variations in uptake of radioactive elements which are to be found in natural ecosystems. Climatic factors may also be important, for there is a good deal of evidence to suggest that rainy areas receive more fallout elements than do arid areas. In a general sense, location is also significant, since due to the fact that the northern hemisphere has so far had a greater proportion of testing than the southern, it may be expected that more fallout will occur here. This view has been supported recently by Kulp,[87] who suggests that, of the

98

approximately 4 million curies* of Sr-90 present in the biosphere in 1960, mean concentrations of 50 mCi/mile^2 were found in the northern hemisphere, as opposed to only 15 mCi/mile^2 in the southern. Most of this was held in the non-biological mineral pool, with only small totals actually circulating within ecosystems.

At our present state of knowledge, in which there are only a few detailed analyses of the rates of uptake of radioactive elements, it is exceedingly difficult to define the more explicit patterns of their accumulation and their pathways of circulation within ecosystems. However, a few preliminary conclusions may be drawn. First, rainy areas of the northern hemisphere may be expected to have relatively large concentrations of fallout elements. Thus, in northern England, Sr-90 uptake is appreciably higher at several trophic levels on acidic moorlands with high rainfall and low calcium and mineral concentrations, than on adjacent more fertile and drier lowlands (see Table 3.7). Both here and elsewhere, Sr-90 may be even more common

Table 3.7   *Amounts of Sr-90 accumulated from fallout in two contrasting ecosystems in the UK, 1956 (after E. P. Odum[125], and F. J. Bryant et al.[19])*

| | Hill pasture Acid peat soil, pH 4·3 Amt Sr-90, Ci/g* | | | Valley pasture, Brown loam soil, pH 6·8 Amt Sr-90, Ci/g* | | |
	Dry material	Calcium	Conc. factor†	Dry material	Calcium	Conc. factor†
Soil (mean, top 4″)	0.112	800	1	0·038	2·6	1
Grass	2·5	2100	21	0·250	41	6·6
Sheep bone	80	160	714	4·4	8·7	115

where matlike vegetation exists, as best exemplified in mosses and lichens, possibly because this has a greater exposed surface area per unit dry weight of tissue than most other communities. In contrast, many angiosperm systems, such as maize field and cattail marsh in the USA,[128] are very inefficient in transferring this and other radioactive materials from the soil. Accumulation rates within aquatic ecosystems are even less well known, although certain

* A curie (named after the discoverer of radium) is a measurement of the activity of radioactive elements; it is equivalent to the amount of material in which $3·4 \times 10^{10}$ atoms disintegrate per second—or the activity of 1 g of radium with its decay products. Following each atomic disintegration, electrons and protons are given off with some radiation. 1 millicurie (mCi) $= 10^{-3}$ curies (Ci); 1 microcurie (μCi) $= 10^{-6}$ Ci; 1 $\mu = 10^{-12}$ Ci.

† The concentration factor is equivalent to the ratio of the amount per g in soil, c.f. the amount per g in biological material.

elements—particularly 1-131—appear to be taken up much more quickly here than on land.[131]

Once assimilated by plants, radioactive elements may be concentrated appreciably as they subsequently pass down the food chain (Table 3.7), although so far, at least among mammals, the average concentrations are over one hundred times lower than the normally accepted danger levels. However, this level may be reached in situations where extreme or prolonged radiation has occurred, as when the average daily intake of Sr-90 in food is in excess of 20 μCi.[33] So far, this has not happened on a worldwide basis, although at the peak fallout periods of 1959 and 1960, the average human dietary intake in the USA reached 15 μCi/day, since when it has fallen to 9. Intake of radioactive materials by specific organisms down the food chain may, of course, be much greater than average in any ecosystem and, as in the case of pesticides, there may be unexpectedly far-reaching side-effects, as when radioactive cobalt-60 (not in itself produced from nuclear fission) was accumulated by clams within two years of their environment having been polluted by fallout elements.[31]

If general irradiation or fallout totals are sufficiently great, more direct ecological consequences may also follow. Coleman and Macfadyen[32] have reported that gamma radiation which kills vegetation also creates intense metabolic activity in soils which could last for several days. This, in turn, stimulates the release of large quantities of carbon dioxide and ammonia, probably from enzymes freed from 'dead' microorganisms,[16] although it is interesting to note that even in excessively irradiated soils, some bacteria and fungi usually stay alive.[76] Even though ammonia totals subsequently remain at a high level for several weeks, small arthropods may begin to reenter irradiated areas within seven days, the exact timing depending on their degree of ammonia tolerance. Eventually, more soil organisms become reestablished and, in a relatively short space of time, cycling processes will recommence.

It is now known that even the most intensive radiation so far created from atomic bomb testing does not necessarily result in the permanent sterilization of the environment, for as in the case of Bikini atoll devastated by a hydrogen bomb explosion in 1947, resettlement by humans has been proved to be practicable following a period of plant and animal recolonization which lasted 21 years. But it may be that the continued exposure of many organisms to sublethal quantities of irradiation, primarily from fallout materials, may be more significant to the biogeochemical environment of eco-systems in the long term, particularly with respect to their carcino-genic or mutagenic qualities. At present, we simply do not know what the ultimate consequences of this might be.

3.12 Air pollution

No other single aspect of environmental pollution is of greater immediate concern than that found in the atmosphere. At its worst, it forms smoke palls and smog banks over our major industrialized conurbations, and its constituent particles may be dispersed by down-wind drift to increase haze levels over large areas. Because of atmo-spheric pollution, the lungs of city-dwellers may be grey in colour as opposed to the normal pink, and the processes of ageing, not only among living organisms, but also on stone and metal work, may be significantly accelerated. Yet this is not a new phenomenon, for the output of industrial coal smoke—with its components of soot, ash, tar, sulphur dioxide, and sulphuric acid—was sufficiently great 200 years ago in the UK to force the enactment of legislation designed to curtail it. It is only the scale which has shown vast increases, particularly since 1945 with the growth of new industries, new fuels, and new technological devices, so much so that, for the first time, pollution is now considered to be a problem of overwhelming inter-national importance. In extreme cases, where pollution elements are concentrated through atmospheric temperature inversions during anticyclones, major disasters may ensue, as in the infamous London smog of 5–9 December 1952 in which 4000 human deaths were directly attributable to the irritant effects of inhaled smog particles.[167] More often, however, the results are less spectacular, if no less tangible, within the ecosystems which they influence.

Three major effects of atmospheric pollution, and probably many minor ones so far undetected, may induce changes in biogeochemical cycling processes. *First*, so much fossil fuel has been burnt in recent years that enormous quantities of carbon, originally withdrawn from biological cycling many thousands of years ago in the formation of sedimentary rock types, have been rereleased within ecosystems. It is now thought that the associated increase of carbon dioxide in the lower layers of the atmosphere may be enough to inhibit outgoing longwave radiation substantially; if present trends continue, this, in turn, could result in an amelioration of mean world temperatures by a factor of 2°C by 2000 AD,[134] which is enough to induce substantial climatic change. Moreover, it is to be expected that the greater amounts of carbon dioxide will give rise to corresponding increases in the rates of photosynthesis, so that dry matter production in green plants at the turn of the next century may theoretically be 20 per cent greater than presentday values.[74] Mattson and Koutler-Andersson[102] have suggested that there might also be a similar augmentation in the rates of nitrogen accumulation in ecosystems, although this is a more dubious assumption based, in part, on present-day figures which are extremely high (see Firbas;[51] Jorgensen;[78] Jensen;[75] and Eriksson[50]), but which result not only from the burning of fossil fuels, but also from the greater intensive use of nitrogenous fertilizers (see Yaalon's work in Israel,[202] and Junge[79]).

Second, the problem of volatile elements in industrial smoke has become greater as new techniques help to make combustion more complete. Today, many types of dust and chemical particles (see Fig. 3.14), including several potential carcinogens, are given off from flues in addition to the basic constituents of water, carbon monoxide, and the oxides of sulphur and nitrogen. Of the last two, conversion of sulphur dioxide to sulphuric acid, and of nitric oxide to nitric acid, takes place as the extracted smoke cools, and these are usually deposited in diluted form a short distance away; but about one-half of the sulphur dioxide will be retained in the atmosphere to fall over large areas as dust containing ammonium and calcium sulphates. In itself, sulphur dioxide is a corrosive gas, which accelerates the weathering of metal and stonework and aggravates lung and other diseases. Despite some attempts to restrict its outflow from industrial sources, it is an increasing problem in many districts since, unlike other combustion elements, its detection and suppression are extremely expensive. In contrast, the carbon monoxide content of industrial fumes is not a major problem at the present time.

Although the precise biogeographical effects of industrial smoke pollution have not yet been worked out, it is already certain that positive correlations exist between the excessive deposition of smoke products and changes in the respiratory systems of many plants and animals. Often, soot may clog stomata in the leaves of plants and reduce the absorption of carbon dioxide from the atmosphere.[175] Plant growth may also be restricted through the increased quantities of acid in rain water which, in the industrial areas of West Yorkshire,

102

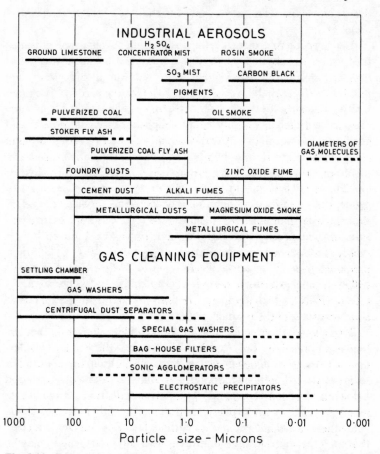

Fig. 3.14 Ranges in particle size of some air pollution materials (from H. P. Munger, The spectrum of particle size and its relation to air pollution, In L. C. McCabe, *Air Pollution*, Copyright McGraw-Hill Book Co. Inc., NY, 1952. Used with permission)

England, may reach concentrations of 5 to 100 ppm.[29] This, in turn, will acidify soil and reduce the amounts of available nutrients, such as calcium, nitrate, and ammonia. Moreover, since bacterial and enzyme activity may also be restrained, many biogeochemical cycling processes will be slowed down, to the detriment of most organic life. Carcinogenic elements are also included in the industrial smoke of many towns, and, to take one example only, it has been suggested that the atmosphere of Liverpool, England, holds 8 to 12 times as many carcinogens as in that of surrounding rural areas.[166]

A *third* group of effects is associated with a specific type of industrial-urban atmospheric pollution, termed *photochemical smog*; this develops over large urban areas when sulphur dioxide levels are relatively low and the use of petrol is relatively high. It is caused mainly through the action of sunlight on a mixture of nitrogen dioxide (emitted in roughly equal proportions by automobiles and gas- and oil-burning industrial premises) and gaseous organic compounds found in automobile fuels, particularly olefins, but also including other products of hydrocarbon chains from which hydrogen atoms have been removed. This creates excessive amounts of ozone, the quantities of which may be raised to between 20 and 30 times greater than normal;* the latter, in turn, can lead to increased rates of oxidation (particularly towards noonday), induced by the relative lack of reducing components in the atmosphere. In addition, peroxides formed by atmospheric reaction with organic materials will eventually give rise to peroxide compounds, such as peroxyacetyl nitrate (PAN), which act as eye-irritants and oxidizing pollutants at concentrations of 0·1 ppm or more.

Even though these conditions are found widely elsewhere, they are best exemplified in Los Angeles, California, where relatively few days are free from smog effects (see Fig. 3.15). Often, the pollution is so great that despite the dry air, visibility is markedly impaired below an inversion layer, and eye-irritation effects are common. As in other types of atmospheric pollution, the full biogeographical consequences are not yet known, although in Los Angeles it is clear that under smog conditions photosynthesis and plant growth may be restricted by as much as 20 to 30 per cent of normal,[83] due to the action of excessive nitrogen dioxide. Leaf-flecking, induced by the unusually high ozone concentrations, may also retard respiratory mechanisms and, therefore, some rates of chemical exchange. Furthermore, the presence of peroxide compounds and gasolene derivatives may give rise to genetic changes,[118] cell collapse,[164] and the interveinal breakdown of tissue. More complicated reactions, stimulated by potentially carcinogenic substances, can result in increased rates of ageing and death among many higher organisms, including man himself.

* Ozone is normally present in unpolluted air at quantities of 1–3 parts per hundred million.[62]

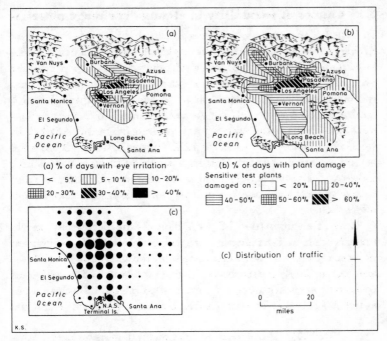

Fig. 3.15 Air pollution in the Los Angeles area for 1956, as compared to the distribution of traffic. In 3.15c, the area of the circles is proportional to the number of vehicles present (from original diagrams in A. J. Haagen-Smit[63] and F. N. Frenkiel, *Scientific Monthly*, Vol. 82, pp. 194–203, 1956).

PATTERNS OF ELEMENT EXCHANGE IN MAJOR WORLD ECOSYSTEMS

Our detailed knowledge of the mechanisms of biogeochemical exchange is still restricted to some extent by the very patchy evidence which is available. In particular, there are very few relevant reports from the southern hemisphere, and from East and South-East Asia; and even in other parts of the world, useful records are still too often concentrated only in very limited areas. However, enough information has been culled already to suggest that rates of element exchange and uptake, both by specific organisms and within whole ecosystems, may be extremely variable (see Lounamaa[95] and Rennie[141]). To date, the most complete analyses of evidence for this have been produced by Soviet scientists, notably Rodin and Basilevic,[148] many of whose conclusions are summarized below.

3.13 Causes of variability in element exchange patterns

It is to be expected that large-scale areal differences in element exchange patterns can be broadly explained by reference to the well-known zonal divisions of incoming solar energy (see chapter 2) which give rise to important regional variations in the quantity and specificity of organic matter (plants and animals), through which most of the cycling of chemical elements takes place. These relationships may perhaps be further particularized by considering briefly some of the factors which induce modifications in the totals and rates of accumulation of plant biomass and litter, for it is these which influence the patterns of element movement most directly, and also help to support the secondary animal populations which can further affect them.

Rates of accumulation of plant biomass are dependent largely upon latitude and the amounts of available moisture, being greatest in those regions with most incoming solar radiation and rainfall (see also section 2.14). Furthermore, in general, warm and humid areas will most commonly support the greatest biomass and litter totals, as noted in Table 3.8. To some extent, however, the precise quantities

Table 3.8   Mean biomass quantities, relative % of green portion, litter fall, and floor litter accumulation in selected major world ecosystems (after L. E. Rodin and N. I. Basilevic, Production and Mineral Cycling in Terrestrial Vegetation, *Oliver and Boyd, Edinburgh, 1967)*

Ecosystem	Biomass	% green part	% roots	Total litter fall, kg/ha	Litter fall, % of biomass	Floor litter kg/ha
Tropical rain forest	> 5000	8	18	250	8	20
Tropical mountain forest (Brazil)	17 000	5	19	—	—	—
Subtrop. decid. forest	4100	3	20	210	5	100
Beech forests	3700	1	26	90	3	150
Taiga spruce forests	1000	8	22	35	4	300
Dry savanna (Ghana)	268	11	42	72	27	—
Meadow steppe (USSR)	250	32	68	137	55	120
Semishrub desert	125	14	87	95	27	—
Harsh desert	16	13	62	6	38	< 1
Arctic tundra	50	30	70	ca. 10	ca. 20	35
Forest sphagnum bogs	370	41	11	25	7	< 1000

of all elements in circulation within a given ecosystem will also vary according to the relative percentages of biomass which are incorporated into the green parts and roots of plants, and according to the total amounts of litter accumulation at the ground surface. Plants and ecosystems whose green-part biomass (i.e., that available to photosynthesis) forms a considerable proportion of the whole may be at a relative advantage when compared to those in which the elements are partly immobilized in new non-photosynthetic woody growth. Immobilization may also be encouraged among ecosystems in which rates of litter decay are slow, and the accumulation of litter elements on the ground surface is greater than normal, as in many of the forests and forest swamps of mid and high latitudes; and, in contrast, element circulation may be enhanced in areas where the processes of decay operate very quickly, as in tropical and equatorial rain forests.

However, despite these reservations, one may look for the maximum rates of movement of chemical elements in those ecosystems with the greatest productive capabilities of organic matter, which in general are forests, and, more specifically, are those forest growing within hot and humid regions (see Table 3.9). This does not necessarily mean that such ecosystems are the most efficient in their uptake of elements, for it has long been suspected that certain subtropical deserts and grasslands, notwithstanding their relatively low biomass and litter totals, may be much more effective in this respect. Such a notion has recently been confirmed by the discovery that relatively large amounts of litter need to be produced by some communities in order to bring about the transfer of small quantities of elements from plant to soil. Thus, taiga spruce forests require between 50 to 100 kg of litter for the movement of 1 kg of ash elements, as opposed to equivalent figures of 35 kg in deciduous forest and tropical rain forest, 30 kg in savanna and steppe grassland, 20 kg in desert communities, and only 10 kg in some harsh deserts.[148] Taken in conjunction with the known variations in biomass production, these figures also serve to emphasize that considerable areal differences in the total quantity of accumulating ash elements in litter may be expected on a world scale. Less is known about other elements, although it is certain that the organic chemicals (Ca, K, P, and S), with Na and Cl, are generally found in the greatest relative abundance among oak and pine forest litter, giving way elsewhere to increasing amounts of Si, Fe, and Al. The abundance of nitrogen

107

Table 3.9 Mean annual quantities of mineral elements circulating in selected major world ecosystems (after L. E. Rodin and N. I. Basilevič, Production and Mineral Cycling in Terrestrial Vegetation, Oliver and Boyd, Edinburgh, 1967)

| Ecosystem | Elements circulating, kg/ha | | | | Nitrogen | | | Ash elements | | | Major elements | |
	Total	% in green part of biomass	Net uptake	Returned in litter	Total	Net uptake	Returned in litter	Total	Net uptake	Returned in litter	In litter fall	On floor
Tropical rain forest	11 081	26	2029	1540	2940	427	261	8141	1602	1279	Si, N, Ca (Al, Fe, S, Mn)	Fe, Si, N (Al, Mn)
Subtropical decid. forest	5283	15	993	795	1359	277	226	3924	716	569	N, Ca, K (Si, Al, Fe)	—
Beech forests	4196	7	492	352	1608	144	82	2588	348	270	Ca, N, Si (Al)	Ca, N, Si (Al, Fe)
Taiga spruce forests	970	22	118	100	350	58	48	620	60	52	N, Ca, K (Si)	N (Si, Ca, Al, Fe)
Dry savanna	978	18	319	312	238	81	80	740	238	232	Si, N, Ca (Fe, Al)	
Meadow steppe (USSR)	1183	36	682	682	274	161	161	909	521	521	Si, N, Ca (K)	Si, N, Ca (Al, Fe)
Semishrub desert	185	8	59	59	61	18	18	124	41	41	N, Ca, K (Na, Cl)	—
Harsh desert	143	33	84	84	31	14	14	112	70	70	Cl, Na, N (S, Mg)	—
Arctic tundra	159	35	38	37	81	21	20	78	17	17	N (K, Si)	N, Ca, Si (Al, Fe)
Forest sphagnum bogs	609	73	109	73	229	40	25	380	69	48	N, Si, Ca (K, Al)	N, Si, Ca (Al, Fe)

in litter also varies considerably, with a maximum of 2·8 per cent by weight in some harsh deserts, and a minimum of 0·6 per cent in certain pine forests. Quantities of individual elements in litter will, of course, usually show a wide range from place to place according to the type of plant community which is present and other variables of the environment.

Taking into account all the available knowledge about circulation patterns of elements within ecosystems, and paying particular attention to the known dominance of certain elements in the annual cycles of exchange, Rodin and Basilevic have concluded that five major terrestrial divisions of element exchange patterns may be delimited on a very rudimentary basis. *First*, it is known that in some areas (e.g., the tundra) the movement of nitrogen greatly predominates. *Second*, there are many ecosystems in which, while nitrogen circulation is less important, it still retains dominance over all other elements; this is the case for certain coniferous and deciduous temperate forests, and also for subtropical forests. *Third*, in several broad-leaved temperate forests in which it forms between 40 to 54 per cent of the ash litter, calcium may be the most common element in circulation. *Fourth*, in steppe and savanna grasslands, many deserts, and tropical rain forests, as much as 50 to 60 per cent of silicon may be present in ash elements, with another 10 to 15 per cent of iron and aluminium. And *fifth*, a few restricted areas such as harsh desert may have a preponderance of chlorine in their elemental patterns of movement.

It is only fair to emphasize that in all these cases, and more especially at a local level, there are many aspects of chemical cycling the significance of which has not yet been determined. This is particularly true with respect to the varied pathways of element movement. Often in present-day calculations, no account is taken of the quantities of nutrients leached out from leaves, or in root exudates.[185] Moreover, not enough is known about the alterations in element exchange patterns which occur as ecosystems develop towards maturity,[161] or about their seasonal variations which, within the northern hemisphere, frequently show a peak rate of exchange in June and July. The relationships between understorey and topstorey cycling in forests, and the roles of animals and birds in both, also need further study, for it may be that, at certain times of the year, more nutrients may be circulating in the lower storeys than elsewhere.[138] Investigations into these and other features may, in time,

109

induce considerable modifications to the generalized areal divisions of elemental exchange patterns which have been outlined above.

REFERENCES

1. ABELSON, P. H., Some aspects of paleobiochemistry, *Ann. N.Y. Acad. Sci.*, Vol. 69, pp. 176–285, 1957.
2. ALEXANDER, P., *Atomic Radiation and Life*, Pelican, Harmondsworth, UK, 2nd edn, 1965.
3. ALLEN, S. E., Chemical aspects of heather burning, *J. App. Ecol.*, Vol. 1, 347–67, 1964.
4. ALLEN, S. E., CARLISLE, A., WHITE, E. J., and EVANS, C. C., The plant nutrient content of rainwater, *J. Ecol.*, Vol. 56, pp. 497–504, 1968.
5. AMES, P. L. and MERSEREAU, G. S., Some factors in the decline of the osprey in Connecticut, *Auk*, Vol. 81, pp. 173–85, 1964.
6. ATTIWILL, P. M., The chemical composition of rainwater in relation to cycling of nutrients in mature *Eucalyptus* forest, *Pl. Soil*, Vol. 24, pp. 390–406, 1966.
7. BAKER, R. J., Notes on some ecological effects of DDT sprayed on elms, *J. Wildl. Mgmt*, Vol. 22, pp. 269–74, 1958.
8. BERTRAND, D., Survey of contemporary knowledge of biochemistry: 2. The biochemistry of vanadium, *Bull. Am. Mus. Nat. Hist.*, Vol. 94, pp. 403–56, 1950.
9. BIRCH, H. F., The effect of soil drying on humus decomposition and nitrate availability, *Pl. Soil*, Vol. 10, pp. 9–31, 1958.
10. BIRCH, L. C. and CLARKE, D. P., Forest soil as an ecological community with special reference to the fauna, *Quart. Rev. Biol.*, Vol. 28, pp. 13–36, 1953.
11. BLAKELY, B. D., COYLE, J. J., and STEELE, J. G., Erosion on cultivated land. In *Soil*, US Dept. of Agriculture, Annual Year Book, pp. 290–307, 1957.
12. BOCOCK, K. L., Changes in the amounts of dry matter, nitrogen, carbon, and energy in decomposing woodland leaf litter in relation to the activities of the soil fauna, *J. Ecol.*, Vol. 52, pp. 273–84, 1964.
13. BOND, G., Evidence for fixation of nitrogen by root nodules of alder (*Alnus*) under field conditions, *New Phytol.*, Vol. 55, pp. 147–53, 1956.
14. BOND, G., MacCONNELL, J. T., and McCULLUM, A. H., The nitrogen-nutrition of *Hippophaë rhamnoides*, L., *Ann. Bot. (n.s.)*, Vol. 20, pp. 501–12, 1956.
15. BOUILLENNE, R., Man, the destroying biotype, *Science*, Vol. 135, p. 706, 1962.
16. BOWEN, H. J. M. and CAWSE, P. A., Effects of ionizing radiation on soils and subsequent crop growth, *Soil Sci.*, Vol. 97, pp. 252–9, 1964.
17. BREIDENBACH, A. W. and LICHTENBERG, J. J., Identification of DDT and dieldrin in rivers, *Science*, Vol. 141, pp. 899–900, 1963.
18. BROWN, A. W. A., Insecticide resistance in arthropods, *W.H.O. Monograph Series*, No. 38, 1958.

19. BRYANT, F. J., CHAMBERLAIN, A. C., MORGAN, A. C., and SPICER, C. S., Radio-strontium in soil, grass, milk and bone in the UK, 1956 results, *J. Nuclear Energy*, Vol. 6, pp. 22–40, 1957.

20. BURGES, A., *Micro-organisms in the Soil*, London, 1958.

21. BURNS, P. Y., Effect of fire on forest soils in the pine-barren regions of New Jersey, *Yale School of Forestry, Bull.*, No. 50, 1952.

22. BUTLER, P. A., Pesticides in the marine environment, *J. Appl. Ecol.* Vol. 3, pp. 253–9, 1966.

23. CARLISLE, A., BROWN, A. H. F., and WHITE, E. J., The organic matter and nutrient elements in the precipitation beneath a sessile oak (*Q. petraea*) canopy, *J. Ecol.*, Vol. 54, pp. 87–98, 1966.

24. CARSON, R., *Silent Spring*, Boston, 1962.

25. CHANDLER, T. J., The changing form of London's heat island, *Geography*, Vol. 46, pp. 295–307, 1961.

26. CHAPMAN, S. B., The ecology of Coom Rigg Moss, Northumberland, II. The chemistry of peat profiles and the development of the bog system, *J. Ecol.*, Vol. 52, pp. 315–21, 1964.

27. CLIFFORD, P. A., Pesticide residues in fluid market milk, *Pub. Health Reports*, Vol. 72, pp. 729–34, 1957.

28. CLOUD, P. E., GRUNER, J. W., and HAGEN, H., Carbonaceous rocks of the Soudan iron formation (early pre-Cambrian), *Science*, Vol. 148, pp. 1713–16, 1965.

29. COHEN, J. B. and RUSHTON, A. L., *Smoke: A Study of Town Air*, London, 1925.

30. COLE, LAMONT, The ecosphere, *Sci. Am.*, Vol. 198, pp. 83–92, 1958.

31. COLE, LAMONT, Man's ecosystem, *Bioscience*, Vol. 1, pp. 243–8, 1966.

32. COLEMAN, D. C. and MACFADYEN, A., The recolonization of gamma-irradiated soil by small arthropods, *Oikos*, Vol. 17, pp. 62–70, 1966.

33. COMAR, C. L., Biological aspects and nuclear weapons, *Am. Scient.*, Vol. 50, pp. 339–53, 1962.

34. COMMONER, B. (Chairman), The integrity of science. A report by the AAAS Committee on Science in the promotion of human welfare, *Am. Scient.*, Vol. 53, pp. 174–98, 1965.

35. COPE, O. B., Contamination of the freshwater ecosystem by pesticides, *J. Appl. Ecol.*, Vol. 3, pp. 33–44, 1966.

36. COTTAM, C., The ecologist's role in problems of pesticide pollution *Bioscience*, Vol. 15, pp. 457–63, 1965.

37. CRICK, F. H. C., Macromolecules and natural selection. In *Growth in Living Systems*, ed. M. X. ZARROW, pp. 3–8, New York, 1961.

38. CRISP, D. T., Input and ouput of minerals for an area of Pennine moorland; the importance of precipitation, drainage, peat erosion and animals, *J. Appl. Ecol.*, Vol. 3, pp. 314–27, 1966.

39. DELWICHE, C. C., Energy relationships in soil biochemistry. In *Soil Biochemistry*, eds. A. D. MCLAREN and G. H. PETERSEN, pp. 173–93, London, 1967.

40. DEWITT, J. B., Effects of chlorinated hydrocarbon insecticides upon quail and pheasants, *J. Agric. Fd. Chem.*, Vol. 3, pp. 672–6, 1955.

41. DEWITT, J. B., Chronic toxicity to quail and pheasants of some chlorinated insecticides, *J. Agric. Fd. Chem.*, Vol. 4, pp. 863–5, 1956.

42. DIMBLEBY, G. W., The historical status of moorland in north-east Yorks., *New Phytol.*, Vol. 51, pp. 349–54, 1952.

43. DONAHUE, R. L., *Soils: An Introduction to Soils and Plant Growth*, Englewood Cliffs, New Jersey, 2nd edn., 1965.

44. DOUGLAS, I., Man, vegetation and the sediment yields of rivers, *Nature*, Vol. 215, pp. 925–8, 1967.

45. DOUGLAS, I., Erosion in the Sungei Gombak catchment, Selangor, Malaysia, *J. Trop. Geogr.*, Vol. 26, pp. 1–16, 1968.

46. DRINKER, P., Air pollution and the public health, *J. R. Inst. Publ. Hth. Hyg.*, 1957.

47. EBERMAYER, E. *Chemie der Pflanzen*, Berlin, 1882.

48. EDWARDS, C. A., Effects of pesticide residues on soil invertebrates and plants. In *Ecology and the Industrial Society*, ed. G. T. GOODMAN, R. W. EDWARDS, and J. M. LAMBERT, pp. 239–62, Oxford, 1965.

49. EGNER, H. and ERIKSSON, E., Current data on the chemical composition of air and precipitation, *Tellus*, Vols. 10–12, 1958–60.

50. ERIKSSON, E., Composition of atmospheric precipitation I. Nitrogen compounds, *Tellus*, Vol. 4, pp. 215–32, 1959.

51. FIRBAS, F., Einige Berechnungen uber die Ernahrung der Hochmoor, *Veröff Geobot. Ints.*, *Zürich*, Vol. 25, pp. 177–200, 1952.

52. FOGG, G. E., *The growth of plants*, Penguin, Harmondsworth, UK, 1963.

53. FOGG, G. E. and WOLFE, M., The nitrogen metabolism of the blue-green algae (Myxophyceae). In *Autotrophic Micro-organisms*, Eds. B. A. FRY and J. L. PEEL, pp. 99–125, Cambridge, 1954.

54. FOWELLS, H. A. and STEPHENSON, R. E., The effect of burning on forest and soils, *Soil Sci.*, Vol. 38, pp. 175–81, 1934.

55. GEORGE, J. L. and FREAR, D. E. H., Pesticides in the Antarctic, *J. Appl. Ecol.*, Vol. 3, pp. 155–67, 1966.

56. GLASSMAN, I., The chemistry of propellants, *Am. Scient.*, Vol. 53, pp. 508–24, 1965.

57. GORE, A. J. P., The supply of six elements by rain to an upland peat area, *J. Ecol.*, Vol. 56, pp. 483–95, 1968.

58. GORE, A. J. P. and URQUHART, C., The effects of waterlogging on the growth of *Molinia caerula* and *Eriophorum vaginatum*, *J. Ecol.*, Vol. 54, pp. 617–34, 1966.

59. GORHAM, E., A comparison of lower and higher plants as accumulators of radioactive fallout, *Can. J. Bot.*, Vol. 37, pp. 327–9, 1959.

60. GRANICK, S., Speculations on the origin and evolution of photosynthesis, *Ann. N.Y. Acad. Sci.*, Vol. 69, pp. 292–308, 1957.

61. GREGORY, F. G., The control of growth and reproduction by external factors, *Rep. 13th Int. Hort. Conf.*, Vol. 1, pp. 96–105, 1952.

62. HAAGEN-SMIT, A. J., Chemistry and physiology of Los Angeles smog, *Ind. Eng. Chem.*, Vol. 44, pp. 1342–6, 1952.

63. HAAGEN-SMIT, A. J., Air conservation, *Science*, Vol. 128, pp. 869–78, 1958.

64. HAWKES, H. E., Principles of geochemical prospecting, *US Geol. Survey*, Bull. 1000F, 130 pp., 1957.
65. HAWKES, H. E. and WEBB, J. S., *Geochemistry in Mineral Exploration*, Harper & Row, New York, 1962.
66. HÉNIN, S. and DUPUIS, M., Essai de bilan de la matière organique du sol, *Annls. Agron.*, Vol. 15, pp. 17–29, 1945.
67. HICKEY, J. J., KEITH, J. A., and COON, F. B., An exploration of pesticides in a lake Michigan ecosystem, *J. Appl. Ecol.*, Vol. 3, pp. 141–51, 1966.
68. HOLDEN, A. V., Organochlorine insecticide residues in salmonid fish, *J. Appl. Ecol.*, Vol. 3, pp. 45–53, 1966.
69. HUTCHINSON, G. E., Nitrogen in the biogeochemistry of the atmosphere, *Ann. N.Y. Acad. Sci.*, Vol. 50, pp. 221–46, 1948.
70. HUTCHINSON, G. E., On living in the biosphere, *Sci. Mon.*, Vol. 67, pp. 393–8, 1948.
71. HUTCHINSON, G. E., The biogeochemistry of the terrestrial atmosphere. In *The earth as a Planet*, ed. G. P. KUIPER, pp. 371–433, University of Chicago Press, Chicago, 1954.
72. HYNES, H. B. N., The effect of sheep-dip containing the insecticide BHC on the fauna of a small stream including *Simulium* and its predators, *Ann. Thop. Med. Parasit.*, Vol. 55, pp. 192–96, 1961.
73. HYNES, H. B. N., A survey of water pollution problems. In *Ecology and the Industrial Society*, eds, G. T. GOODMAN, R. W. EDWARDS, and J. M. LAMBERT, pp. 49–64, 1965.
74. JACKSON, R. M. and RAW, F., Life in the soil, *Studies in Biology* No. 2, Institute of Biology, London, 1966.
75. JENSEN, J., Undersøgelser over nedbørens indhold af plantenoeringstoffer, *Tidsskr. Plav. 1*, Vol. 65, pp. 894–906, 1962.
76. JOHNSON, L. F. and OSBORNE, T. S., Survival of fungi in soil exposed to gamma irradiation, *Can. J. Bot.*, Vol. 42, pp. 105–13, 1964.
77. JONES, E., Contribution of rainwater to the nutrient economy of soil in northern Nigeria, *Nature*, Vol. 188, p. 432, 1960.
78. JORGENSEN, C. A., Kraelstofproblemet paa Maglemose oq andre Hojmoser, *Bot. Tidsk.*, Vol. 29, pp. 463–87, 1927.
79. JUNGE, C. E., *Air chemistry and Radio-activity*, New York, 1963.
80. KEARNEY, P. C., KAUFMAN, D. D., and ALEXANDER, M., Biochemistry of herbicide decomposition in soils. In *Soil Biochemistry*, eds. A. D. MCLAREN and G. H. PETERSEN, pp. 318–42, Arnold, London, 1967.
81. KEITH, J. A., Reproduction in a population of herring gulls (*Larno argentatus*) contaminated by DDT, *J. Appl. Ecol.*, Vol. 3, pp. 57–70, 1966.
82. KONONOVA, M. M., *Soil Organic Matter*, New York, 1961.
83. KORITZ, H. G. and WENT, F. W., Physiological action of smog on plants. I. Initial growth and transpiration studies. *Pl. Physiol.*, Vol. 28, pp. 50–62, 1956.
84. KORTLEVEN, J., Kwantitatieve aspecten van humusopbouw en humusafbraak, *Versl. landbouwk. Onderz. Rijkslandb. Proet Stn.*, Vol. 69, 1963.

85. KRASILNIKOV, N. A., *Soil micro-organisms and higher plants*, Acad. of Sci., USSR, Moscow 1958.

86. KRULWICH, T. A. and FUNK, H. B., Stimulation of *Nitrobacter agilis* by biotin, *J. Bact.*, Vol. 90. pp. 729–33, 1965.

87. KULP, J. L., Radionuclides in man from nuclear tests, *J. Agric. Fd. Chem.*, Vol. 9, pp. 122–6, 1961.

88. LEYTON, L., The growth and mineral nutrition of spruce and pine in heathland plantations, *Inst. Pap. Commonw. For. Dist.*, pp. 31, 1954.

89. LICHTENSTEIN, E. P. and SCHULTZ, K. R., Persistence of some chlorinated hydro-carbon insecticides as influenced by soil types, rate of application and temperature, *J. Econ. Ent.*, Vol. 52, pp. 124–31, 1959.

90. LIEBIG, J., *Die organische Chemie in ihrer Anvendung vor Agricultur und Physiologie*, Braunschweig, 1840.

91. LITAV, M. Mycorrhizal association in dwarf-shrub species growing in soft cretaceous rocks, *J. Ecol.*, Vol. 53, pp. 147–51, 1965.

92. LOACH, K., Relations between soil nutrients and vegetation in wet-heaths. I. Soil nutrient content and moisture conditions, *J. Ecol.*, Vol. 54, pp. 597–608, 1966.

93. LOACH, K., Relations between soil nutrients and vegetation in wet heaths. II. Nutrient uptake by the major species in the field and under controlled conditions, *J. Ecol.*, Vol. 56, pp. 117–27, 1968.

94. LOACH, K., Seasonal growth and nutrient uptake in a Molinietum, *J. Ecol.*, Vol. 56, pp. 433–44, 1968.

95. LOUNAMAA, J., Trace elements in plants growing wild on different rocks in Finland, *Ann. Bot. Soc. Vanama*, Vol. 29, pp. 1–196, 1956.

96. MCLAREN, A. D. and PETERSEN, G. H., Introduction to the biochemistry of terrestrial soils. In *Soil Biochemistry*, eds. A. D. MCLAREN and G. H. PETERSEN, pp. 1–18, Arnold, London, 1967.

97. MCVEAN, D. N., Growth and mineral nutrition of Scots Pine seedlings on some common peat types, *J. Ecol.*, Vol. 51, pp. 657–70, 1963.

98. MADGEWICK, H. A. I. and OVINGTON, J. D., The chemical composition and precipitation in adjacent forest and open plots, *Forestry*, Vol. 32, pp. 14–22, 1959.

99. MALMER, N., Notes on the relation between the chemical composition of mire plants and peat. *Bot Notiser*, Vol. 111, pp. 274–88, 1958.

100. MARTIN, J. P. and PRATT, P. F., Fumigants, fungicides and the soil, *J. Agric. Fd. Chem.*, Vol. 6, pp. 345–8, 1958.

101. MASON, B., *Principles of Geochemistry*, New York, 1960.

102. MATTSON, S. and KOUTLER-ANDERSSON, E., Geochemistry of a raised bog. II. Some nitrogen relationships, *Lantbr. Högsk. Ann.*, Vol. 22, pp. 219–24, 1955.

103. MEIKLEJOHN, J., The nitrifying bacteria: a review, *J. Soil Sci.*, Vol. 4, pp. 59–68, 1953.

104. MEIKLEJOHN, J., Microbiology of the nitrogen cycle in some Ghana Soils, *Emp. J. Exp. Agric.*, Vol. 30, pp. 115–26, 1962.

105. MEIKLEJOHN, J., Microbiological studies of large termite mounds, *Rhod. Zamb. Mal. J. Agric. Res.*, vol. 3, pp. 67–79, 1965.

106. MEIKLEJOHN, J. Numbers of nitrifying bacteria in some Rhodesian soils under natural grass and improved pastures, *J. Appl. Ecol.*, Vol. 5, pp. 291–9, 1968.

107. MELIN, E. and NILSSON, H., Transfer of radioactive phosphorus to pine seedlings by means of mycorrhical hyphae, *Physiol. Plant.*, Vol. 3, pp. 88–92, 1950.

108. MELLANBY, K., *Pesticides and Pollution*, New Naturalist Series, London, 1967.

109. MILLER, R. B., The chemical composition of rainwater at Taita, New Zealand, 1956–58, *N.Z. Jl. Sci.*, Vol. 4, pp. 844–53, 1961.

110. MINA, V. N., Cycle of N and ash elements in mixed oakwoods of the forest-steppe, *Pochvovedenie*, Vol. 6, pp. 34–44, 1955.

111. MINDERMAN, G., Addition, decomposition and accumulation of organic matter in forests, *J. Ecol.*, Vol. 56, pp. 355–62, 1968.

112. MINDERMAN, G. and LEEFLANG, K. W. F., The amounts of drainage water and solutes from lysimeters planted with either oak, pine, or natural dune vegetation, or without any vegetation cover, *Pl. Soil*, Vol. 28, pp. 61–80, 1968.

113. MOORE, N. W., Environmental contamination by pesticides, In *Ecology and the Industrial Society*, eds. G. T. GOODMAN, R. W. EDWARDS, and J. M. LAMBERT, pp. 219–37, Oxford, 1965.

114. MOORE, N. W. and TATTON, J. O'G., Organochlorine insecticides in the eggs of seabirds, *Nature*, Vol. 207, pp. 42–3, 1965.

115. MOORE, N. W. and WALKER, C. H., Organic chlorine insecticide residues in wild birds, *Nature*, Vol. 201, pp. 1072–3, 1964.

116. MORRISON, T. M., Uptake of sulphur by excised beech mycorrhizas, *New Phytol.*, Vol. 62, pp. 44–9, 1963.

117. MULLA, M. S., Vector control technology and its relationship to the environment and wild life, *J. Appl. Ecol.*, Vol. 3, pp. 21–8, 1966.

118. MULLER, H. J., Do air pollutants act as mutagents? *Am. Rev. Resp. Dis.*, Vol. 83, pp. 571–2, 1961.

119. MUNRO, P. E., Inhibition of nitrite oxidizers by roots of grass., *J. Appl. Ecol.*, Vol. 3, pp. 227–9, 1966.

120. MUNRO, P. E., Inhibition of nitrifiers by grass root extracts, *J. Appl. Ecol.*, Vol. 3, pp. 231–8, 1966.

121. NEWBOULD, P. J., Ecology of Cranesmoor, a New Forest valley bog. I. The present vegetation, *J. Ecol.*, Vol. 48, pp. 361–83, 1960.

122. NEWMAN, A. S. and DOWNING, C. R., Herbicides and the soil, *J. Agric. Fd. and Chem.*, Vol. 6, pp. 352–3, 1958.

123. NIKLEWSKI, B., Über die Bedingungen der Nitrifikation im Stallmist. *Zentbl. Bakt. Parasitkde. Abt. II.*, Vol. 26, pp. 388–442, 1910.

124. NYE, P. H. and GREENLAND, D. J., The soil under shifting cultivation, *Comm. Bureau of Soils, Harpenden*, Tech. Comm. No. 51, 1960.

125. ODUM, E. P., *Fundamentals of Ecology*, Saunders, Philadelphia, 2nd edn., 1959.

126. ODUM, E. P., *Ecology*, New York, 1963.

127. OVINGTON, J. D., Quantitative ecology and the woodland ecosystem concept, *Adv. Ecol. Res.*, Vol. 1, pp. 103–92, 1962.

128. Ovington, J. D. and Lawrence, D. B., Strontium-90 in maize-field, cattail marsh and oakwood ecosystems, *J. Appl. Ecol.*, Vol. 1, pp. 175–81, 1965.
129. Park, K. J. F., Rawes, M., and Allen, S. E., Grassland studies on the Moor House National Nature Reserve, *J. Ecol.*, Vol. 50, pp. 53–62, 1962.
130. Pearsall, W. H., *Mountains and Moorlands*, London, 1950.
131. Pendleton, R. C. and Hanson, W. C. Absorption of caesium-137 by components of an acquatic community, *Proc. 2nd U.N. Int. Conf. Peaceful Uses Atom. Energ.*, Vol. 18, pp. 419–22, 1958.
132. Perel'man, A. I., *Geochemistry of Epigenesis*, Plenum, New York, 1967.
133. Piper, C. S., *Soil and Plant Analysis*, New York, 1950.
134. Plass, G. N., The carbon dioxide theory of climatic change, *Tellus*, Vol. 8, 1956.
135. Pokrovskii, V. A., The lower limit of the biosphere in the European USSR, as defined by new geothermal investigations, *Tr. Inst. Mikrobiol. Akad. Nank SSSR*, Vol. 9, 1961.
136. Prestt, I., Studies of recent changes in the status of some birds of prey, and fish feeding birds in Britain, *J. Appl. Ecol.*, Vol. 3, pp. 107-12, 1966.
137. Putnam, P. C., *Energy in the Future*, New York, 1953.
138. P'Yavchenko, N. E., The biological cycle of nitrogen and mineral substances in bog forests, *Pochvovedenie*, Vol. 6, pp. 21–32, 1960.
139. Ramann, E., Mineralstoff-Bedarf und Stickstoff-Bedarf zur Holzerengung von Kiefer. Fichte und Rothbuche im Hochwaldbetriebe, *Z. Forest—U. Jadgw.*, Vol. 19, No. 10, 1887.
140. Rankama, K. and Sahama, T. H., *Geochemistry*, New York, 1950.
141. Rennie, P. J., The uptake of nutrients by mature forest growth, *Pl. Soil*, Vol. 7, pp. 49–95, 1955.
142. Reynolds, E. R. C., Ecological and physiological investigations into conifer growth on wet-heath soils, *Ph.D. Thesis, U. of London*, 1956.
143. Rice, E. L., Inhibition of nitrogen-fixing and nitrifying bacteria by seed plants, *Ecology*, Vol. 45, pp. 824–37, 1964.
144. Rice, E. L., Inhibition of nitrogen-fixing and nitrifying bacteria by seed plants. II. Characterization and identification of inhibitors. *Physiologia Pl.*, Vol. 18, pp. 255–68, 1965.
145. Robertson, R. A. and Davies, G. E., Quantities of plant nutrients in heather ecosystems, *J. Appl. Ecol.*, Vol. 2, pp. 211–19, 1965.
146. Robinson, G. W., *Soils, Their Origin, Constitution and Classification*, London, 1932.
147. Robinson, J. B., Nitrification in a New Zealand grassland soil, *Pl. Soil*, Vol. 19, pp. 173–83, 1963.
148. Rodin, L. E. and Basilevic, N. I., *Production and Mineral Cycling in Terrestrial Vegetation*, Oliver and Boyd, Edinburgh, 1967.
149. Romney, E. M., Lindberg, R. G., Hawthorne, H. A., Bystrom, B. G., and Larson, K. H., Contamination of plant foliage with radioactive fall-out, *Ecology*, Vol. 44, pp. 343–9, 1963.
150. Rosedahl, R. O., The effect of mycorrhizal and non-mycorrhizal fungi

on the availability of soluble potash and phosphate minerals, *Proc. Soil. Sc. Soc. Am.*, Vol. 7, pp. 477–9, 1942.

151. ROUTIEN, J. B. and DAWSON, F. R., Some interrelationships of growth, salt absorption, respiration and mycorrhizal development in *Pinus echinata, Am. J. Bot.*, Vol. 30, pp. 440–51, 1943.

152. ROVIRA, A. D. and McDOUGALL, B. M., Microbiological and biochemical aspects of the rhizosphere. In *Soil Biochemistry*, eds. A. D. McLAREN and G. H. PETERSEN, pp. 417–63, Arnold, London, 1967.

153. RUDD, R. L., *Pesticides and the Living Landscape*, Madison, 1964.

154. RUSSELL, E. J., *Soil Conditions and Plant Growth*, 9th edn., London, 1961.

155. RUSSELL, E. J. and RICHARDS, E. H., The amounts and composition of rain and snow falling at Rothamsted, *J. Agric. Sci., Cambridge*, Vol. 9, pp. 309–37, 1919.

156. RUSSELL, R. S., Radio-isotopes and environmental circumstances: the passage of fission products through food chains. In *Symposium on Radio-isotopes in the Biosphere*, eds. R. S. CALDECOTT and L. A. SNYDER, pp. 269–92, University of Minnesota, 1960.

157. SCHNEIDER, F., Some pesticide-wildlife problems in Switzerland, *J. Appl. Ecol.*, Vol. 3, pp. 15–20, 1966.

158. SHUTT, F. T., The influence of grain growing on the nitrogen and organic matter content of the western prairie soils of Canada, *J. Agric. Sci.*, Vol. 15, pp. 162–77, 1925.

159. SIMPSON, J. R., Mineral nitrogen fluctuations in soils under improved pasture in Southern New South Wales, *Aust. J. Agric. Res.*, Vol. 13, pp. 1059–72, 1962.

160. SKUJINS, J. J., Enzymes in soil. In *Soil Biochemistry*, eds. A. D. McLAREN and G. H. PETERSEN, pp. 371–416, Arnold, London, 1967.

161. SMIRNOVA, K. M. and GORODENTSEVA, G. A., The consumption and rotation of nutritive elements in birch woods, *Bull. Soc. Nat. Moscow (biol.)*, Vol. 62, pp. 135–47, 1958.

162. SMITH, P. F., Mineral analyses of plant tissues, *A. Rev. Pl. Physiol.*, Vol. 13, pp. 81–108, 1962.

163. STARKEY, R. L., Some influences of the development of higher plants upon the microorganisms of the soil, *Soil Sci.*, Vol. 45, pp. 207–49, 1938.

164. STEPHENS, E. R., SCOTT, W. E., HANST, P. L., and DOERR, R. C., Recent developments in the study of the organic chemistry of the atmosphere, *J. Air Poll. Control Assoc.*, Vol. 6, pp. 159–65, 1956.

165. STERN, V. M., SMITH, R. F., BOSCH, R. VAN DEN, and HAGEN, K. S., The integrated control concept, *Hilgardia*, Vol. 29, pp. 81–101, 1959.

166. STOCKS, P. and CAMPBELL, J. M., Lung cancer death rates among non-smokers and pipe and cigarette smokers. Evaluation in relation to air pollution by benspyrene and other substances, *Br. Med. J.*, Vol. 2, pp. 923–9, 1955.

167. STOKINGER, H. E., Effects of air pollution on animals. In *Air Pollution*, ed. C. S. STERN, Academic Press, New York, 1962.

168. STRICKLAND, A. H., Some estimates of insecticide and fungicide usage in agriculture and horticulture in England and Wales, 1960–64, *J. Appl. Ecol.*, Vol. 3, pp. 3–13, 1966.

169. Tamm, C. F., Some observations on the nutrient turnover in a bog community dominated by *Eriophorum vaginatum*, *Oikos*, Vol. 5, pp. 189–94, 1954.

170. Tamm, C. O., The atmosphere, *Handb. Pfl. Physiol.*, Vol. 4, pp. 233–42, 1958.

171. Tamm, O., *Northern Coniferous Soils*, Oxford, 1950.

172. Tarzwell, C. M., The toxicity of synthetic pesticides to aquatic organisms, and suggestions for meeting the problem. In *Ecology and the Industrial Society*, eds. G. T. Goodman, R. W. Edwards, and J. M. Lambert, pp. 197–218, Oxford, 1965.

173. Templeton, W. L., Ecological aspects of the disposal of radioactive wastes to the sea. In *Ecology and the Industrial Society*, eds. G. T. Goodman, R. W. Edwards and J. M. Lambert, pp. 65–89, Oxford, 1965.

174. Theron, J. J., The influence of plants on the mineralization of nitrogen and the maintenance of organic matter in the soil, *J. Agric. Sci.*, *Cambridge*, Vol. 41, pp. 289–96, 1951.

175. Thomas, M. D., The effects of air pollution on plants and animals. In *Ecology and the Industrial Society*, eds. G. T. Goodman, R. W. Edwards, and J. M. Lambert, pp. 11–33, Oxford, 1965.

176. Thornton, I., Nutrient content of rainwater in the Gambia, *Nature*, Vol. 205, p. 1025, 1965.

177. Timmermans, J. A., Lutte contre la végétation aquatique envahissante, *Trav. Sta. Rech. Groenendaal*, D, No. 31, 1961.

178. Ulianova, O. M., Isolation of pure *Nitrosomas* cultures from various substances, and their characteristics, *Mikrobiologiya*, Vol. 29, pp. 813–19, 1960.

179. US Dept. of Health, Education and Welfare, Pesticide residues, facts for consumers, *Food and Drug Administration*, Pub. No. 18, Washington DC, 1963.

180. Urey, H. C., On the early chemical history of the earth, and the origin of life, *Proc. Nat. Acad. Sci. US*, Vol. 38, pp. 351–63, 1952.

181. Van der Drift, J. and Witkamp, M., The significance of the breakdown of oak litter by *Enoicylla pusilla*, *Arch. néerl. Zool.*, Vol. 13, pp. 486–92, 1959.

182. Van Klingeren, B., Koeman, J. H., and Van Haaften, J. L., A study of the hare (*Lepus europeus*) in relation to the use of pesticides in a polder in the Netherlands, *J. Appl. Ecol.*, Vol. 3, pp. 125–31, 1966.

183. Vernadskii, V. I., *Studies in geochemistry*, 2nd Russ. edn, Leningrad, 1934.

184. Vinogradov, A. P., Biogeochemical provinces and their role in organic evolution, *Geokhimiya*, Vol. 3, 1963.

185. Vinokurov, M. A. and Tyurmenko, A. N., Materials in the forest's biological cycle of nitrogen and phosphorus, *Pochvovedenie*, Vol. 7, pp. 787–91, 1958.

186. Waksman, S. A., *Soil Microbiology*, New York, 1952.

187. Way, M. J., The natural environment and integrated methods of pest control, *J. Appl. Ecol.*, Vol. 3, pp. 29–32, 1966.

188. Weatherell, J., The checking of forest trees by heather, *Forestry*, Vol. 26, pp. 37–40, 1953.

189. WEBER, R., Vergleichende Untersuchungen über die Auspüche der Weisetanne und Fichte an die mineralischen Nährstoffe der Bodens, *Allg. Forst-u. Jagdztg.*, Vol. 57, No. 1, 1881.

190. WEISS, H. V. and SHIPMAN, W. H., Biological concentration by killer clams of cobalt-60 from radio-active fallout, *Science*, Vol. 125, p. 695, 1957.

191. WELLS, H. G., HUXLEY, J. S., and WELLS, G. P., *The Sciences of Life*, Garden City, New York, 1939.

192. WETSELAAR, R. and HUTTON, J. T., The ionic composition of rainwater at Katherine, N. T., and its part in the cycling of plant nutrients, *Aust. J. Agric. Res.*, Vol. 14, pp. 319–29, 1963.

193. WHEATLEY, G. A. and HARDMAN, J. A., Indications of the presence of organochlorine insecticides in rainwater in Central England, *Nature*, Vol. 207, pp. 486–7, 1965.

194. WHITE, G., *The Natural History of Selborne*, London, 1789.

195. WHYTE, R. O., The myth of tropical grasslands, *Trop. Agric.*, Vol. 39, pp. 1–11, 1962.

196. WILKINSON, H. R., Man and the natural environment, *Occasional Papers in Geography No. 1*, University of Hull, UK, 1963.

197. WILL, G. M., Nutrient return in litter and rainfall under some exotic conifer stands in New Zealand, *N.Z. Jl. Agric. Res.*, Vol. 2, pp. 719–34, 1959.

198. WILLRICH, T. L., Management of agricultural resources to minimize pollution of natural waters. In *Proceedings of the National Symposium on Quality Standards for Natural Waters*, Univ. of Michigan, pp. 303–14, 1966.

199. WITKAMP, M., FRANK, M. L., and SHOOPMAN, J. L., Accumulation and biota in a pioneer ecosystem of kudzu vine at Copperhill, Tennessee, *J. Appl. Ecol.*, Vol. 3, pp. 383–91, 1966.

200. WITTICH, W., Untersuchungen über den Verlauf der Streuzersetzung auf einem Boden mit starker Regenwurmtätigkeit, *Sch. Reihe forstl. Fak*, Göttingen, Vol. 9, pp. 5–33, 1953.

201. WOLFF, E., *Aschen-analysen*, Berlin, 1871, 1880.

202. YAALON, D. H., The concentration of ammonia and nitrate in rainwater over Israel in relation to environmental factors, *Tellus*, Vol. 16, pp. 200–04, 1964.

203. YOUNG, L. A. and NICHOLSON, H. P., Stream pollution resulting from the use of organic insecticides, *Progressive Fish Culturist*, Vol. 13, p. 193, 1951.

4

Environmental Limitations in Ecosystem Development

4.1 Introduction

In addition to the primary controls of energy availability and the subsidiary effects of biogeochemical cycling, all ecosystems have inherent environmental limitations which may restrict the numerical or physiological growth of organisms within them. It is these which are to be examined in this chapter.

As used in biogeography, the concept of *environment* or *habitat* is essentially holistic in application, embodying aspects of physical phenomena such as climate, the water balance, rock structure, soil, and topography, as well as interorganism reactions of a biotic nature. Moreover, it must include a time factor, since it is often the case that the influences of environmental parameters such as these only become apparent when timelag effects are taken into consideration. To a very large extent, the physiology of all organisms reflects their environment, even though certain growth processes may, at times, follow a rhythmic *endogenous* pattern which is strictly genetically controlled, as when the leaf attitudes of some trees vary according to a circadian cycle of approximately 24 hours in length (the biological clock). Most other physiological interactions are, to some extent, either directly or indirectly affected by the relative availability of such basic necessities as water, heat, oxygen, and carbon dioxide. This is most clearly seen in the case of vegetation communities since, unlike the vast majority of animals, plants cannot quickly escape from their immediate habitat, so that their morphogenetic response to external change has often to be extremely rapid. This chapter will therefore be concerned mainly with an examination of relationships between the plant components of ecosystems and their environment, and the reader is referred to chapter 5 for a more detailed study of restraints among animal populations. In both chapters, only those influences which (to take Mason and Langenheim's phrase[78]) are 'operationally significant' to organisms at some stage of their life cycle are to be considered.

Before the turn of this century, studies of environmental limitations on plant growth were few in number due to a general lack of information. Even in occidental countries, such important features as temperature, precipitation, and windspeed were not recorded on a consistent and continuous basis until towards the end of the last century; in other parts of the world, their collation may date back only for a score of years. Scientific analyses of the soil-vegetation complex have a slightly longer history in that they can be traced to Justus Liebig's report[74] on the minimal nutrient requirements of certain plants, produced in the first half of the nineteenth century (see also chapter 3). But the more general study of interactions between living organisms and their environment was not formally conceived until 1869, when the German zoologist Haeckel[50] introduced the term *ecology*. Even then, this notion was not used widely until the appearance of Warming's book on ecological plant geography in 1895,[145] despite the fact that isolated attempts to explain such interrelationships continued to appear from time to time in the later years of the nineteenth century, as in the case of Semper's *Animal Life*, published in 1881.[115]

At first, investigations within this field were largely concerned with the analysis of single, or relatively narrow groups of environmental parameters: thus, patterns of temperature and rainfall were often assessed separately with little attention paid to the way in which they might act reciprocally with other influences. The inprofitability of attempting to prove too close a connection between plant formations and their environment in this way was soon recognized, notably by Livingston and Shreve[75] in their extensive review of vegetation in the USA, and it is now generally agreed (see Went[152]) that, as yet, we simply do not know what most of the detailed physiological responses of organisms to particular parameters will be. Indeed, in view of the fact that all environments are changing continuously, often according to diurnal, seasonal, annual,

121

and permanent trends at one and the same time, these are often much more complex than might appear at first sight. Consequently, recorded mean values of data for any area may not be of significance for prolonged periods, except where they are incorporated into advanced computer-channelled systems of data analysis involving trend-forecasting. Surveying the field as a whole, it is still true to say that many more measurements need to be taken and examined before the full significance of each and every environmental factor is realized. It is the purpose of this chapter, in summarizing the results of our accumulated knowledge to date, to indicate our current progress towards this aim.

THE CONCEPT OF TOLERANCE

Within a given environment, all organisms will have *limits of tolerance* beyond which they will not grow. For individual components, these may prove to be exceedingly difficult to define, since all are interrelated to some extent, and modifications in one may lead to changes in tolerance levels for all of the others, as when the presence of one greatly in excess of normal helps to compensate for deficiencies in others, a process known as *factor compensation*. But despite this interaction, it is still often the case that the influence of only one or two components (termed respectively the *master limiting factor* or *paired limiting factors*) can lead to restraints upon successful growth, though the relative importance of these may be modified at different times in an organism's life history, or with changing population pressures. Thus, Larsen[69] has noted that among certain moth groups the level of food availability may be the controlling influence on population size in times of food scarcity, even though tighter restrictions may be exercised by the amounts of local rainfall and the relative incidence of ultraviolet light in times of food abundance. On occasions, limiting factors which operate at the margins of tolerance may become critical to the survival of a species, in which case they are often termed *regulatory factors* instead. Difficulties also arise in the satisfactory definition of the term *tolerance*, which is closely associated among plant communities with the related notions of *avoidance* and *evasion*. Most commonly plant physiologists refer to *tolerance* as a hardiness factor, which enables an organism to withstand harmful conditions inside its cell tissues, whereas *avoidance* and *evasion* imply that the organism has successfully prevented injurious factors from

122

entering the tissues; thus, in xerophytic plants, extreme drought may be evaded through dormancy, avoided through the development of a hard and thick cuticle on leaves which reduces transpiration, or

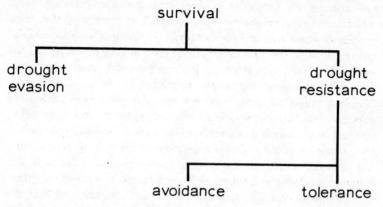

Fig. 4.1 The relationships between drought resistance, tolerance, avoidance, and evasion (from an original diagram in J. Levitt,[73] copyright Academic Press, New York. Reproduced by permission)

tolerated through a wide range of other controls. Summary illustrations of these concepts are given in Table 4.1 and Fig. 4.1. At times, combinations of all three effects can operate within the same plant, when the specific physiological reactions to each may be heavily obscured.

Table 4.1   The nature of resistance towards an injurious factor (from J. Levitt,[73] copyright Academic Press, New York. Reproduced by permission)

	Conditions of living plant cells possessing—	
Environmental factor	*(a) avoidance*	*(b) tolerance, hardiness*
Low temperature	(warm)*	cold
Freezing	(unfrozen)*	frozen
High temperature	(cool)*	hot
Water deficiency	high vapour pressure	low vapour pressure

* Condition rare or non-existent.

Some of the problems associated with these concepts were outlined initially by Shelford in 1911 and 1913.[117,118] It was he who was the first to accept the idea that while all organisms have maximum, minimum, and optimum requirements for growth which are related to their physical environment, these are not invariable and, indeed, may be appreciably modified should external circumstances change.

123

In other words, organisms will always eventually adapt to alterations in their habitat if they are able to do so. He also put forward the now generally adopted view that limiting factors within an environment will affect many species most severely at the reproductive period, which is the most critical time for growth, and recognized that even within the same species wide ranges in the levels of tolerance might be found, as in the case of several prairie grasses in the USA, the reactions of which to seasonal differences in day length show considerable latitudinal variations (section 4.3). As a general rule, Shelford suggested that those species with the greatest tolerance ranges would be most widely distributed, although this supposition has had to be modified as a result of later research which established that *reverse compensatory factors* may prevent this taking place: thus, the drought tolerance of certain grasses is greatly reduced in the event of available soil nitrogen being present in subnormal quantities.[102] More recently, too, the curious fact has emerged that the general optimum conditions for growth often depend on suboptimal relationships for specific functions within organisms, a feature which is especially noticeable among long-living pine trees in California and Nevada.[112] It is now also known that many species apparently prefer to live in natural environments which are suboptimal for growth, a pattern first noted by Kinne[64] for the marine coelenterate *Cordylophora caspia*.

4.2 Techniques for measuring environmental conditions

Environmental conditions may be determined through the use of a wide range of instrumentation, the technology of which is continually being revised. Through this, data may be collected at specific intervals of time (daily, twice-daily, hourly, etc.) or, as is more frequently the case today, from continuous recording devices whose readings are penned mechanically on long paper strips. Recording stations may be manned or largely automatic, and the data obtained may refer either to the *macroenvironment*, which lies largely outside the influence of vegetation and its roots, or the *microenvironment* within the vegetated layer.*

* It should be noted that the two terms *macroenvironment* and *microenvironment* may be used differently in climatology and meteorology, where they refer respectively to data recorded at, or below, a height approximately 1·5 m above ground surface, where 'standard' temperature, wind and humidity measurements are usually taken.

Usually, most environmental parameters are measured individually by means of increasingly intricate and sensitive equipment. Thus, radiation income, the intensity of which has been assessed in the past with pyrheliometers, is now being detailed through the use of radiometers which indicate the net energy *flux* (i.e., its exchange and movement) at or near to the ground surface to within very fine limits. Wavelength quality may be determined by reference to photoelectric cells and, as already noted in chapter 2, these are especially valuable for examining the inflow of visible light waves required by green plants in the process of photosynthesis. Infrared and ultraviolet sensors are also available. Temperatures of air, soil, and plants, which traditionally have been ascertained by a range of standard or maximum-minimum thermometers and their associated automatic recorders, are at present being evaluated through three main types of sensor: the resistance thermometer for soil and leaves, in which variations in the electrical resistance of metal show a linear relationship with modifications in temperature; the thermistor, which shows similar, but non-linear, changes in resistance; and the thermocouple, in which metals help to record temperature disparities between the environment and a known standard. Atmospheric humidity is a more difficult parameter to appraise and, although new sensors can be utilized, the inexpensive wet-dry bulb thermometer (or psychrometer) is often still accurate enough to calculate even minor differences in absolute humidity, dew point, or vapour pressure (section 4.5). Precipitation may be collected in a variety of recording or non-recording rain gauges, and evaporation and transpiration are determined from evaporating pans, evapotranspirometers, and atmometers. Soil moisture quantities, analysed in the past through the tedious procedure of oven-drying field samples so as to obtain a percentage weight of water in dry soil, can now be obtained through using accurately-balanced weighing lysimeters and neutron-scattering probe techniques. The measurement of soil chemical factors has already been discussed in chapter 3. Wind speeds and directions are collated mainly from self-recording vanes and anemometers. Additional details of these and other means of delimiting environmental parameters may be gleaned from an examination of standard works, such as the British Meteorological Society's Handbook,[82] and more technical summaries, such as those incorporated into the British Ecological Society's recent publication on the measurement of environmental factors.[139]

In addition to the precise appraisal of individual components, attempts have recently been made to measure the general holistic effect of the environment on plants, either through the establishment of large-scale lysimeters (e.g., at Davis, California), in which water exchanges in ecosystems of up to an acre in size may be determined, or through the concept of a *phytotron*,[152] a large-scale laboratory which enables the growth of plants to be examined in an artificial environment in which the parameters are closely controlled. Although, so far, the technical difficulties and the cost of undertaking such experiments have made these relatively unprofitable undertakings, subsequent developments along these lines may be expected to add substantially to our knowledge of the wider influences of environment upon the organisms which live within it.

LIMITS OF TOLERANCE IN TERRESTRIAL ECOSYSTEMS

Granted that some difficulties exist in both the interpretation of the tolerance concept and the measurement and analysis of environmental components, the consequences arising from the existence of limiting factors may be seen everywhere. For land-based plant communities, with which this chapter is most immediately concerned, four overriding divisions of potential restraints on growth result from them, and may be categorized as follows: *first*, a climatic group, which is usually most important, involving conditions of light, heat, and temperature, moisture availability, and wind; *second*, a series of topographic influences which can assume significance especially where effects of altitude, aspect, and slope help to modify vegetation formations on high mountains or in high latitudes; *third*, edaphic checks induced by the presence of certain physico-chemical and moisture conditions within the soil; and *fourth*, biotic restrictions which occur notably when animal populations overgraze their ranges. More complete details as to the full effects of each of these may be found in subsequent sections of this chapter.

4.3 The light factor

It has already been established in chapter 2 that the energy resources of all ecosystems are largely dictated by the quantity of incoming solar radiation received by them at the earth's surface. Certain characteristics of this radiant energy may also in themselves be responsible for shaping (and sometimes limiting) the growth of organisms within ecosystems. Thus, the presence and intensity of visible light are the most important single factors which stimulate the process of photosynthesis and the production of chlorophyll, the opening and closing of *stomata* (leaf pores), and the formation of *auxins* (growth-forming substances) in green plants. In the latter case, variations in light intensity may eventually produce altered forms either of whole organisms or their component parts, so that leaves in the shaded centre of a tree may be greatly modified in shape as compared to those fully exposed to the sun, and even large trees, such as oak (e.g., the English common oak, *Quercus robur*), may assume a different appearance in forests as opposed to open parkland. Recent experiments have made it abundantly clear that many plants can vary their physiology and morphology very quickly to adapt to changing light conditions. Hughes[59] has reported that seedlings of small balsam (*Impatiens parviflora*), coffee, and several temperate deciduous forest species can all respond to falling light intensities by increasing their leaf area, so that rates of photosynthesis remain the same; in very strong light, Cooper[30] has noted that even cell structures within leaves can be reoriented relatively quickly to become aligned parallel to the beams of light, a feature which is characteristic of much of the scrub vegetation of central California.

Taken as a whole, three aspects of light income—intensity, quality, and photoperiod (or daylength)—are significant to plant growth, and each of these may vary diurnally and seasonally according to latitude. Intensity of light tends generally to be greatest towards the Equator and in arid climates, where the amounts lost by diffusion within and reflection from the atmosphere are least; in contrast, very low light intensities may be found in high latitudes, areas with a high percentage of cloud cover, and urban regions with major atmospheric pollution problems. Despite the ease with which many species can adapt successfully to a wide range of different light intensities, it is also the case that both too much and too little light may at times restrict growth considerably. Direct exposure of protoplasm to light can lead to death, but even where this does not occur excessive light often has other deleterious effects, as when it induces the photooxidation of enzymes in certain species, thus reducing chemical synthesis, particularly with respect to protein formation.[131] The consequences of too little light may be equally marked, causing adaptive features to form in many seedlings (as noted in Table 4.2),

127

Table 4.2 Light climate, regenerating species and adaptive features of seedlings in selected shaded habitats of temperate regions (modified from J. P. Grime[48])

Type of habitat	Intensity of illumination 0–40 cm above ground surface	Species capable of regeneration Herbaceous	Species capable of regeneration Woody	Adaptive features of seedling
1. Closed grassland	Marked increase in intensity with small increase in height above ground surface	Plantago lanceolata† Rumex acetosa†	Quercus rubra* Acer rumbrum*	(a) Large seeds, (b) tall stature, (c) rapid extension of growth on shading.
2. Open woodland and scrub	High intensity before canopy expanded; often moderately high later in growing season. Where ground veg., increases in intensity with small increases in height.	Impatiens parviflora† Endymion non-scriptus* Urtica dioica†	Quercus rubra* Acer rubrum*	(a) Large leaf area if shaded, (b) tall stature, (c) horizontal laminae, (d) low diffusion resistance in leaves, (e) rapid extension of growth, on shading.
3. Dense woodland and forest	Very low intensity over most of growing season. Only small changes in intensity with change in height above ground.	Deschampsia flexuosa†	Tsuga canadensis* Acer saccharum*	(a) Resistance to fungal attack, (b) low respiration rates, (c) horizontal laminae, (d) limited extension of growth, on shading.

* American species.
† European species.

or even at times preventing the growth of all but the most shade tolerant species. Observed and distinctive differences in light tolerance among plants have in the past given rise to a broad division between those which are *heliophytic* and grow best in strong sunlight, and those *sciophytes* which normally develop under lower light intensities, but these terms may often be used only in a very general sense, particularly in the humid mid-latitudes, where for many plants and in many microenvironments light intensities are constantly changing according to the time of the day or season, the passing of clouds, the direction of wind and the swaying movement of surrounding species, the number and size of holes in a forest canopy[5] and many other habitat factors.

Usually, however, it is accepted that most plants will have optimum mean light conditions for growth, to which they are genetically best suited, and which they will strive to reach in competition with their neighbours. This struggle for light is responsible, in part, for the development of *vertical stratification* in many communities, in which distinctive layers of vegetation occur, each of which has a different mean tolerance to light and, thus, markedly dissimilar component species. But the vertical extent and the number of these layers can vary considerably, as indicated in Fig. 4.2. In relatively restricted environments, such as grassland, desert, or arctic-alpine zones, only one layer may be found, but elsewhere several are present, as in most forests: here, the topmost layer of tall trees is termed the *canopy*, and is made up of species which benefit from strong light at maturity. Although it is customary to find only one canopy in temperate mid-latitudes, two or three may develop in energy-rich tropical rain forests, where strongly heliophytic plants, such as palms, may even protrude above them as *emergents*. Beneath the canopy everywhere, shade tolerant shrubs usually form another stratum, while herb and ground layers can often also be distinguished; in non-evergreen plant communities of mid-latitudes, the last two frequently consist of species whose flowering and germination times are well in advance of those of others in the same community, possibly so as to take the maximum advantage of light which penetrates to the ground before the leafing period of the canopy and shrub plants. Thus, wood sorrel (*Oxalis acetosella*), violet (*Viola* spp.), and bluebell (*Endymion non-scriptus*) all follow this pattern in British oakwoods. Generally, of course, once leafing has taken place in deciduous forests, and throughout the year in evergreen communities, the

129

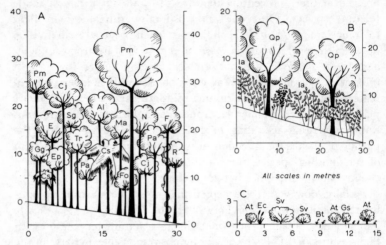

Fig. 4.2 Vegetation stratification in selected communities: (a) tropical rain forest in Jamaica; (b) oakwoods at Killarney, Eire; (c) cold desert vegetation in Washington State, USA. (From original diagrams, respectively in G. F. Asprey and R. G. Robbins, *Ecol. Mon.*, Vol. 27, pp. 251–65, 1952; A. G. Tansley;[129] and W. D. Billings, *Plants and the ecosystem*, © 1964 by Wadsworth Publishing Co., Belmont, California, 94002. Reprinted by permission of the publishers.) Key to species:

Rain forest: *Al, Alchornia latifolia* (jumba); *Cj, Calophyllum jacquinii* (Santa Maria); *Cs, Calyptronoma swartzii*; *E, Eugenia sp.* (redwood); *Ep, Exothea paniculata* (wild genip); *F, Ficus* sp.; *Fo, Faramea occidentalis* (wild coffee); *Gg, Gaurea glabra* (wild akee); *Ma, Matayba apetala* (wannika); *N, Nectandra sp.* (sweetwood); *Pa, Pithe-collobium alexandrii* (shadbark); *Pm, Psidium montanum* (mountain guava); *R, Rubia-*ceous species; *Sg, Symphonia globulifera* (hog gum); *Tr, Trophis racemosa* (ramoon). Oakwoods: *Qp, Quercus petraea* (sessile oak); *Ia, Ilex aquifolium* (holly); *Sa, Sorbus aucuparia* (rowan).

Cold desert: *At, Artemesia tridentata* (sage brush); *Bt, Bromus tectorum* (cheatgrass); *Ec, Elymus cinereus* (giant wild rye); *Gs, Grayia spinosa* (hopsage); *St, Sarcobatus vermiculatus* (greasewood)

effect of stratification is to develop substantially increased levels of shade, so that light intensities at ground level may be only 15 per cent of those at the canopy top in mid-latitudes, and as low as 1 per cent in the more complex tropical rain forests. This means that helio-phytic species within or outside these communities may have diffi-culty in regenerating or colonizing on the forest floor unless a break in the canopy occurs, so allowing greater amounts of light than nor-mal to break through to ground level, though this problem is often partly relieved by the fact that the seeds of some light-seeking species, particularly in tropical rain forests, are able to remain viable in the ground for many years before they germinate, while others are capable of surviving at the seedling stage for periods of up to 50 years, before they are reached by sufficient light to stimulate further growth.[136]

These individual and community adaptations to shading and light intensity must, however, be regarded as broad generalizations when applied to ecosystems as a whole, for not only are the morphogenetic responses to shade extremely complicated, but the whole question of shading is often also closely associated with other potential environmental restrictions, such as root competition for available water nutrients. Thus, Murray and Nichols[90] have indicated that for certain tropical plants, if nutrition is poor, an increased tolerance for shade becomes a necessary preliminary to successful growth; and Oosting and Kramer[98] have observed similar features in more northerly species. Loach[76] has also pointed out the curious fact that in many ecosystems seedlings of shade tolerant species may be slower to adapt to increased shade than those which are usually heliophytic in nature, even though rates of photosynthesis remain close to identical for the two groups, and that in these instances it may not be the tolerance to shade, but rather the rates of respiration under shade which are critical to the survival of a plant. In all ecosystems, ground shading effects which are restrictive to growth become more critical when snow is lying for much of the year, even though, as Daubenmire[33] has suggested, photosynthesis may still be possible underneath as much as 40 cm of snow cover. In water, light can penetrate much further to stimulate vegetation growth, although its intensity decreases geometrically as the depth of water increases arithmetically, so that even in clear and still water only 50 per cent of surface light is capable of reaching a depth of 18 m, with much smaller quotients in moving water, especially when it carries a large silt load. Under oceanic conditions, photosynthesis virtually ceases at depths of over 50 m, when darkness is almost complete.

It frequently happens that not only the intensity, but also the *quality* of light can affect plant growth. Probably the most important restrictions so far noted in this respect refer to those imposed by relative differences in penetration to the ground surface of ultra-violet radiation. It has already been made clear in chapter 2 that, as incoming light energy passes through the atmosphere, greater amounts are lost from the ultra-violet segment of the spectrum than elsewhere, since this is particularly vulnerable to interception by ozone layers approximately 16 miles above sea level. Ultraviolet losses continue to remain substantial as one approaches lower altitudes, so that usually more is taken into plant systems in high mountain regions than on other parts of the earth's surface; accordingly, Billings[11] has estimated that the ultraviolet income at altitudes of 4000 m in the Rocky Mountains of Colorado may be twice that at sea level. This fact may have important repercussions on the development and distribution of plants at high altitudes, especially since ultraviolet is known to be potentially among the most lethal of all incoming radiation, having the greatest photochemical effect in plant cells. The main consequence of this is that, while high-altitude plants as a group have adapted themselves to survive excessive ultraviolet, low-altitude plants frequently have not, so that they are often unable to disperse into major mountain regions due to this one feature alone, even though all other environmental conditions may be favourable. However, it must not be assumed that the effects of ultraviolet radiation are always wholly harmful, for moderate increases in income have encouraged egg production in poultry and may possibly raise the rates of reproduction in other species too.

Differences in the quality of light can also vary appreciably within vegetation communities. Ultraviolet rays may continue to be strongly intercepted once they have reached a canopy, so that much smaller amounts are received at the ground surface of forests than in more open communities. Considerable variations in the seasonal patterns of visible light are also known in forests, as noted initially by Salisbury[110] in detailing changes in the ratio of direct to diffuse light in the *light phase* (leaf fall to leaf expansion) and the *shade phase* (mid-May to mid-autumn) of oak (*Quercus robur*) woodlands in Britain; more recent estimates have suggested that up to 95 per cent of light penetrating to the forest floor may be direct in the shade phase, as opposed to only 40 to 70 per cent in the light phase.[14] Equivalent figures for direct light income in mid-latitudinal pine forests may reach 50 per cent in the summer, and close to zero in the winter months, when stems and twigs may be very important interceptors of direct light.[37] However, the final effects of these features on plant growth have not yet been determined.

For many species the photoperiod (or length of day) is a further critical factor in successful growth. Even though physiological reactions among some seedlings are sufficiently sensitive to be stimulated by moonlight[9,114], most plants require either direct or diffuse sunlight for their healthy development, so that differences in the length of day according to latitude, ranging from the constant 12 hours of the equator to the 6 month daylight period at the Poles, may be expected to produce some modification to growth patterns. At times, also, the length of the nightime period may be equally significant, as Nitsch[92] has pointed out with regard to the initiation of leaf fall in the western plane (*Platanus occidentalis*). It was Garner and Allard[42] who first observed the response of plants to these variations, and Allard[1] who first suggested that the relatively constant photo-period in the tropics may explain why flowering occurs there through-out the year while it is restricted to a shorter season with long photo-periods in extra-tropical areas. It is now an accepted fact that *short-day* and *long-day* species and varieties of many plants exist,* and that each of these may not be capable of growing in areas where the photoperiod differs substantially from that to which they are accustomed, a feature which helps to explain why northern plants, such as wheat, do not grow well in the tropics, despite the fact that temperature and moisture conditions are often suitable. However, these relationships may become somewhat blurred at high latitudes in which the summer photoperiods are long, where for some species (e.g., the sycamore, *Acer pseudo-platanus*) endogenous controls may stop growth mechanisms after the passage of a certain length of time as measured from daybreak, regardless of the length of the photo-period. In addition to the direct stimulus of growth patterns and other physiological reactions within plants, differences in photo-periodism may also account for the instigation of hibernation and reproduction seasons among certain mammals and rest phases in insect populations, the establishment of nesting and migration habits among birds,[107,19] and, possibly, even the numbers of some nitrogen-fixing bacteria present in legume nodules.[94] As expected, the in-terruption of normal photoperiodism often induces abnormal reactions among populations, and in the case of an eclipse of the sun, may even lead to the temporary roosting of birds and the closing of flowers at any time during the day.

Despite these comments, caution should always be exercised in any attempt to suggest too close a degree of interaction between the growth and distribution of plants and the light factor in ecosystems,

* These terms are usually used to describe the characteristics of their flowering period.

for, as already noted with respect to shading, it is often the combination of light and other environmental parameters which is more meaningful, in this context, than the direct effects of light in themselves. Thus, Blackman[13] has suggested that, for leaf growth in *Salvinia natans*, light conditions may only become limiting when the leaf temperature is in excess of 20°C, and in laboratory experiments Steward[128] has concluded that metabolism in some plants (mint, banana, tulip, Jerusalem artichoke, peas, and conifers) seems to be largely controlled jointly by the three associated influences of light intensity, photoperiodism, and the range of atmospheric temperatures, rather than by any one of these in particular.

4.4 Heat and temperature

An adequate supply of heat is essential to the growth of organisms, all of which have their own maximum, minimum, and optimum temperature requirements. For plants, as for most animals, these are often not clear-cut, for the temperatures necessary to the efficient functioning of a wide range of internal mechanisms can vary appreciably. Thus, those most favourable for photosynthesis are usually lower than those for respiration, a feature which may give rise to some physiological problems since the rate of photosynthesis will fall appreciably should the temperatures rise to the optimum for respiration. Critical temperatures for any function can also be altered with changes in age, state of health, and the amount of water present in a plant; furthermore, they may be modified seasonally in a few cases.

Heat contrasts within terrestrial ecosystems may be analysed either by reference to the *absolute temperatures* on which physiological functions depend, or relatively, as an indicator of the *flux* of heat moving between soil, plants and animals, and atmosphere. Differential patterns of absolute temperature can be accounted for largely in terms of variations in latitude, altitude, and distance from the sea. Since the annual receipt of energy in the tropics is much greater than towards the Poles, it is to be expected that temperatures will also differ considerably on a latitudinal basis. This means that plant species may be broadly divided into those which can withstand the *megathermal* conditions of the tropics, where mean monthly temperatures do not fall below 20°C and are usually above 25°C; those which are *microthermal*, occurring within high latitudes in areas where the mean monthly temperature does not exceed 10°C; and those *mesothermal* plants in mid-latitudes with a physiology adapted to strongly marked seasonal rhythms of life (see also Table 4.7). Since high mountain ranges have distinctively colder climates, and vegetation communities more tolerant of cold than one would normally expect for the latitude in which they occur (section 4.8), the effects of altitude may disrupt this general pattern; and relative continenality (or distance from the sea) can lead to further modifications, if only because more extreme ranges of mean monthly and diurnal temperatures develop as one moves away from open oceans. Temperature conditions may also be ameliorated or depressed by the persistent presence of warm or cold currents of air or water: thus, the western coasts of Scotland, Ireland, and Norway have much warmer winters than those of similar latitudes in Labrador and Newfoundland, due to the predominance of warm south-westerly air currents and the proximity of the Gulf Stream ocean current in the first three regions. Moreover, it should be remembered that the state of the surface can lower or raise temperatures from those anticipated, at times on a relatively large scale, as when soil temperatures are reduced over extensive areas of the northern hemisphere by the presence of *permafrost* (permanently frozen ground, a relic of the Pleistocene ice ages, the current distribution of which can be noted in Fig. 4.3). However, apart from this, such surface effects are usually more local than regional in importance, and result predominantly from relatively minor convolutions in topography, or from changing edaphic conditions.

Over the world as a whole, the known extremes of air temperature in land ecosystems, as recorded under a screen at standard heights above the ground (ca. 1·5 m), range from −88·2°C (−126°F) in Vostok, Antarctica, and −70°C (−94°F) in Verkhoyansk, Siberia, to 57°C (134°F) in Death Valley, California, and 58°C (136°F) in the northern Sahara.* Mean diurnal, mean monthly, and mean annual screen temperatures naturally show much smaller variations, from maxima in each case of approximately 27°C (80°F) in equatorial regions to minima of below freezing point on the polar icecaps. But except as a general guide to the absolute limits of megathermal, microthermal, and mesothermal species, mean annual temperatures

---

* At times, temperatures on non-organic surfaces exposed to direct sunlight may approach 100°C; other than these, the highest normally recorded in natural ecosystems are found in hot springs, at 70°C.

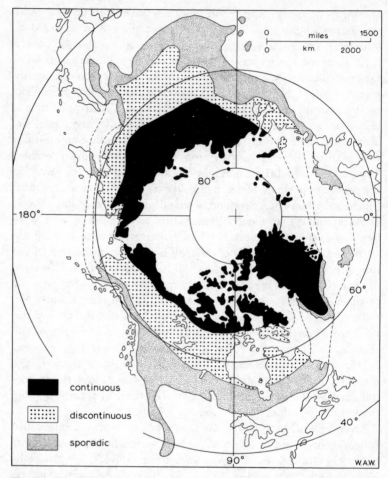

Fig. 4.3 The extent of permafrost in the northern hemisphere (from an original diagram in R. F. Black[12])

are only rarely of significance to the biogeographer; for particular terrestrial ecosystems, mean monthly temperatures are much more important. However, it is usually the case that the daily ranges in temperature, especially those near to the ground where germination and sprouting occur, are the most critical of all to plant growth, these often being much greater than would be inferred from analyses of mean diurnal temperatures at screen level (see Fig. 4.4). Where stratified communities, such as forests and woodlands, have become

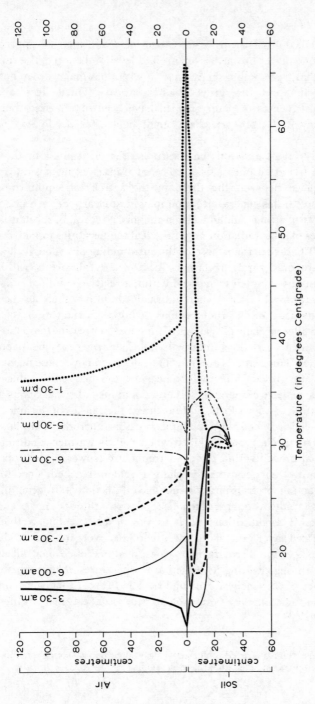

Fig. 4.4 The march of temperature in air and soil, 31 July 1953, Nevada desert (from W. D. Billings, *Plants and the ecosystem,*[11] © 1964 by Wadsworth Publishing Co., Belmont, California. Reprinted by permission of the publisher)

established, diurnal temperature ranges may also increase appreciably as one moves upwards from ground level to the top of the canopy, so that tree seedlings may have to withstand major differences in microclimate as they grow towards maturity. These effects, and their consequences to plant growth, have been summarized more extensively in Geiger's excellent small book, *The climate near the ground*.[43]

Usually, plant and soil temperatures are very similar to those found in the atmosphere, for a constant transfer of heat is always taking place between the three in the search for equilibrium. Exceptions to this general situation may most clearly be seen under conditions of strong sunlight, when the direct receipt of substantial quantities of solar radiation may raise leaf temperatures to between $20°$ to $30°C$ higher than those of the surrounding air, though these values are only rarely recorded. As a whole, all plants within ecosystems are adapted to succeed within a relatively limited range of temperatures, so that when unusual conditions such as these prevail, damage to cell structures may be initiated, and this, in turn, can often affect specific physiological functions, so creating temporary or permanent restrictions in growth. Thus, enzymes may be inactivated when temperatures are only a few degrees above those needed for optimum growth, a fact which accounts for the greatly reduced metabolic activity of many mesothermal and megathermal plants at temperatures of above $45°C$, and (at times) their death or dormancy when $55°C$ is reached; and for similar reasons microthermal plants (e.g., arctic species) may be killed when slightly warmer conditions than usual prevail within their environment.* However, the upper lethal limit of temperatures for plants is seldom reached, since the cooling action of transpiration on leaves (section 4.5) generally increases rapidly as temperatures rise; or where this does not occur, physiological modifications, such as the development of a thick cuticularized layer on leaf surfaces, help effectively to insulate the living parts of the plant. In most cases, death is much more likely to occur through wilting at sublethal temperatures or through the persistence of extremely warm conditions in the root zone, for many roots tend to become less permeable to water intake at temperatures of over $40°C$[58].

---

* In temperatures above $55°C$, blue-green algae adapted to hot spring conditions may be the only form of life, these often developing best at $70°C$.

Excessively cold temperatures may also result in death, as when freezing induces tissue dessication or the precipitation of proteins. Similar reactions may also occur above freezing point if, for example, abnormally low temperatures prevail within megathermal communities. Nevertheless, the consequences of excessive cold are often reduced or negated through seasonal adaptations of a physiological or morphological nature, whose effect is to reduce the metabolic activity within species while at the same time ensuring their survival. Thus, in many mesothermal plants, rates of metabolism may fall appreciably when the *leaf* temperatures drop to below 6°C (43°F), and the cell-sap concentration of many conifers is known to increase during the autumn months so as to effectively reduce its freezing point. Morphological adaptations include the adoption of dwarf forms which hug the ground surface, so as to be covered by an insulating layer of snow in winter, or the restriction of life to the subsurface root layer of plants during the severest months. It is not only the degree of cold, but also the length of the cold period which may determine whether or not a species survives, a point which has been recognized in many parts of the world. To take one example only, succulents such as the giant cactus (*Cereus giganteus*) in the deserts of the American south-west can withstand up to but no more than 18 hours of continuously freezing temperatures, so that, in consequence, their distribution is restricted to the warmer lower parts of this region.

Despite the very narrow tolerances found in most plants to atypical changes in temperature, some will grow well in widely contrasting régimes of heat availability; thus species may have broad or restricted distributions within megathermal, mesothermal, and microthermal limits, or at times may even overlap all three. Nevertheless, the ultimate temperature controls for all plants seem to be clear. *First*, limits are set by extremes of heat or cold; these may be isolated extremes, or those resulting from abnormally low or high mean monthly temperature conditions which indicate prolonged periods of severe weather, or a combination of both. Thus, the poleward limit for growth and reproduction of ivy (*Hedera helix*) in northern Europe is set by the extreme winter minimum temperatures in areas where the mean temperature of the coldest month is −1·5°C or below, as intimated in Fig. 4.5. *Second*, restraints on growth may be set by the amounts of *accumulated heat* absorbed by plants in the growing season, and these, of course, vary widely according to

139

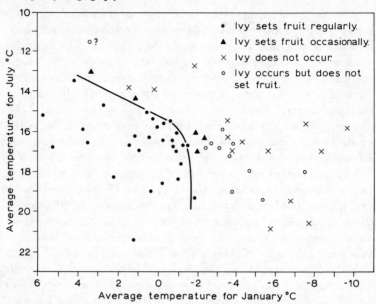

Fig. 4.5 Temperature conditions controlling the reproduction of ivy (*Hedera helix*) (from an original diagram in J. Iversen[61])

latitude, as shown in Fig. 4.6. This is an extremely complex factor, the needs of plants with respect to accumulated heat being very diverse; some mesothermal species may need continuous summer daytime temperatures of between 20°C to 40°C in order to maintain effective rates of photosynthesis and respiration, while others in the same ecosystem can be much more flexible in their heat demands. Certain restrictions caused by the relative availability of accumulated heat may, however, be very clear, as when trees are customarily confined to regions in which the mean monthly temperatures are in excess of 10°C for at least part of the year, except where they grow in a very stunted form, because there is not sufficient heat to keep them alive in colder climates than this. A *third* temperature limitation derives from the fact that there is a need for considerable winter chilling during the rest phase of growth for many microthermal and mesothermal plants. This has been particularly well documented in the case of mid-latitudinal fruit trees and shrubs, among which blueberries need 800 hours and peaches 400 hours of temperatures of below 7°C before their dormant buds are capable of development, but it appears to be true also of many other species.

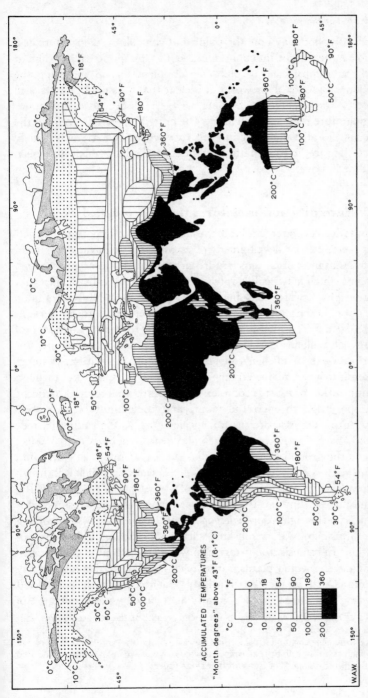

Fig. 4.6 Mean accumulated temperatures on a worldwide basis (from an original diagram in A. A. Miller, Three new climatic maps, *Institute of British Geographers, Transactions and Papers*, 17, pp. 15–20, 1951)

ACCUMULATED TEMPERATURES
"Month degrees" above 43° F (6.1°C)

°C	°F
0	0
10	18
30	54
50	90
100	180
200	360

W.A.W.

Finally, one may note the point that controls of temperature as a whole appear to be much more marked at the species rather than at the community level of plant development for, in surveying the major ecosystems of the world, it is clear that grasslands, forests, and deserts are all able to become well established in every major temperature régime which allows the growth of plants, although the specific floral component in each case will vary extremely widely according to the basic patterns of heat availability which have been outlined above.

4.5 Humidity and moisture: the hydrological cycle

Limitations of water availability may often be an important restricting factor in the development of ecosystems. It has already been noted that most organisms require large amounts of water to survive, a point which may be illustrated in the case of plants by the fact that the weight of water may be 70 to 90 per cent of the fresh organic weight in herbaceous species, and 40 to 60 per cent in many trees; in both instances, water is essential not only to the maintenance of many physiological and chemical processes within the plants, but also as a carrier of their nutrient food supply in solution. Water is taken into green plants largely through root hairs, and lost through transpiration processes; complex mechanisms are required to aid its transportation through the plant, and these must be extremely efficient in the case of very tall species, such as the California redwood (*Sequoia sempervirens*), which may reach heights of over 350 feet above the ground surface. Usually, the initial intake of water into plant roots is achieved either through *imbibition*, in which it is drawn into cell walls, and then into protoplasm, at pressures of up to 100 atmospheres* through the natural attraction of solid substances for liquids; or through *passive absorption*;† or by means of *osmosis*, involving the diffusion of watery substances from areas of high to low concentration. In osmosis, the pressure of water movement in most terrestrial and freshwater plants is usually close to 2 atm, though this may be raised with an increase in aridity;[142] moreover, in the case of salt tolerant species (*halophytes*), greatly augmented osmotic pressures of

* 1 atmosphere is equivalent to the mean weight of the atmosphere per unit area at sea level.

† *Passive absorption* takes place when unusually high transpiration rates prevail, in which case root cells are passively subjected to a massive inflow of water, induced by the direct demands of transpiration from above.

between 35 to 40 atm are often required to move the saline solutions. Once admitted into plant systems, water is then drawn upwards through trunks and stems, stimulated by the difference in *vapour pressure* (the pressure that water vapour exerts in the atmosphere) between leaf surfaces and root hairs; this may be referred to as the *diffusion pressure gradient*, and, at times, can reach very great proportions. There is no doubt that the up-pulling of water is also maintained partially by the natural tensile strength of a column of water held within an enclosed rigid structure, such as a stem or trunk, which when free from air bubbles can attain pressures of up to 300 atm.[25] Eventually, upon reaching the leaf surface, water is transpired back into the atmosphere by processes which are to be discussed in detail later in this section.

Although some of the water available for plant growth can originate from deep aquifers or other ground sources, most is derived fairly directly from atmospheric moisture (or water vapour) precipitated in the form of liquid water—rain, fog, and dew—or snow and ice. For any given temperature, the atmosphere will only be able to hold a certain amount of water vapour, the exact quantities being greater in warm air than cold (Table 4.3), so that if air which holds the

Table 4.3   The capacity of the atmosphere to hold moisture

Temperature °C	Capacity g/litre	Rate of change g/litre/10°C
40	40·06	15·84
30	24·22	10·36
20	13·86	6·33
10	7·53	3·68
0	3·85	

maximum amount of vapour for a given temperature (*saturated air*) is then cooled, it will be forced to lose some through condensation in adjusting to a new level of saturation; when this happens, the point on the temperature scale at which condensation begins is termed the *dew point*. Usually, the atmosphere requires rather a large amount of cooling before the dew point is reached since, in stable conditions, it is far from being saturated. This cooling may be achieved in freely moving air through uplift either by means of *convection* currents when land surfaces are strongly heated, or through *cyclonic* disturbances when warm moist air is pushed over colder and denser air in frontal situations, or as a result of *orographic* effects when air is forced over major mountain ranges. In all three instances, substantial

143

amounts of precipitation may result in the form of mist, fog, rain, ice, and snow. Direct radiative cooling at night can also induce condensation on land surfaces as dew, rime, or fog. Although they may be mentioned in passing, the precise mechanisms involved in such cooling processes are more a matter for meteorological analysis than biogeographical discussion, and the reader is therefore referred to any standard text in physical meteorology for a more complete explanation of these.

All forms of atmospheric precipitation may be of some importance to the growth and development of organisms within ecosystems. On a worldwide basis, the significance of *rainfall* to moisture income can vary appreciably, both on an annual basis, as depicted in Fig. 4.7, and seasonally. In low latitudes, regions with onshore winds (e.g., tradewind coasts on the western sides of tropical oceans, or monsoon coasts), and areas on the windward sides of mountains everywhere, rainfall totals may be exceptionally high, whereas, in contrast, very low falls can be anticipated in high latitudes, regions with offshore winds (e.g., the western sides of tropical deserts) or prevailing anti-cyclonic conditions (the *horse latitudes*, at 20° to 30° north and south of the Equator), or on the lee side of mountains, where the capacity of the air to hold moisture is rapidly raised as its temperature increases in moving downslope. The effects of seasonal variations, particularly in mid-latitudes, may often be felt most acutely in terms of differences in the intensity of rainfall, a factor which may considerably affect its rate of interception by vegetation and the general rates of runoff (pp. 147 to 150). In mid- and high-latitudinal areas, and on the higher parts of tropical mountains, considerable quantities of precipitation may also fall as *snow* which acts as a very efficient water reservoir should it lie on the ground for long periods of time.

Fog and *rime* can also contribute substantially to moisture income in certain ecosystems, especially when they are caused by advective or orographic cooling over weather fronts or mountains, for in these situations not only is precipitation increased, but transpiration is correspondingly reduced in the saturated air. In the summer fog zone of the central California coast close to San Francisco, the amounts of *fog-drip* reaching the ground surface under these conditions may be so great as to account for 50 per cent of the total precipitation, a fact which has been used in attempts to explain the very restricted distribution of the moisture-loving California redwoods. In turn, Oberlander[93] has suggested that rates of fog drip in

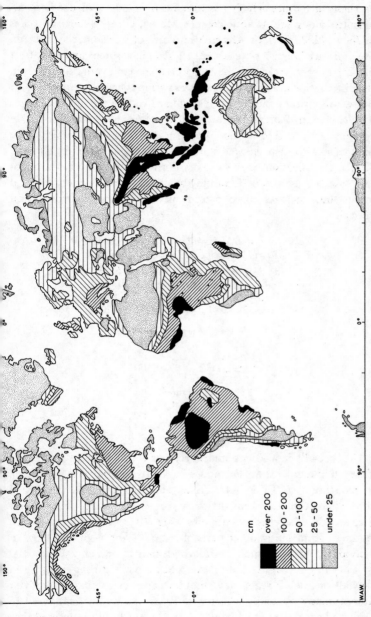

Fig. 4.7 The world distribution of mean annual rainfall (from an original diagram in W. G. Kendrew, *Climate*, 3rd edn, Clarendon Press, Oxford, 1949)

this area are certainly increased by the presence of forest trees such as these, a feature which has also been noted in spruce-fir communities along the Atlantic coast of Maine, USA.[34] Above altitudes of 760 m (2500 ft) in the Green Mountains of Vermont, the effects on the moisture balance of persistent fog drip in summer and hoar frost in winter are so great that they have created significant differences in the distribution of major plant species within the mountain range, for below this altitude the New England hardwoods predominate (sugar maple, *Acer saccharum;* beech, *Fagus grandifolia*); while above only conifers such as red spruce (*Picea rubens*) and balsam fir (*Abies balsamifer*) appear to be well suited to the excessively moist climate. Similar modifications to species composition as compared to adjacent rain forests may also be found in the cloud forests of misty tropical mountains, such as those present in Vera Cruz (Mexico) and Hawaii.

Table 4.4   Rates of dewfall in selected areas, mm/night
(after J. L. Monteith,[87] quoting G. Hofmann,[55] and R. L. Jones[63])

(a) Maximum, on crops at Rothamsted, UK[87]

Spring wheat, 20 May 1957:	0·26
Sugar beet, 8 August 1958:	0·47
Grass (60 cm), 1 August 1960:	0·20

(b) Maximum, on artificial surfaces[55, 63]

Israel	0·45
Jamaica	0·43
S. England	0·43
Munich	0·43
Baltic coast, Germany	0·37
Montpellier, France	0·22
Moscow	0·22
Romania	0·17

Especially in arid or semiarid regions with a considerable night-time fall in temperature, *dew* may serve as an additional major source of moisture (see Table 4.4). Heavy dewfalls have been recorded throughout the year from the drier tropical lowlands, deserts, and areas with *Mediterranean* climates (Fig. 4.10), and seasonally, in summer, from many other parts of the mid- and high latitudes. In arid areas, such as Israel, dewfall totals may approach values of 100 to 150 mm/yr,[7] and recent research has indicated that exposure to dewfall on such a scale may stimulate an increase in productivity among many plants by a factor of 50 per cent when compared to undewed specimens.[35] Although in the past it has been argued that dew can be absorbed directly into plants through leaf surfaces, recently both Waisel,[140] and Cloudsley-Thompson and Chadwick[26] have come out strongly against this notion; their view is that, as with other forms of precipitation, water from dew must enter into the ground before it can be taken into plants through root hairs.

Not all of the water which reaches the vegetation canopy or surface ultimately becomes available to plants, and in order to explain this point one must seek an understanding of the mechanisms of the *hydrological cycle*, a term used to describe and evaluate the circulation of water and moisture between atmosphere, earth, and living organisms (see Fig. 4.8 for a diagrammatic representation of this).

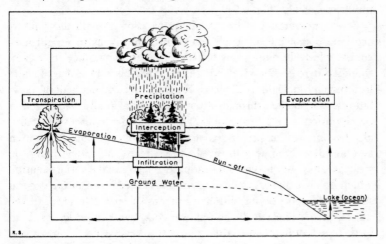

Fig. 4.8 The hydrological cycle

First, it is important to realize that some precipitation may be *intercepted* by vegetation communities before it reaches the ground, to be returned to the atmosphere as water vapour by means of evaporation from the intercepting plant surfaces. The degree of interception will vary according to the conditions of precipitation, the nature of the vegetation, and time, but on occasions will be extremely high, and when the foliage is wet, but not saturated, may even exceed the rates of evaporation expected from an open water surface in the same area. Usually, interception is greatest in forest formations, in which 90 per cent of rainfall may be lost from light summer showers in mid-latitudes.[144] For mean conditions, including periods of more prolonged rainfall, Kittredge[65, 66] has suggested that an interception

147

loss of 40 per cent may be more typical for the summer in Douglas fir forests of the Pacific Northwest (USA), though for other communities the value may be less, at 10 to 25 per cent; at first sight, the considerably higher percentage loss of the former may be unexpected, though it is true generally that interception is greater under conifers than elsewhere during the leafing period, possibly because air circulation is freer, or the presence of smaller water droplets on the needle leaves substantially encourages evaporation rates. On a seasonal basis, the differences may be much larger, for though summer losses are approximately equal to those of winter in coniferous forests, they may increase by a factor of two to three times in deciduous communities (see also Fig. 4.9).

Fewer measurements are available to assess the effects of interception on snowfall. Geiger[43] has argued that its extra weight may enable relatively more snow to reach the ground surface than rain, though Kittredge[67] has indicated that, at least in some areas of white fir forests in the California Sierra Nevada, up to 40 per cent of snowfall may still be lost through interception. But in all cases, whether precipitation is in the form of rain or snow, interception will gradually become less as prolonged storms develop, for once vegetation communities become saturated, the weight of water lying on stems, trunks, and leaves will gradually overcome the surface tension which holds it there, so allowing more to move down to the ground surface through gravity, either by means of *direct throughfall* in the atmosphere, or *indirect flow* along branches, trunks, and leaves. Both of these are difficult to evaluate in natural ecosystems, though it is known that the degree of indirect flow may be determined to some extent by the type of stem or bark: thus, Rowe[108] has estimated that up to 15 per cent of precipitation may move down the trunks of smooth-barked species, such as beech (*Fagus* spp.), while Kittredge[67] is of the opinion that 2 to 3 per cent may be a more realistic figure for rough-barked trees. Very few detailed examinations of throughfall and stem flow are available for study, and usually interception values are estimated by means of a crude formula:

$$C = et, \qquad (4.1a)$$

where C is the estimated interception capacity of the vegetation cover, e is the average rate of evaporation during a storm, and t is the duration of the storm.[56] This is, however, only applicable for storms

148

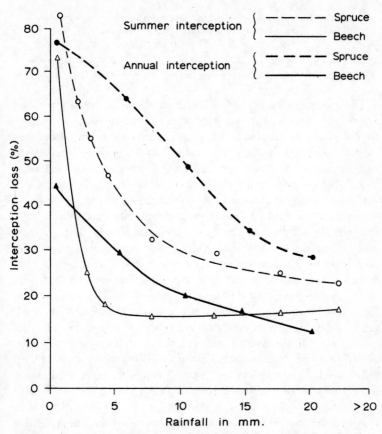

Fig. 4.9 Interception losses from spruce and beech forests. (Based on data from F. E. Eidmann quoted by H. L. Penman,[103] and from E. Hoppe quoted by R. Geiger,[43] compiled in an original diagram in R. C. Ward, *Principles of Hydrology*, London, 1967, copyright © McGraw-Hill Publishing Co., London, UK. Used with permission)

in which precipitation does not reach the interception capacity; for longer periods of rain a different formula is required, using data similar to that obtained in Table 4.5:

$$I = R - Rg - S, \qquad (4.1b)$$

in which the interception (I) is calculated with reference to R, the precipitation above the vegetated layer; Rg, the precipitation beneath the vegetated layer; and S, the stem flow.

149

*Table 4.5   Stem-flow, throughfall, and interception in a 17-year-old
stand of Norway spruce (Picea abies) July 24–November 18, 1959
(data from E. R. C. Reynolds and L. Leyton[104])*

Stem-flow	Through catch	Total throughfall	Gross precip.	Interception
0·27 ± 0·046	4·20 ± 0·230	4·46 ± 0·240	6·09	1·64

The patterns of interception outlined above do not occur in all areas. Where well-stratified vegetation communities are absent, or bare surfaces predominate, interception may be minimal or non-existent; and it may even be negative where fog drip is persistent. But for most ecosystems, interception in forested and wooded areas is a viable phenomenon which affects considerably the amount of water reaching the ground surface from precipitation. Once it has arrived on the ground, water may either enter the soil through infiltration, or leave ecosystems through runoff or further evaporation. *Runoff* totals often show considerable variability, depending largely on the climate and the type of vegetation; usually, they become greater as the intensity of precipitation increases, so that more water may be lost in this way from heavy summer showers than from light rainfall in winter. Since the process of interception itself reduces the intensity of precipitation, in that it delays the arrival of rain or snow on the ground surface, one may expect less runoff from completely vegetated surfaces than from bare ground or areas with clean-weeded crops. Thus, in Texas, Jarvis[62] has observed that for days with rainfall totals of between 1 and 4 in. (ca. 2·5 to 10 cm), runoff values of between 40 to 80 per cent of the rainfall actually reaching the ground surface could be expected on land under maize or cotton with a vetch cover, as opposed to values of only 1 to 2 per cent, even on steep hillsides, when thickly matted Bermuda grass (*Cynodon dactylon*) was the predominant vegetation. Even lower rates of runoff may be found in communities with a thick litter accumulation which serves as a further interceptive layer. Local variations in runoff which depart from these general patterns may at times be stimulated by soil conditions, for if the soil is compacted or frozen or close to saturation, the chances of its acceptance of water are correspondingly reduced, so that runoff rates which are close to the maximum are encouraged. Especially on soils which do not support a well-developed vegetation cover, runoff may also be substantially increased if the humus or litter layers are absent or, of course, if steep slopes prevail.

150

Direct *evaporation* from wet soil surfaces can also be a significant factor in reducing rates of infiltration into the soil, particularly within the tropics, or in areas where summer temperatures are consistently high. Providing that there is no limit to the amount of available water, evaporation from soil surfaces may often be broadly comparable with that from water, or as Lee[70] has noted, may be even greater, though it should be remembered that once a surface has become less than saturated, evaporation rates will begin to fall away rapidly. Evaporation from soil may also at times be increased to much above normal values through the effects of wind, which can raise the evaporative power of an area by bringing in considerable amounts of advective energy.

Water which remains at the soil surface, and which has not been lost to plants through runoff or evaporation, may enter the soil through *infiltration*, the precise rates of which may be affected by the temperature, the quality of the water, the texture, compactness and wetness of the soils, and the number of colloids present. Gaertner[41] has argued that the quantity of water in a soil may, for many plant species, have a greater influence on their final patterns of distribution than any other single environmental factor, though Meinzer[80] has qualified this in suggesting that its influence may become progressively less important as one moves from arid to humid areas. The idea has been mooted by Rice[105] that differences in soil water availability may also lead to major modifications within animal populations of certain deciduous forests. Yet the processes of input, retention, and movement of soil water, and their effects on growth patterns, are still by no means fully understood. In general, once water has penetrated into a soil, it continues to move downwards, by means of gravity, towards a *water table* beneath which the soil is saturated. Alternatively, it may be held above this level by hygroscopic forces in which soil particles (and especially colloids) adsorb the water, or by capillary forces in which water is trapped by surface tension in the air spaces between fine-grained particles of soil. Particularly during dry summers in mid-latitudes, and in the tropics, these capillary forces may be great enough to encourage water to move upwards within the soil, so helping to keep the surface layer moist and raising the rates of evaporation from it. However, the effectiveness of capillary action falls away as soil depth increases, so that in most cases moisture held at depths of greater than four feet below ground level (a figure which may be modified according to

151

soil texture) will not be able to reach the surface in this way, being lost only through transpiration processes.[57]

It follows from this that one of the most important factors in determining the availability of soil moisture to plants is the level of the water table. Under bog, fen, or marsh this may reach the ground surface, so that anaerobic waterlogged conditions prevail in which free oxygen is lacking, so limiting the cycling of nutrients and restricting plant growth and the numbers of organisms in the soil. But, more often, it will lie some distance beneath the surface, below a *zone of intermittent saturation* which dries out periodically or seasonally and in which water may then be found only in hygroscopic and capillary forms; within this aerobic zone, the vertical extent of which will vary according to season and soil texture, nutrients normally circulate freely, plants and their roots grow well, and organic life as a whole is encouraged. Usually in humid areas the water table lies close to the surface, whereas under more arid conditions it may be located at great depths, thus creating very contrasting situations in terms of the mean annual availability of water, the frequency and length of the periods of saturation, and the relative restriction or stimulus of organic growth. However, it is difficult to make generalizations about the expected location of water tables on a climatic basis alone, for local environental components may be equally important in helping to determine this. Thus, the presence of trees can in itself have the effect of depressing water tables to below normal depths,[159,161] and should hard pans develop within the soil (section 4.9) or moisture-attractive clays be present (as in the case of montmorillonitic clays in many parts of Barbados[148]), climatically abnormal *perched water tables* are often formed very close to the surface. Nevertheless, no matter what the position of the water table may be, in extremely wet periods all soil layers can eventually become saturated so that downward movement of water ceases, or becomes only barely perceptible. At this point, the levels of soil water have reached a point which is close to the *field capacity*,* beyond which no more can be absorbed, and for as long as this condition remains, all water reaching the ground surface from precipitation will be removed *in toto* through runoff (floods) or evaporation. Field capacity is usually attained when the amounts of water in a soil form between 5 and 40 per cent of its oven-dry weight, the large percentage variation being

* Field capacity is usually defined as the percentage of water retained in the soil after gravitational water has drained away.

explained in terms of differences in soil texture. Soil moisture-holding qualities may also be measured by means of *moisture equivalents* (Table 4.6) which refer to the percentage weight of water which an oven-dried specimen of soil will retain against a centrifugal force equal to 1000 times that of gravity; this also varies according to texture from about 2 per cent in coarse sands to over 40 per cent in heavy clays.[83]

Thus, for any given ecosystem and within known time limits, the amounts of water theoretically available for plant growth can be broadly and empirically determined by reference to the following equation:

$$Pa = P - (I + Eg + R + Ma + Md) + G, \qquad (4.2)$$

where Pa is the amount of available water; P is the recorded precipitation above the vegetated layer; I is the amount intercepted by the vegetation, according to equations (4.1a) or (4.1b); Eg is the evaporation from the ground surface; R is the runoff; Ma is the absorption of water within the soil by particles and colloids; Md is the net amount of water percolating downwards to deep water lying beyond the root zone; and G is the inflow from ground water or other sources which may add to the water income. Once water has become available for plants, it may then be absorbed as previously described, chiefly through root hairs and largely within the zone of intermittent saturation, some being taken in as it moves downwards through gravity, but most coming from capillary water.

Following absorption, the hydrological cycle is completed when water moves through the plant, helping to modify internal and leaf temperature régimes, and ensuring the constant flow of nutrients in solution before it is transpired as vapour from leaf surfaces back into the atmosphere. Daubenmire[33] has estimated that as much as 99 per cent of absorbed water may pass through the plant system in this way, the remainder going to form new chemical compounds within it. Rates of transpiration, which help to control the internal water movement of plants, may vary appreciably for individual species, though, as Alway[4] pointed out as early as 1913, communities of plants appear to differ only slightly in their ability to move water from a soil within a given climatic area. In order to explain this apparent paradox, one must first consider the possible mechanisms which initiate and govern the rate of transpiration, some of which are

153

physiologically and some climatically induced (see also pp. 160 to 163). Willis and Jeffries[160] have noted that transpiration may be effected in three ways from plant surfaces. First, water can be evaporated directly from moist membranes into the atmosphere through the cuticle, in a process termed *cuticular or direct transpiration*, which accounts for up to 10 per cent of the total transpiration of some species. Second, in a relatively few plants, water may be lost in liquid form from an uninjured stem or leaf through *water stomata* or *hydathodes*, by means of *guttation*.

Third, and the most important mechanism of water loss, is *stomatal transpiration*, in which moisture is diffused through lenticular or stomatal pores within leaf surfaces. This is responsible for over 90 per cent of the water loss in many plants, though the exact values may vary according to the numbers, size, and spacing of the stomata. Stomatal density is known to range from 50 000 to 800 000/in.^{2} (ca. 7750 to 124 000/cm.^{2}) in many plant leaves, often being distributed in a ratio of 3:1 between the lower and upper sides of the leaf, though there are a few plants with no stomata facing upwards. Conditions of light may induce modifications in the size of stomata and in the length of the period in which they are open, and since most stomata are closed during the hours of darkness, it has been argued by Monteith[85] that transpiration must then be negligible. In daylight, stomatal transpiration rates are controlled directly by the size of the stomatal aperture until they are more than half open, following which general meteorological parameters assume a greater importance as stimuli. In particular, increases in light intensity, temperature, and wind speed may all be responsible for raising the rates of water loss, sometimes very rapidly and very considerably indeed. Thus, transpiration is augmented by factors of 20 per cent over still-air values when wind speeds reach 8km/hour, 35 per cent at 16 km/hour, and 50 per cent at 24 km/hour, due to the increased amounts of advective energy brought into localities in these situations.

So far within this argument, it has been assumed that the water supply within the soil has at all times been adequate for the plant's needs. Frequently, however, it is not, particularly within the zone of intermittent saturation, so that modifications to the expected patterns of transpiration may occur from time to time. In general, transpiration rates will stay at near normal values until the reservoir of available water (and particularly capillary water) begins to near exhaustion, when the tensional forces holding the water in the soil increase slowly until they reach a pressure of between 15 to 20 atm;[36] beyond this point they grow sharply and out of all proportion to the ability of a plant to absorb water through the diffusion pressure gradient. Once this pattern has set in, the plant may react physiologically through closing stomata, or reducing stomatal activity, thus also minimizing the movement of carbon dioxide in the leaves and lowering the rates of photosynthesis.[60] If the availability of water is then further reduced, moisture stresses can develop in the plant, followed by wilting either on a temporary basis or as a permanent state, when the leaves become so flaccid due to moisture deficiencies that they will not recover normal turgidity if placed in an approximately saturated atmosphere, without addition of water to the soil.[70] In mid-latitudes, the effects of wilting may most clearly be seen in the leaf fall or stem wilt of plants on desiccated, cracked clay soils during summer droughts. However, they may also be found in the *physiological droughts* of winter or early spring, when air temperatures are often much warmer than soil temperatures, thus stimulating transpiration to a much greater extent than can be met by the rate of water absorption, especially since in winter the viscosity of most plant water has been increased so as to effectively reduce its freezing point and offset the effects of cold. If the desiccation is severe enough, or lasts for a sufficiently long period of time, plants may then become dormant or eventually die if the metabolic and mechanical damage within cells is too great.[72] However, it must be remembered that minor moisture stresses are not always detrimental to the plant; indeed, they may be necessary to stimulate flowering in some species (e.g., *Caffea* spp.).

Once permanent wilting has begun, the amount of water still present in a soil is referred to as the *permanent wilting percentage*, which ranges from approximately 1 to 12 per cent in coarse sand, 10 to 15 per cent in clay loams, and 25 to 30 per cent in absorbent heavy clays[83] (see also Table 4.6). On occasions, these values may prove exceedingly difficult to determine, since many species (e.g., the perennial *Bouteloua gracilis*) have the ability to reduce transpiration rates so sharply at wilting point that they may be able to postpone permanent wilting indefinitely, but once this condition has been reached, only the most resistant capillary and hygroscopic water remains in the rhizosphere. Even then, beyond the state of permanent wilting, some water may still be absorbed by root hairs in a wicklike action,[16] if the attraction of root cells for water exceeds the back-pull of the soil.*

Table 4.6   *Interrelationships among certain soil moisture constants of a group of soils (after F. J. Veihmayer[137])*

Moisture constant	Clay	Loam	Silt loam	Sandy loam	Fine sand
Moisture equivalent	28·4	21·7	16·1	9·5	2·3
Permanent wilting %	13·4	10·3	7·5	2·9	1·0

It is clear from all these comments that not only can plants influence indirectly the availability of water in a soil, but they themselves must adapt to a wide range of different soil moisture conditions, both regionally and locally, and on a seasonal and permanent timescale. It was this realization which led Warming[145] to distinguish broad and, to some extent, ill-defined groups of plants on the basis of their tolerance to water. Thus, moisture-loving *hydrophytes* have their roots in water, or are present in soil with above-normal quantities of water within it (e.g., alders, *Alnus* spp.). In contrast, *xerophytes* are able to withstand long periods, and even years, of atmospheric and soil drought by means of specific physiological and morphological adaptations†: in the deserts of the American south-west, they may be present as ephemeral annuals, as succulents in which water is stored in stems or leaves (e.g., Cactaceae), or as non-succulent perennials including grasses, whose growth periods may be extremely limited. Species with extensive root systems designed to reach deep soil water, such as mesquite (*Prosopsis* spp.) and alfalfa, whose roots have been known to extend downwards for 20 to 40 m respectively, may also be classed as xerophytes, in addition to those which are widely spaced in order to increase the supply of water to individual plants (e.g., the creosote bush *Larrea tridentata*); those with the very high osmotic pressures which are needed to combat increasing salinity; those with very low rates of transpiration which may lead to a postponement of the permanent wilting phase, such as the ocotillo (*Fouqeria splendens*), whose minute

* The back-pull of the soil for water is often measured by a $-pF$ factor, this referring to the log of the tension (F), in cm of water, by which the water is held in the soil.

† It is important to make a distinction between xerophytes, which live in dry places, and other *xeromorphs*, which also have adaptations to prevent water loss, but are not necessarily confined to dry areas. An example of the latter is the rush family (*Juncus*) which, although living in wet places, has protected stomata and no leaves.

leaves spring up only after rain showers, or the sage brush (*Artemesia tridentata*) whose transpiration is reduced by means of a grey pigment present in its leaves, or a wide variety of species which possess thorny leaves or leaves with a hard cuticularized layer (*sclerophylly*); and *phreatophytes*, such as willows (*Salix* spp.), whose roots can penetrate across the zone of intermittent saturation to the water table below. Between these two extremes, *mesophytes* show a wide range of adaptation to life in environments which are neither too wet nor too dry, and form a group which include most of the plant species to be found in terrestrial ecosystems.

4.6 Energy/temperature—water relationships

Thus far within this chapter, it has been implicit that one of the most significant physical relationships within ecosystems is that between the energy and water balances. Indeed, as Slayter[124] has recently emphasized, the exchanges of water and energy between plants and habitat determine to a considerable extent not only the characteristics of the physical micro-environment, but also the distinctive nature and functions of the plants themselves. It was the early tentative recognition of this fact which led to the first attempts to correlate parameters of relative temperature, available moisture, and major vegetation groups. These were crystallized initially by de Candolle[20] in the mid-nineteenth century, in his application of the regionally descriptive terms, *megathermal*, *mesothermal*, *microthermal*, and *xerophilous*, to plant communities. Later, Grisebach[49] showed the way towards a more accurate system of analysing temperature and precipitation patterns as controls of vegetation, a point more specifically taken up by the now famous Russian-trained biologist-climatologist, Wladimir Köppen,[68] whose system of climatic classification is still widely used today.

Köppen's main interest in this field lay in the definition of broad climatic limits which would fit in with de Candolle's plant regions. In order to achieve this aim, he divided the world into five major climatic zones, according to the criteria presented in Table 4.7. Three of these (A, C, D) supported forest, one (E) was a snow climate, and the last (B) a dry climate. Delimitation of the A, C, D, and E climates was accomplished by reference to mean annual and mean monthly temperatures, and the B climate by a critical rainfall

Table 4.7 *Regional climates, based on the Köppen system of classification, and modified by G. T. Trewartha, in* An Introduction to Weather and Climate, *copyright McGraw-Hill, New York, 2nd edn, 1943, used with permission*

Major divisions	Characteristics
A	*Tropical forest* climate; mean monthly temperature of the coldest month above 18°C (64·4°F).
B	*Dry* climates (for limits, see graph, Fig. 4.10). BS—steppe or semiarid; BW—desert or arid.
C	*Mesothermal forest* climate: mean monthly temperature of the coldest month above 0°C (32°F); that of the warmest month above 10°C (50°F).
D	*Microthermal snow-forest* climate; mean monthly temperature of the coldest month below 0°C (32°F); that of warmest month above 10°C (50°F).
E	*Polar* climates: mean monthly temperature of warmest month below 10°C (50°F); if above 0°C (32°F), then ET, *tundra* climate; if below, then an EF, *frost* climate.

Subdivisions Use

a	With C + D climates	Mean monthly temperature of the warmest month above 22°C (71·6°F)
b		Mean monthly temperature of the warmest month below 22°C (71·6°F)
c		Less than 4 months with a mean monthly temperature of over 10°C (50°F)
d		Same as *c*, but mean monthly temperature of coldest month below −38°C (−36·4°F).
f		Constantly moist.
m	(With A climates)	Monsoon climate.
s		Summer dry season } Precipitation of driest month must be below 4 cm in C climates, 6 cm in A climates.
w		Winter dry season }
h	With B climates only	Hot and dry, mean monthly temperature of all months over 0°C (32°F)
k		Cold and dry, mean monthly temperature of at least one month below 0°C (32°F)
n		Frequent, or infrequent (n') fog.

factor which compared existing precipitation with a temperature-evaporation component calculated from the following formula:

$$R = 0.44(T - k), \tag{4.3}$$

in which R is the mean annual precipitation, T is the mean annual temperature, and k is a constant related to variations in the seasonal concentration of rainfall. Once these had been determined, secondary and tertiary characteristics of climate were evaluated from mean monthly temperatures and mean monthly precipitation. The world-wide distribution of these climatic zones is depicted in Fig. 4.10.

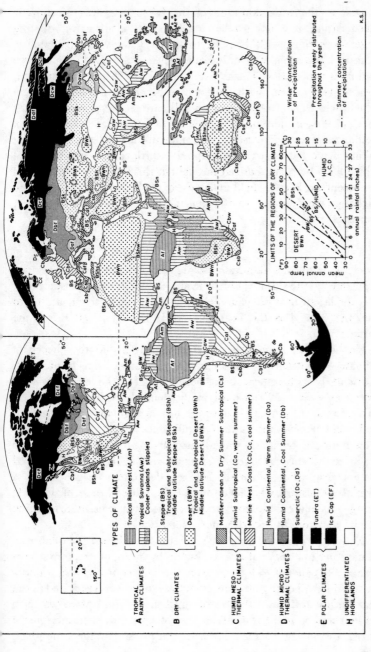

Fig. 4.10 The distribution of climates (modified from the Köppen classification by G. T. Trewartha, in *An Introduction to Weather and Climate*, copyright McGraw-Hill Book Co., New York, 2nd edn, 1943. Used with permission)

Despite the wide adoption of the Köppen system of climatic classification, it is not wholly satisfactory as a means of equating vegetation groups with the patterns of energy and water availability, since its bases are always only crude representations of temperature and moisture and, as Thornthwaite and Hare[134] have pointed out, it never gets to grips with the essential processes of the atmosphere which help to control weather and climate. In other words, it fails in that it was developed too early to appreciate the complex exchanges of energy, moisture, and momentum which are continually taking place between the atmosphere, earth, and vegetation, and which have been emphasized throughout this book. Indeed, most of the recent work in this field makes use of the knowledge that all these exchanges are subject to the general laws of radiation balance, expressed thus:

$$R = Le + A + Q, \qquad (4.4)$$

where R is the net radiation (solar, plus net infrared fluxes after reflection) at the surface; e is the evaporation or evapotranspiration; L is the latent heat of evaporation; Q is the convective heat flux to the air; and A is the heat flux to or from the soil below the surface.

Furthermore, some conceptual modifications to the theory of climate-vegetation relationships have recently been formulated, particularly with respect to the development of the idea of *potential evapotranspiration* (*PE*): as originally defined by Thornthwaite,[132] this refers to the potential water loss from a vegetated surface which will occur if soil water is always sufficient to meet plant demands. This notion is obviously somewhat different to those of water loss from ecosystems by means of evaporation, transpiration, and evapotranspiration, since all these are dependent on conditions of available moisture, whereas limits to *PE* are strictly those of available energy. In view of its importance, it will perhaps be useful to examine the implications of this idea more closely.

As a preliminary, it should be recalled that the precise quantities of moisture lost from natural ecosystems by means of evapotranspiration are dependent to a large extent on four controls: *first*, an adequate source of energy, mainly from the sun, which is required at the leaf surface to induce the transformation of liquid plant water to vapour and so make it available for absorption by the atmosphere; *second*, air movement, and a lower vapour pressure in the atmosphere than on the evaporating plant surface; *third*, the type of vegetation, particularly with regard to the extent of its root system and the reflectivity of its surface parts (i.e., its albedo); and *fourth*, the amount of water present in the root zone. Of these, the climatic controls of energy and air movement are both closely related, as are also the soil-vegetation factors, though it is the former which will usually be dominant since, in the long run, evapotranspiration rates must be determined by the amounts of available energy, which will remain fairly constant for given latitudes (eq. 4.4). Further to this, van Wijk and de Vries[158] have suggested that, in theory, the windspeed factor cannot be of too great a significance, an argument based on the observation that on bright sunny days in mid-latitudes, when the wind speed is low, as much as 60 per cent of the available energy goes to provide the latent heat of evaporation; but despite this, evapotranspiration rates can still increase substantially when hot dry winds are blowing.

Probably the most important controlling variable other than energy income is that of reflectivity, for it is known that albedo values range widely from about 25 per cent for most green crops,[86] to 15 per cent or less in coniferous forests.[103] Detailed records of these have been published by Angström[6], with values of 26 per cent for grassland, 17·5 per cent for oak woodland, and 14 per cent for pine forests. Translated into more meaningful terms, these indicate that pine forests may absorb up to 12 per cent more energy than grasslands. However, it is not the case that these differences can be transcribed directly into equivalent variations in the rates of evapotranspiration except in the case of some individual plants, for observations have confirmed that within similar climates, dissimilar terrestrial ecosystems tend to evapotranspire *in toto* at comparable rates, no matter what their vegetational components may be, always assuming that the plant cover of the ground surface is complete—a situation which confirms the primary importance of the meteorological controls, and the secondary nature of the vegetation type. Often, the only exceptions to this rule are found under conditions of limited water availability within the soil, when ecosystems comprised of deep-rooted species, or of species with high osmotic pressures, may transpire more water than others. It is clear that the idea of PE, in which water availability is always adequate, excludes this latter possibility and ensures that the theoretical rates of evapotranspiration must be broadly correlated with patterns of temperature, and controlled by the radiation balance equation and also by the laws of the conservation of energy, as presented in chapter 2.

Thus, the PE concept was developed partly in response to the need to obtain an accurate estimate of the atmosphere's evaporative power, a parameter which is difficult to measure even with sophisticated present-day techniques of instrumentation.* The final determination of the formulae through which PE could be calculated was worked out empirically by Thornthwaite,[133] by estimating mean runoff and precipitation from and within small catchments and watersheds, and then equating these with the easily recorded values of mean monthly temperature and daylength, both of which were taken to indicate indirectly the state of the radiation balance at the earth's surface. It was argued at a later stage by Mather[79] that these values might also reflect important variations in the condition of other climatic parameters, such as wind and humidity, though Thornthwaite himself always regarded the latter as being insignificant as an influence on PE.

The essentials of Thornthwaite's 1948 formulae may be found in the two following equations:

$$i = (t/5)^{1 \cdot 514} \tag{4.5}$$

$$e = 1 \cdot 6b(10t/I)^a \tag{4.6}$$

From the first of these, one can derive a *monthly* heat index i by first calculating t, the mean screen temperature of the month (°C); then, from each monthly heat index recorded throughout the year, an annual heat index (I) is obtained through simple addition. Monthly PE may subsequently be determined by referring to eq. 4.6, in which e is the monthly PE in cm for a 30 day month of 12 hour photoperiods, t is the mean monthly temperature (°C), a is a constant which is a cubic function of I, and b is a factor to correct for unequal daylength on a monthly basis.† As Hare[52] indicates, these equations give a complicated and non-linear relationship between e and t, in which e is made to depend logarithmically on t for the period concerned, and also in which the actual value of e for a given t relies to some extent on a heat index for a *climatologically normal year*. This

---

* This has also been attempted by Budyko[18] in the USSR, whose idea of an evaporability factor in land environments does not, however, take into account the secondary effects of vegetation as precisely as does the notion of PE.

† Further details of this method, including the necessary tables of conversion for constants, may be found in Thornthwaite and Mather.[135]

latter premise has been criticized by Serra[116] on the grounds that it may be unrealistic to calculate monthly PE from the mean temperatures of all other months of the year, and also by van Wijk and de Vries[158] with respect to the idea that, since temperatures tend to lag behind energy income, there may be seasonal discrepancies of estimated as opposed to actual *e*, a supposition which has been confirmed by Green[46] and Ward[143]. In turn, replies to these criticisms have pointed out that the annual heat index is an essential part of Thornthwaite's formula, in that it forms a means of evaluating the annual radiation balance, a component which is of great significance to the water balance; and that the discrepancies caused by seasonal lags in temperature may to some extent be offset by the daylength index *b*.

Despite the problems inherent in all empirically derived formulae, Thornthwaite's method of calculating rates of PE has been well received in every part of the world, being used especially widely by climatologists, soil scientists, and hydrologists. Sibbons[123] has commented that it can certainly be employed effectively under a wide range of climatic conditions, particularly in a general sense as an indicator of relative balance between temperature, radiation income, and evaporation. Indeed, it appears to be relatively unsuccessful only where specific environmental circumstances interrupt this balance, as when a rapid interplay of very different weather patterns prevails in mid-latitudes; or when an *oasis effect*, first described by Penman[103] and since defined more explicitly for tradewind islands such as Hawaii[21] and Barbados,[106] causes considerable variations between the computed and observed values of PE due to the large amounts of incoming advective energy found in these areas; or when the observed PE values are reduced by the presence of vegetation communities which are comprised largely of non-vascular plants, as in the lichen-woodlands of Central Labrador[91,149] and other northern regions.

However, in most cases, the advantages of using the Thornthwaite formulae to assess the relationships between climatic parameters and terrestrial ecosystems are clear. Through it, annual values of PE have been equated with thermal regions, as in Table 4.8, and the broad zones of crop growth. Moreover, although it has not yet been proved that differences in the rates of PE can be used to explain more precise changes in the patterns of natural vegetation, Hare[51] has noted some degree of correlation between the two over large

Table 4.8   Relationships between annual PE and thermal province
(after C. W. Thornthwaite and F. K. Hare[134])

	Thermal province	Annual PE, cm
E	Frost	0–14.2
D	Tundra	14·3–28·5
C_1	} Microthermal	28·6–42·7
C_2		42·8–57·0
B_1	}	57·1–71·2
B_2	} Mesothermal	71·3–85·5
B_3		85·6–99·7
B_4	}	99·8–114·0
A	Megathermal	over 114·0

areas in the boreal forest zone of Labrador, as indicated in Table 4.9. If one knows the field capacity of water within a soil, and also

Table 4.9   Relationships between annual PE and vegetation in Labrador
(after C. W. Thornthwaite and F. K. Hare[134])

Forest subzone	Annual PE, cm	Vegetation
Tundra		Tundra
	.... 31	
Forest-tundra		Tundra on interfluves; woodland in valleys
	.... 35	
Woodland		Predominantly *Cladonia*-rich open woodland; closed-crown forest in isolated groves
	.... 42	
Forest		Closed-crown forest occupying most mesic sites
	.... 52	
Mixed temperate forests		Forest dominated by non-boreal species, typically deciduous broad-leaved

its *root constant*, a term first coined by Penman to describe the maximum moisture deficiency which can be built up for any soil without checking transpiration (i.e., the equivalent of 7·5 to 12·5 cm of soil moisture under grass and, possibly, 25 to 30 cm under certain trees), the annual water budget may also be worked out in detail using appropriate tables for the soil moisture storage change as given in Thornthwaite and Mather.[135] Two contrasting examples of this are given in Table 4.10 and Fig. 4.11. The nature of the water budget may also have a marked bearing on the degree of restriction imposed upon growth in ecosystems. Finally, *humidity and aridity indices* (I_h and I_a) may be calculated from Table 4.10 by comparing annual

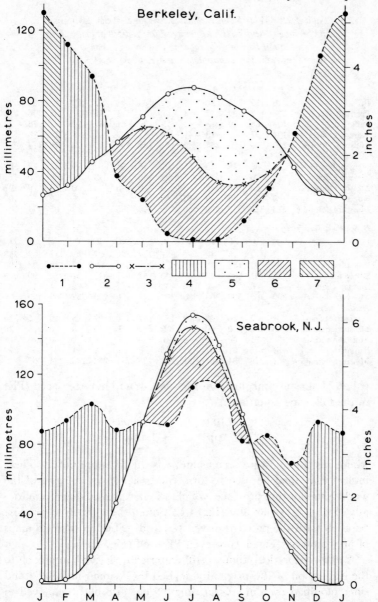

Fig. 4.11 The average march of precipitation (1), PE (2) and actual evapo-transpiration (3) at Berkeley, California and Seabrook, New Jersey. Periods of water surplus (4), water deficit (5), soil moisture utilization (6), and soil water recharge (7) are also shown (from an original diagram in C. W. Thornthwaite and J. R. Mather[135])

*Table 4.10   Waterbalance computations for Berkeley, California,
and Seabrook, New Jersey, assuming 300 mm depth of water stored
in soil layer at field capacity (all values in mm)
(after C. W. Thornthwaite and J. R. Mather[135])*

A. Berkeley, California

	J	F	M	A	M	J	J	A	S	O	N	D	Yr
PE	26	32	45	56	71	84	88	82	75	63	43	28	693
Precipitation	130	112	94	37	24	5	1	1	13	31	62	106	616
Difference	104	80	49	−19	−47	−79	−87	−81	−62	−32	19	78	−77
Storage change	104	22	0	−19	−41	−56	−47	−33	−19	−8	19	78	
Moisture storage	278	300	300	281	240	184	137	104	85	77	96	174	
Actual evapo-transpiration	26	32	45	56	65	61	48	34	32	39	43	28	509
Moisture deficit	0	0	0	0	6	23	40	48	43	24	0	0	184
Moisture surplus	0	58	49	0	0	0	0	0	0	0	0	0	107
Runoff*	0	29	39	19	10	5	3	1	1	0	0	0	107
Moisture detention	278	329	339	300	250	189	140	105	86	77	96	174	

B. Seabrook, New Jersey

	J	F	M	A	M	J	J	A	S	O	N	D	Yr
PE	1	2	16	46	92	131	154	136	97	53	19	3	750
Precipitation	87	93	102	88	92	91	112	113	82	85	70	93	1108
Difference	86	91	86	42	0	−40	−42	−23	−15	32	51	90	358
Storage change	0	0	0	0	0	−38	−35	−17	−10	32	51	17	
Moisture storage	300	300	300	300	300	262	227	210	200	232	283	300	
Actual evapo-transpiration	1	2	16	46	92	129	147	130	92	53	19	3	730
Moisture deficit	0	0	0	0	0	2	7	6	5	0	0	0	20
Moisture surplus	86	91	86	42	0	0	0	0	0	0	0	0	378
Runoff*	61	76	81	61	31	15	8	4	2	1	1	37	378
Moisture detention	361	376	381	361	331	277	235	214	202	233	284	337	

values of moisture surplus, moisture deficit, and moisture need (PE) in the following equations:

$$I_h = \frac{100s}{\text{PE}}; \qquad I_a = \frac{100d}{\text{PE}}$$

where s is the moisture surplus and d is the moisture deficit. These indices may then be used to form the basis for the division of the world into moisture provinces which, when expressed diagrammatically (Fig. 4.12) or areally (Fig. 4.13), often show much more close parallels with the patterns of vegetation distribution than do maps of precipitation (Fig. 4.7) or even PE itself (Fig. 4.14).

A more theoretical, though still empirically derived, approach to the estimation of PE, originally devised by Penman in 1948[101] and since modified by him several times,[103] has also been employed in many parts of the world. Penman thought that the best indication of

---

* Assuming 50 per cent of the water available for runoff in any month is held over until the following month.

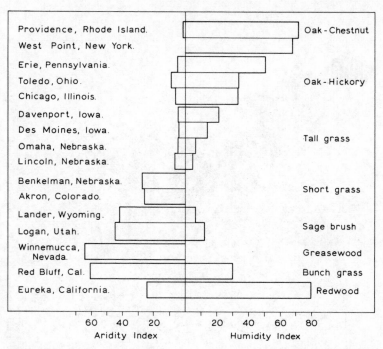

Fig. 4.12 Humidity and aridity indices at selected stations near latitude 41°N across the USA (from an original diagram in C. W. Thornthwaite and J. R. Mather[135])

moisture loss from an evapotranspiring surface could be obtained from a combination of the radiation balance approach and a means of taking into account the effects of atmospheric turbulence. Accordingly, he produced three equations, the first of which is:

$$Ea = 0{\cdot}35(e_a - e_d)(1 + U/100) \text{ mm/day}, \qquad (4.7)$$

in which e_a is the saturation vapour pressure of water and e_d is the measured vapour pressure at the mean screen level temperature in mm of mercury; and U is a factor representing the mean wind speed at screen level, in miles per day. In effect, this equation measures the drying power of the atmosphere (Ea) which naturally increases with an increase in the saturation deficit and wind speed. His second equation is more concerned with the availability of radiation for heating and evaporation at the earth's surface:

$$H = A - B \text{ mm/day}, \qquad (4.8)$$

167

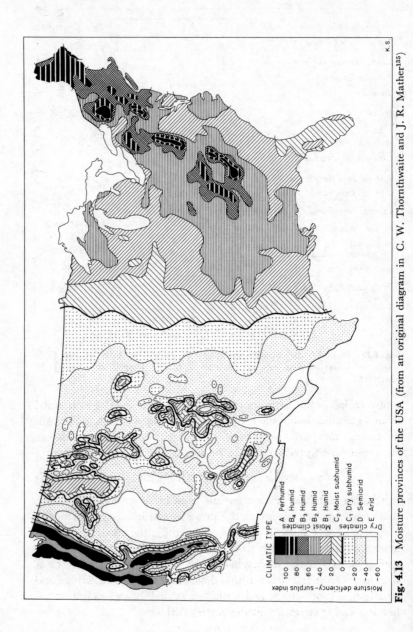

Fig. 4.13 Moisture provinces of the USA (from an original diagram in C. W. Thornthwaite and J. R. Mather[135])

K.S

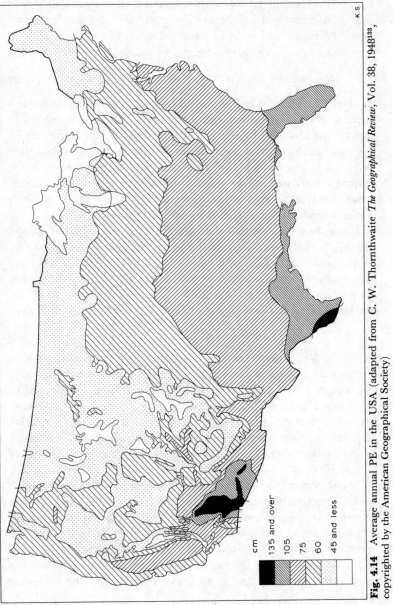

cm

135 and over
105
75
60
45 and less

Fig. 4.14 Average annual PE in the USA (adapted from C. W. Thornthwaite *The Geographical Review*, Vol. 38, 1948[133], copyrighted by the American Geographical Society)

in which A is the incoming shortwave, and B the outgoing longwave radiation, expressed in evaporation equivalents of mm/day, and derived from the following subsidiary equations:

$$A = (1 - r)Ra(0.18 + 0.55n/N) \text{ mm/day} \tag{4.8a}$$

$$B = \sigma Ta^4(0.56 - 0.09e_d)(0.10 + 0.90n/N) \text{ mm/day} \tag{4.8b}$$

In these, Ra refers to the intensity of radiation income at the ground surface, as measured directly in evaporation units and corrected to assume the absence of an atmosphere; r is an albedo coefficient for the evaporating surface; n/N is the relationship of actual/possible hours of bright sunlight; σTa^4 is the back radiation leaving the surface, also corrected to assume the absence of an atmosphere, when Ta is the mean air temperature at screen level and σ is Stefan's constant;* and e_d is as in eq. 4.7.

If one then assumes that the heat flux into and out from the soil (usually about 2 per cent of the incoming energy) is sufficiently insignificant not to affect the final totals of PE, as Pearl *et al*[100] have suggested, then one may divide net radiation income between that used to heat the air, and that used in evapotranspiration processes, the latter being estimated by combining eq. (4.7) and (4.8) to give:

$$E = (\Delta/\gamma H + Ea)/(\Delta/\gamma + 1) \text{ mm/day,} \tag{4.9}$$

in which Δ represents the rate of change of the saturation vapour pressure for water at mean screen air temperatures in mm mercury/°C, and γ is the psychrometric constant for wet and dry bulbs (0.27 mm Hg/°C). The dimensionless component Δ/γ is consequently a factor which takes into account the changing relative importance of net radiation and the evaporative potential in the total rates of evaporation, so that, as Penman has noted,[103] the values of this at 10°C, 20°C, and 30°C will be respectively 1.3, 2.3, and 3.9. This component also assumes that the coefficients of water vapour and convective heat transfer are equal, a situation which is best exemplified in conditions of turbulence,[28] but which is less complete when insulation is strong and winds are light,[84] as in the summer months of temperate latitudes; in these latter conditions, the net radiation will be given greater importance in the equation than the evaporative potential, and in any case H tends to be the dominant

* Stefan's law states that a black body (in this case, the earth) will emit radiation from its surface constantly in proportion to the fourth power of its absolute temperature.

factor generally, as opposed to *Ea*, particularly in humid climates. Final evaporation totals (*E*) will also depend to some extent on the albedo of the evaporating surface, as indicated by the albedo coefficient *r* in eq. (4.8a); thus, a turf cover will have a coefficient of 0·25, and open water 0·05. If the evaporation is measured over open water, then the total *potential transpiration* (Penman's term, equivalent to PE) may be computed with an empirical weighting factor *f*, as follows:

$$\text{PE} = fE_0 \qquad (4.10)$$

in which *f*, representative of a photoperiod component, varies seasonally (0·6 to 0·8 at Rothamsted, England), and E_0 is the evaporation from an open water surface. But while the *f* factor has been applied successfully throughout western Europe, it is not so generally acceptable elsewhere, especially in the tropics where little seasonal variation in photoperiod occurs and where Smith,[125] among others, has found it hard to use.

4.7 Wind

At times, the presence of strong or persistent winds may be responsible not only for inducing local modifications to the patterns of energy flux, but also for restricting growth within ecosystems, either by means of abrasive action and mechanical injury or through altering the rates of transpiration or the effective temperature. Cases in which physical damage to plant structure results from wind action are most often seen on exposed mountain ridges, open plains, or land close to the sea. Frequently, it is the exceptional single gust which causes this, though persistent strong winds in mid-latitudinal cyclones may result in the *windthrow* of insecurely rooted forest trees, and vegetation may be similarly devastated at heights of over a few inches above the ground layer during tropical hurricanes or typhoons, in which constantly high wind speeds may last for 24 hours or more. Moreover, at times, the combined effects of wind and freezing rain (or freezing fog) in mid- and high latitudes are such that ice may build up on branches and trunks, generally in winter or early spring, to so great an extent that the plant can no longer support the additional weight on its structure, in which case it may either partially or wholly collapse.

Usually, however, it is more common for vegetation growth to be restrained by wind less directly than this, as when wind speeds coupled with high temperatures increase the transpiration rates so steeply that many species can no longer extract sufficient water from the soil to keep pace with their output, a situation which may lead to visible effects, such as the browning of needle leaves in conifers, and eventual wilting, cell damage, or, at times, death (*physiological drought*, see also section 4.5). Tall trees and shrubs are most commonly affected by these conditions, so that in especially exposed places they may adopt low and xeromorphic forms which hug the ground surface, as in the elfin woodlands (*krummholz*) of Alaska, in which dwarf pines, willows, alders, and birches grow slowly to maturity at heights of only a few inches. Occasionally, deformed individuals may also become established, in which most of the living parts of the plant lie to the leeward of the main stems, so being sheltered to some extent from the consequences of the drying wind. On high mountain ridges or in exposed situations at high latitudes, where wind effects may induce not only physiological drought, but also an above normal degree of cold penetration into plant tissues (*windchill*), plants may eventually be killed off completely (*windkill*), or restricted to much lower altitudes or to more sheltered valleys than would usually be the case. Similar restrictions to growth, found in many coastal areas, are incurred when the joint influence of strong winds and salt spray thrown up from the sea surface reaches a critical level. Here, the height of plants is often conspicuously reduced by *wind-bevelling*, a process which smoothly moulds the top of shrub or tree canopies at their upper limits as they rise away from windward shores, in response to the gradual fall of local transpiration rates inland from coasts which are extremely rich in solar and advective energy. Here, too, the number of species present may also be significantly low when compared to other terrestrial ecosystems, mainly because few plants can withstand the relatively large amounts of salt deposited in these localities. Indeed, many authors (e.g., Chapman;[23] Oosting and Billings[97]) have pointed out that distinctive patterns of vegetation zonation may be present on coastlands as a direct reaction to the differing degrees of tolerance to salt spray which are found among dissimilar plant species. Sometimes, less obvious effects of wind-blown salt spray may also be felt over a much wider area than the immediate coastal zone, as on the tradewind island of Barbados where they can be detected everywhere,[58] or in parts of Britain where they have been observed at distances of up to 40 miles inland.[88]

4.8 Topography

Three major topographic influences may sometimes act as limiting factors. First, the direct *effects of altitude* may be extremely important where the relief is sufficient to induce marked vertical changes of temperature and humidity: usually, temperatures decrease with an increase in altitude either at the dry adiabatic lapse rate (1°C/100 m) or, more frequently, at rates somewhat lower than this (ca. 0·65°C/100 m), and, since humidity is also partly dependent on temperature, moisture conditions may also be appreciably modified. These conditions are responsible for the well-known development of

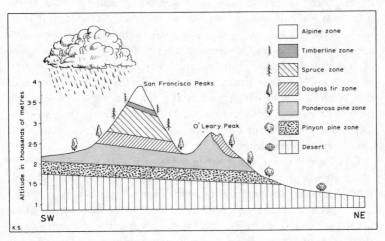

Fig. 4.15 Vegetation zonation on San Francisco Peak, Arizona (modified from an original diagram in C. H. Merriam[81])

altitudinal zonation patterns in animal and plant communities of major mountain ranges, as when rain forest gives way upslope to cloud forest, dwarfed forest, and, finally, alpine scrub and grasses in the tropics; or when arid communities merge into more humid formations in mid-latitudes, as shown in Fig. 4.15; or when spruce-fir forests die out in favour of tundra mosses and lichens in subarctic regions. On some of the highest tropical mountains, as in the alpine zone of Mount Kenya, East Africa, the general lowering of temperature coupled with the effects of a relatively small diurnal range creates a situation at certain elevations in which daily temperatures at screen level fluctuate above and below freezing point on almost

173

every day of the year, thus inducing further consequential changes in relative humidity from values of close to 90 per cent in the early morning to below 20 per cent towards midday; however, at the ground surface itself, relative humidity may remain surprisingly constant at approximately 70 per cent. Apparently as a response to these unique conditions,[27] which make it difficult for plants to survive, many species in this locality (e.g., *Senecio* spp., *Lobelia* spp.) have adopted a genetically new rosette leaf form designed to function efficiently close to the immediate ground surface in the zone of constant humidity and to maintain plant temperatures at night. Further general modifications to the genetic and morphological development of plants may be triggered by differences in light quality at high altitudes as opposed to sea level (section 4.3). Moreover mountains may act as major barriers to the spread of organisms (see section 6.3) or as well-defined lines of division between regions of broadly dissimilar climates and ecosystems; one of the best examples of the latter situation is that of the Cascade Mountains in the Western USA, which within the space of 40 miles separate very wet forests of Douglas fir (*Pseudotsuga taxifolia*) on their western slopes from sage-brush (*Artemesia tridentata*) communities on their semidesert, eastern slopes.

Local variations in *aspect* may also give rise to some restrictions on plant growth in mountain and hilly regions of the mid- and high latitudes, where the angle of incidence of solar radiation is relatively low, especially in the winter months. Because of this, the accumulated year-round intensity, duration, and quality of insolation in these areas may be so dissimilar on different sides of certain valleys as to give rise to major alterations in the general vegetation mosaic; this is best seen in east-west valleys of strong relief, where the shaded, cold, damp, and often forested and floristically-limited north-facing slopes tend to contrast appreciably with south-facing slopes* which are warmer and drier, and can therefore support much more diverse communities. Similar, though less noticeable, features may also at times be detected in north-south valleys, as between east- and west-facing slopes, for the warmth received from the afternoon sun by the latter is usually much more intense than that gained in the morning by the former. Furthermore, if on a sufficiently large scale, effects such as these can lead to local adjustments within the

---

* Often, the French alpine terms *adret* and *ubac* are used to signify the sunny and shaded slopes respectively.

altitudinally induced vegetation zonation structure of large mountain masses, as in the San Francisco ranges of Arizona (see Fig. 4.15), in which all zones are displaced upwards on south-facing slopes looking towards the sun.

On occasions, the relative *steepness of slope* can form a third very important topographic limiting factor. As a general rule, steeper slopes in mountain areas are drier than more gentle ones, so that the vegetation communities developed upon them can be expected to be of a more xerophytic nature than elsewhere. Moreover, while there are very few slopes too steep to support vegetation of any kind, some which lie close to or beyond the normal angle of rest for loose material (ca. 35°) may be sufficiently unstable to have numerous rock slips (screes), snowslides, or avalanches, all of which can restrain the growth of communities, or even destroy them along certain specified tracks. In addition, the deposition of moving material at the base of the scree or avalanche track may remove most of the vegetation which is there, so creating more new open sites. Rather similar features may also be noted in and at the mouth of steep-sided mountain valleys with a narrow valley floor, in which case plants may be uprooted by flood water carrying a large sediment load.

4.9 Edaphic considerations

Those environmental components which are dependent in some measure on the soil may be referred to as edaphic in nature. The influence of some of these on vegetation growth has already been noted in passing, particularly with respect to those which affect the exchanges of energy and moisture within ecosystems, the breakdown of organic matter in the energy chain, and the biogeochemical cycling of elements. It may be inferred from this that it is difficult to isolate edaphic influences from others present in any habitat; moreover, in many cases, it is not easy to discern whether soil characteristics are being moulded more by (say) vegetation than vice versa since interactions between the two are inextricably interwoven; however, for given environments and within specific time limits, some broad distinctions as to the relative effect of soil on vegetation may be drawn, as indicated below.

As an environmental medium, soil must be capable of providing plants with four essentials: an anchorage for roots, a supply of water, adequate amounts of chemical elements in nutrient form, and enough space for air circulation. As far as plants are concerned, this means that all soils must, in turn, have four basic constituents: mineral elements derived from the breakdown of parent rock, organic material which may be added in several ways (see chapter 3), and the ample presence of water and air. Oosting[96] has pointed out that variations in the supply of each of these may lead to restrictions in the development both of individual plants and whole communities, though Billings[11] rightly qualifies this in stating that such edaphic controls tend to have their greatest effect in climates which are dry or cold, or in regions where there are prominent relief forms, in which instances soils may develop only very slowly, often reflecting the qualities of the underlying rock more sharply than elsewhere. However, most soils in terrestrial ecosystems have been formed much more in equilibrium with the overlying conditions of climate and vegetation than with the underlying rock. These *zonal* soils, so named because of their affinities with the major latitudinally based divisions of climate on the earth's surface, show their response to these through the development of stratified vertical layers or *horizons*, three of which may usually be detected: an *A* horizon which is greatly influenced by climate and surface weathering processes; a *C* horizon, often consisting of only slightly weathered fragments of the under-lying rock; and an intermediate *B* horizon which may show charac-teristics of both the *A* and *C* layers. Each of these may be further subdivided, as shown in Figs. 4.16 and 4.17. The overall soil structure as represented by the horizons is termed the *soil profile*, good examples of which may be found under any relatively undisturbed vegetation community, for it is probable that they take only a comparatively few years to form.

The particular environmental circumstances which encourage the formation of strikingly different soil profiles may be illustrated by briefly considering two contrasting examples of mature zonal soils, a *podsol* and a *chernozem*, the former located underneath many northern coniferous forests, and the latter on the semiarid steppe grasslands of southern Russia. Annual precipitation in podsol regions is in excess of evapotranspiration, so that there is a predominantly down-ward movement of water and soluble nutrients from the surface layers of the soil through gravity (as shown in Fig. 4.16) in a process termed *leaching*. This gives rise to an infertile *A* horizon which is often extremely acidic, and if leaching has been of sufficient magni-tude to cause the removal of almost all the mineral elements except

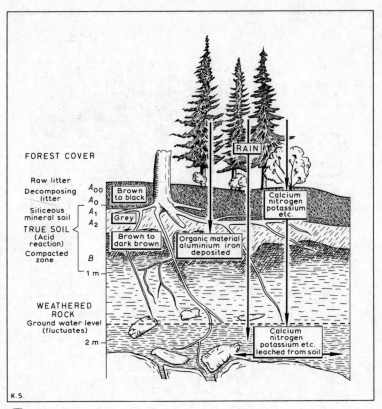

FOREST COVER

Raw litter — A_{00}
Decomposing litter — A_0
Siliceous mineral soil — A_1
A_2
TRUE SOIL (Acid reaction)
Compacted zone — B
1 m

WEATHERED ROCK
Ground water level (fluctuates)
2 m

RAIN

Brown to black
Grey
Brown to dark brown
Organic material aluminium iron deposited
Calcium nitrogen potassium etc.
Calcium nitrogen potassium etc. leached from soil

K.S.

Fig 4.16. A generalized vertical section through a forest podsol (modified from an original diagram in G. L. Clarke[25])

clay, an excessively white, or *bleached* subhorizon may also form. Usually, the downward movement of water in these soils becomes less pronounced a few inches beneath the surface, so that an accumulation of the leached minerals begins to take place in the *B* horizon, sometimes as an iron-rich hardpan; below this the *C* horizon is normally little affected by weathering and is closely related to the underlying rock. Such a profile may sometimes be further modified if the soil is excessively cool and wet throughout the year, for then a sticky, compact, structureless, blue-grey layer, termed a *gley*, may form at its base; this is characterized by poor aeration and a strong reduction of iron compounds, so that streaky reddish-brown stripes of ferric oxide may often be found scattered within the gley.

177

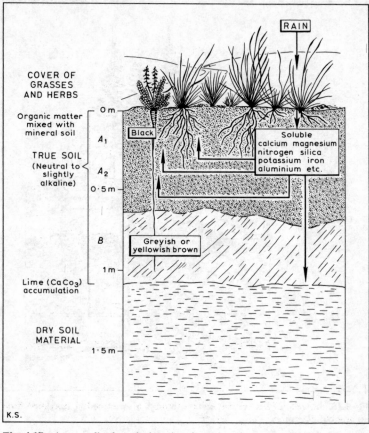

Fig. 4.17 A generalized vertical section through a chernozem (modified from an original diagram in G. L. Clarke[25])

Very different conditions to these prevail in the semiarid steppe-grassland region of the USSR, where evapotranspiration exceeds precipitation and the year-round movement of nutrients and minerals in soil solution is predominantly upwards through capillary action (pp. 151 to 152). Consequently, fertile elements tend always to accumulate in an expanded A horizon, so creating a rich, black, neutral humus soil (the chernozem) which encourages the growth of grass and its roots; in turn, the decaying roots will themselves eventually add more organically derived nutrients to the A horizon, increasing its fertility still further (see Fig. 4.17). In this situation,

178

the extent of the *A* horizon often becomes so great that the *B* horizon is reduced in scale, although it too will be fertile with further strong accumulations of organic material. In contrast, as in other soil profiles, the *C* horizon is generally little affected by these processes and often still retains its strong affinities with the underlying rock.

Several other zonal soil groups may also be delimited. These were first detailed by Glinka[44] in 1927 and are summarized in Fig. 4.18,

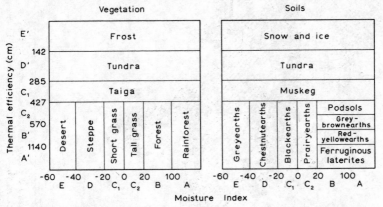

Fig. 4.18 A diagrammatic representation of the relationships between climate, vegetation, and soil (from an original diagram in C. W. Thornthwaite and F. K. Hare[134])

which also shows their general association with world climates, as calculated from the indices of humidity, aridity, and thermal efficiency. It will be noted that, taken as a whole, these groups fall into two major overriding divisions: first, those humid soils, such as podsols, which have a predominantly downward movement of soil water and relatively strong accumulations of aluminium and iron in their elemental structure (*pedalfers*); and second, those *pedocals* (e.g., chernozems), found mainly in arid and semiarid regions, which are characterized by the presence of calcium in relatively large quantities and an upward movement of water. The worldwide distribution of zonal soil groups is presented in Fig. 4.19. For a detailed description of their essential characteristics, the reader is referred to any one of several relevant textbooks.*

* See, for example, G. W. Robinson, *Soils: their Origin, Constitution and Classification*, 3rd edit., London, 1949.

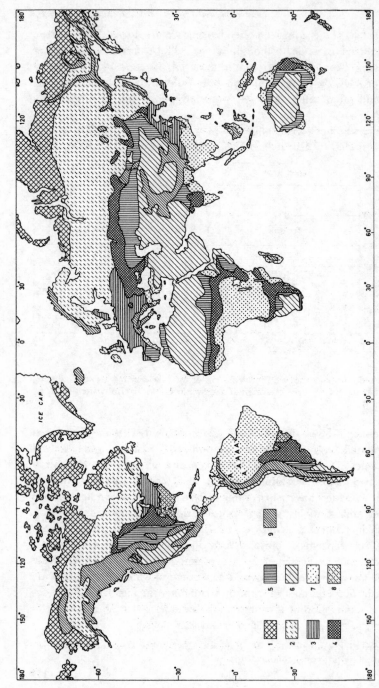

Fig. 4.19 A generalized world map showing the distribution of the major zonal soil types. 1, Tundra soils; 2, Podsols and muskeg; 3, Grey-brown earths; 4, Prairie soils and chernozems (black earths); 5, Chestnut earths; 6, Desert grey earths; 7, Ferrugineous laterites; 8, Red-yellow earths; 9, Mountain soils; A, Alluvial soils.

ICE CAP

Superimposed upon this general pattern in all parts of the world are more locally developed *intrazonal* soils, which form when the chemical or mineral composition of the underlying rocks, or the conditions of drainage or topography, begin to exert a particularly strong influence upon them. In these situations some modification to the expected composition and structure of vegetation communities can take place. Some of the best examples of the effects of drainage and topography on soil and vegetation are seen in *catenas*, a term used to describe a sequence of closely related soil types which, nevertheless, show certain basic differences because of their position on a hillslope. Often, catena soils are better aerated, deeper, have a richer accumulation of nutrients and usually support more complex vegetation assemblages on the lower slopes than on the upper. Thus, within the North Yorkshire Moors region of England bracken-dominated assemblages of 20 to 30 species on the lower slopes of hillsides can be contrasted with the floristically impoverished *Calluna* heath of the upper slopes and the moorland tops, where *Calluna* is in certain districts the only species present. Similar local variations in vegetation can also be expected when underlying rocks give rise to the above-normal accumulation of sand or clay fragments in soil, so modifying its texture. Thus, over large parts of the humid coastal plain of Georgia and Alabama, USA, the presence of sandy, coarse textured and freely draining soils increases apparent aridity, and results in the replacement of the climatologically normal warm temperate deciduous broad-leaf forest of the region by drought-tolerant southern pines. Elsewhere, coarse textured soils may produce an opposite effect, particularly in semiarid environments, where Patton[99] has indicated that their greater rates of infiltration may reduce the surface runoff so much that tongues of forest are able to maintain themselves, even under very dry climatic conditions, much more easily than on more compact soils. Too great quantities of the fine-textured clay component in soils will also change mosaics of vegetation, either through generally restricting aeration or root penetration, or by means of reducing infiltration and encouraging runoff, if it should develop a concretelike consistency during dry weather;[150] if these conditions occur in arid areas, they can also lead to the excessive concentration of salts in the topmost layer of soil, so further limiting plant growth to a few salt-tolerant halophytes, which have exceptionally high osmotic pressures (section 4.5). Both Chapman[22] and Choudhuri[24] have suggested that excessive salinity

181

in such soils is particularly harmful to plants in the germination stage, though Evans[38] has pointed out that poor aeration may be equally damaging to their growth at this time. It should also be noted that the effects of differing soil texture can, in themselves, be further modified by the conditions of soil structure.* Thus, in Texas, black 'waxy' soils, with a plastic clay content of 60 per cent, would have very poor aeration and extremely limited patterns of movement of water and nutrients were it not for the grouping of their component particles into larger granules, with considerable air space between them.

Intrazonal restraints on ecosystems can also arise when the underlying rock exerts an exceptional influence on soil chemistry. Normally, most soils have a wide range of mineral nutrients available for plant growth, the richest supply often being found in alluvium, but in some areas these favourable conditions are replaced by an imbalanced situation. Thus, soils developed on limestones, dolomites, and gypsum may have above-normal levels of $CaCO_3$ held within them; up to a point, this can encourage the activities of soil organisms and the development of plants, but once $CaCO_3$ totals comprise three per cent of all soil constituents by weight, vegetation communities begin to become more and more restricted, due to their inability to tolerate such high levels of lime accumulation, until eventually very few plants survive. Similarly, very acidic and poorly aerated soils formed from granites, or on bog or peat lands, may have limiting deficiencies of Ca, K, Mg, and other elements, as in the cotton grass (*Eriophorum vaginatum*) moorland communities of north-west Europe, where K is the limiting factor. They may also have restraints associated either with toxic effects which result from the modified chemical reactions of Mn, Zn, Cu, some soluble compounds of Al and Fe, and even H and OH ions themselves at the very low levels of pH found in these circumstances, or with the generally reduced intake of nutrients in these areas. At times, the explanation of some puzzling vegetation patterns may be clarified

* Soil *texture* refers to the coarseness or fineness (i.e., the size) of individual soil particles, whereas soil *structure* refers to the degree of aggregation of individual soil particles into clusters. Although some soils are structureless, most have reasonably well-defined patterns of aggregation. Those structures which best encourage the movement of water and nutrients (and therefore make the best agricultural soils) are granular or crumby, the latter having somewhat smaller aggregations than the former. Others, which are less advantageous, are platy structures, which often impair soil permeability; and larger blocky structures, which have sharply-angular faces.

by an examination of nutrient deficiencies induced by pH differences, as in parts of the semidesert of Utah, USA; here, Salisbury[111] suggests that the presence of islands of yellow pine (*Pinus ponderosa*) within more extensive areas of sage brush (*Artemesia tridentata*) may be accounted for by the existence of hydrothermally altered rocks with a strikingly low pH beneath the yellow pine. It is now known that dissimilarities in pH may substantially affect respiration and enzyme activity within plants, and that the associated potential nutrient insufficiency may, in turn, reduce the ability of seeds to germinate, and change the size and morphology of species and their root systems, their vigour and flowering times, and their susceptibility to other restricting factors. Local nutrient deficiency such as this may also be accentuated if the amounts of fragmented parent material in the soil are greater than normal, for this is usually lacking in N, P, and S, so that plants in these environments must require only small quantities of these elements if they are to survive.[47]

Some of the best defined limitations to growth induced by soil chemical factors are to be found in areas underlain by serpentine rock. Soils here are derived from magnesium-iron silicates which are low in major nutrients, but high in magnesium, chromium, and nickel, the last two being close to toxic in concentration for most plants. It is therefore not surprising to find that vegetation communities developed on serpentine are quite distinct from those on adjacent soils within the same climatic zone, and that the line of division between the two may be extremely clear-cut. Rune[109] in northern Sweden, West[153] in the Shetlands, and Steele[127] in parts of Scotland, have all emphasized the relatively poor plant cover in serpentine areas, through which many bare rock patches protrude; this cover is also usually characterized by a marked poverty of species, though this is not true everywhere, as Coombe and Frost[29] have shown on the Lizard peninsula of south-west England. The apparent need for many serpentine plants to adhere to xeromorphic forms[156] can create additional restrictions on growth and on the number of species present, and, in several instances, new morphological varieties which are serpentine-tolerant have been evolved, leading at times to strong endemism (section 6.3). Even where these adaptations do not occur, many plants may still show a marked competitive advantage over others only on serpentine, as is the case with the digger pine (*Pinus sabiniana*) in parts of north-western California.[147]

Although it is clear from these examples that certain intrazonal soils may exert considerable restraints on the growth of plants and vegetation communities, it is now also becoming increasingly appreciated that in some circumstances the type of vegetation present in an area can, in turn, be as important a *direct* influence on the formation of soil type as any other factor; indeed, it may in itself paradoxically lead to restrictions on the development of specific plants by inducing complicated reactions within the soil. Thus, humus products from the litter of species of walnut (*Juglans nigra*[17]), sunflower (*Helianthus rigidus*[32]), and *Encelia farinosa*[89] can often prove toxic to other species which are normally able to compete successfully with them. On a broader scale, the environmental deterioration of the North Yorkshire Moors (see section 3.9, p. 87) has given rise not only to modifications in the dominant plant communities, involving a change from oak-birch forests to heather (*Calluna vulgaris*), but also a parallel alteration of soils from brown earths to very acidic peaty podsols; indeed, traces of the former soils may still be seen as fossil remnants underneath burial mounds erected during the bronze age. Similar fossilized forest soils, which date back to the early agricultural colonization of southern Russia, have also been observed beneath the present chernozem of the steppe grassland, and it is again clear that the formation of chernozem here did not take place until the original forest had been cleared by man. These two examples serve to emphasize the point that soil profiles are rarely static and, indeed, unless intrazonal restrictions are sufficiently strong, may change quickly and substantially should the patterns of vegetation and/or climate themselves be altered. However, timelag effects may, for a while, help to preserve a soil which is more closely related to conditions of the immediate past, rather than to the present, a point which is particularly relevant to analyses of soil-vegetation relationships in the fluctuating climates of today, and the recent past.

4.10 The biotic factor

Thus far, the effects of animals on the development of plant communities in ecosystems have not been considered. It should be noted at the outset that these may be extremely intricate, involving a range of environmental, physiological, and demographic controls, all of which are examined in greater detail in chapter 5. Discussion in this section will therefore be limited to a review of some of the more direct

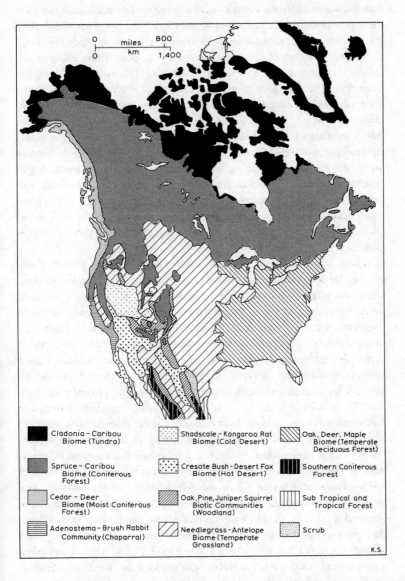

Fig. 4.20 The biomes of North America (from an original diagram in V. E. Shelford, *The ecology of North America*, Univ. of Illinois Press, 1963)

relationships between individuals or groups of animals and the particular vegetation communities in which they live or with which they are associated. Although more is known of these in mid- and high latitudes than in the tropics, it would appear that, in general, mammals have a much more precise ecological *rapport* with vegetation types than do most other animal groups, except, perhaps in the case of some insects and birds. Accordingly, mammals are usually given primacy in studies of the *biome* (or *biotic area*), a concept which aims to extend the ideas of community among vegetation and animal populations to cover the patterns of life within both. Biomes may be defined as major regions in which distinctive plant and animal groups usually live in harmony with each other, and are also well adapted to the external conditions of environment, so that one may make tentative, but meaningful, correlations between all three. A relatively simple diagrammatic map of the biomes of North America is presented in Fig. 4.20.

These idealized situations are interrupted if the numbers of one particular animal group become excessively high in relation to their environmental resources of food: in this event, the group concerned may eventually begin to act as an effective biotic limiting factor to vegetation growth. In parts of Britain, this has clearly happened among rabbit populations, whose numbers have been considerable since the sixteenth century and which in extreme situations have been responsible for clearing off almost all vegetation in their warren area. The mechanisms which give rise to this may perhaps be clarified by considering first the case of individual rabbits, which are known always to consume food resources in the immediate vicinity of their burrows before moving farther away to search for new supplies.[40] This means that, in time, a marked biotically induced zonation of vegetation may be formed around each burrow, in which all plants have been eaten off *immediately* adjacent to the burrow, and selective chewing effects are noticeable up to a height of 50 cm above ground level (i.e., that to which a rabbit can reach when standing on its hind legs) and to a distance of approximately 100 m away, beyond which there is little disturbance (see Fig. 4.21). Within the chewed area, several of the less palatable species, such as bracken (*Pteridium aquilinum*), elder (*Sambucus nigra*), and bramble (*Rubus fruticosus*) may gain a competitive advantage, this being more clearly seen as one approaches the burrow. If one then multiplies the effects of rabbit grazing as outlined above by a factor of several thousand, or

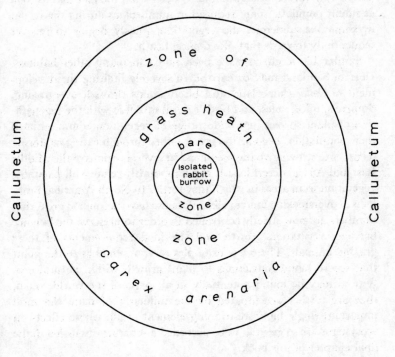

Fig. 4.21 Zonation of vegetation round an isolated rabbit burrow (modified from an original diagram in E. P. Farrow[40])

several tens of thousand times, to cover the size of a typical large warren, one may begin to appreciate the reasons as to why rabbit populations tend to have such a devastating effect on plant growth. In some areas, indeed, the final results of grazing have been disastrous, as in the sandstone Breckland region of west Norfolk, UK[146], where it has been responsible not only for inducing a general change in vegetation from *Calluna* heath to a much more restricted grassland formation dominated by common bent (*Agrostis* sp.) and sheep's fescue (*Festuca ovina*), but also for giving rise to a severe local removal of the grassland itself, to create bare ground and surface instability in the topmost layers of the soil, which, in turn, has led to the subsequent erosion of soil and sand by wind action.[130] Proof that responses

187

such as these are essentially of a biotic nature may be obtained partly from direct observation and partly from the fact that as soon as rabbit populations are reduced in numbers, as during the recent myxomatosis epidemic, the vegetation quickly begins to recover sequentially towards that of a *Calluna* heath.

Similar biotic effects have been noted in many other habitats. Deer in Scotland may be capable of severely limiting the development of *Calluna* moorland and birch woods through overgrazing. Squirrels, mice, voles, and birds can all interfere with the regeneration of plant species in a wide range of vegetation communities if their populations become too great, either through eating or storing seeds, or destroying young seedlings. In several regions of the Middle East and Africa, desert locusts can temporarily remove all palatable vegetation from areas in which they settle. In North America, problems of overstocking among elk and moose have become so great that numbers are now strictly controlled in order to preserve the balance between vegetation growth and the food requirements of these grazing animals. These few examples serve to emphasize the point that severe biotic restrictions to plant growth within natural ecosystems may be found potentially in all parts of the world; often, they are further accentuated by the influence of man, the most important single biotic restraining element of all, whose effects on ecosystems are so great as to be accorded a separate discussion in the final chapter of this book.

REFERENCES

1. ALLARD, H. A., Length of day in relation to the natural and artificial distribution of plants, *Ecology*, Vol. 13, pp. 221–34, 1932.
2. ALLARD, H. A., Lack of available phosphorus, preventing normal succession on small areas on Bull Run Mountain in Virginia, *Ecology*, Vol. 23, pp. 345–53, 1942.
3. ALVIN, P. de T., Moisture stress as a requirement for flowering coffee, *Science*, Vol. 132, No. 3423, p. 345, 1960.
4. ALWAY, F. J., Studies of the relation of non-available water of the soil to the hygroscopic coefficient, *Nebraska State Research Bull.*, Vol. 3, pp. 5–122, 1913.
5. ANDERSON, M. C., Some problems of simple characterization of the light climate in plant communities. In *Light as an Ecological Factor*, eds. R. BAINBRIDGE, G. C. EVANS, and O. RACKHAM, pp. 77–90, Oxford, 1966.
6. ANGSTRÖM, A., The albedo of various surfaces of the ground, *Geografiska Annaler*, Vol. H. 4, pp. 323–42, 1925.

7. ASHBEL, D., Frequency and distribution of dew in Palestine, *Geog. Rev.*, Vol. 39, pp. 291–7, 1949.

8. BAILEY, F., Ultraviolet boosts egg production, *Poultry Trib.*, Vol. 56, p. 34, 1950.

9. BEESON, C. F., Forestry, horticulture and the moon, *For. Abstr.*, Vol. 8, pp. 191–8, 1946.

10. BERNSTEIN, L., The needs and uses of water by plants. In *Water*, *U.S.D.A. Yearbook*, pp. 18–25, 1955.

11. BILLINGS, W. D., *Plants and the Ecosystem*, London, 1964.

12. BLACK, R. F., Permafrost—a review, *Bull. Geol. Soc. Am.*, Vol. 65, pp. 839–56, 1954.

13. BLACKMAN, G. E., Responses to environmental factors by plants in the vegetative phase. In *Growth in Living Systems*, ed. M. X. ZARROW, pp. 525–56, New York, 1961.

14. BLACKMAN, G. E. and RUTTER, A. J., Physiological and ecological studies in the analysis of plant environment. I. The light factor and the distribution of the bluebell (*Scilla non-scripta*) in woodland communities, *Ann. Bot.*, *N.S.*, Vol. 10, pp. 361–90, 1946.

15. BOSIAN, G., Assimilations- und Transpiration-bestimmungen an Pflanzen des Zentralkaiserstuhls, *Z. Bot.*, Vol. 26, pp. 209–84.

16. BRIGGS, L. J. and SCHANTZ, H. L., The relative wilting coefficient for different plants, *Bot. Gaz.*, Vol. 53, pp. 229–45, 1912.

17. BROOKS, M. G., Effect of black walnut trees and their products on other vegetation, *W. Virginia Agric. Exper. Sta. Bull.* No. 347, 1951.

18. BUDYKO, M., *The heat balance of the earth's surface*, Gidrometeoizdat Leningrad, 1956, trans. by N. A. STEPANOVA, PB 121692, Office of Technical Services, US Department of Commerce, Washington DC, 1958.

19. BURGER, W., Review of experimental investigations on seasonal reproduction in birds, *Wilson Bull.*, Vol. 61, pp. 221–30, 1949.

20. CANDOLLE, A. L. de, *Géographie Botanique Raisonée*, Paris, 1855.

21. CHANG, J., Microclimate of sugar cane, *Hawaiian Planters' Record*, Vol. 56, pp. 195–214, 1961.

22. CHAPMAN, V. J., The new perspective in the halophytes, *Quart. Rev. Biol.*, Vol. 17, pp. 291–311, 1942.

23. CHAPMAN, V. J., *Coastal vegetation*, London, 1964.

24. CHOUDHURI, G. N., Effect of soil salinity on germination and survival of some steppe plants in Washington, *Ecology*, Vol. 49, pp. 465–71, 1968.

25. CLARKE, G. L., *Elements of ecology*, Wiley, New York, 1954.

26. CLOUDSLEY-THOMPSON, J. L. and CHADWICK, M. J., *Life in Deserts*, London, 1964.

27. COE, M. J., *The ecology of the Alpine Zone of Mt. Kenya*, The Hague, 1967.

28. COLE, J. A. and GREEN, M. J., Measurements of net radiation over vegetation, and of other climatic factors affecting transpiration losses in water catchments, *I.A.S.H., Cttee. for Evaporation Publ.*, 62, pp. 190–202, 1963.

29. COOMBE, D. E. and FROST, L. C., The heaths of the Cornish serpentine, *J. Ecol.*, Vol. 44, pp. 226–56, 1954.

30. COOPER, W. S., The broad-sclerophyll vegetation of California, *Carnegie Inst.*, Pub. No. 319, Washington DC, 1922.
31. CRAFTS, A. S., CURRIER, H. B., and STOCKLING, C. R., *Water in the Physiology of Plants*, Chronica Botanica, Waltham, Mass., 1949.
32. CURTIS, J. T. and COTTAM, G., Antibiotic and antitoxic effects in prairie sunflower, *Torrey Bot. Club Bull.*, No. 77, pp. 187–91, 1950.
33. DAUBENMIRE, R. F., *Plants and Environment*, New York, 2nd edn, 1959.
34. DAVIS, R. B., Spruce-fir forests of the coast of Maine, *Ecol. Mon.*, Vol. 36, pp. 79–94, 1966.
35. DUVDEVANI, S., Dew research for arid agriculture, *Discovery*, Vol. 18, pp. 330–4, 1957.
36. EDLEFSEN, N. E. and ANDERSON, A. B. C., Thermodynamics of soil moisture, *Hilgardia*, Vol. 15, pp. 31–298, 1943.
37. EVANS, G. C., Model and measurement in the study of woodland light climates. In *Light as an ecological Factor*, eds. R. BAINBRIDGE, G. C. EVANS, and O. RACKHAM, pp. 53–76, Oxford, 1966.
38. EVANS, L. T., The ecology of the halophytic vegetation at Lake Ellesmere, New Zealand, *J. Ecol.*, Vol. 41, pp. 106–22, 1953.
39. EVENARI, M. and RICHTER, R., Physiological-ecological investigations in the Wilderness of Judaea, *J. Linn. Soc. (Bot.)*, Vol. 51, pp. 333–81, 1937.
40. FARROW, E. P., On the ecology of the vegetation of Breckland. III. General effects of rabbits on the vegetation, *J. Ecol.*, Vol. 5, pp. 1–18, 1917.
41. GAERTNER, E. E., Water relations of forest trees. In *The water relations of plants*, eds. A. J. RUTTER and F. H. WHITEHEAD, pp. 366–78, London, 1963.
42. GARNER, W. W. and ALLARD, H. A., Effect of the relative length of day and night and other factors of the environment on growth and reproduction in plants, *J. Agric. Res.*, Vol. 18, pp. 553–606, 1920.
43. GEIGER, R., *The Climate Near the Ground*, 2nd edn, Harvard U.P., 1957.
44. GLINKA, K. D., *The Great Soil Groups of the World and their Development*, (trans. by C. F. MARBUT), Ann Arbor, Michigan, 1927.
45. GOODMAN, G. T. and PERKINS, D. F., The role of mineral nutrients in *Eriophorum* communities, IV. Potassium supply as a limiting factor in an *E. vaginatum* community, *J. Ecol.*, Vol. 56, pp. 685–96, 1968.
46. GREEN, F. H. W., Some observations of potential evapotranspiration, 1955–1957, *Quart. J. R. Met. Soc.*, Vol. 85, pp. 152–8, 1959.
47. GRIGGS, R. F., The colonization of Katmai ash, a new inorganic 'soil', *Am. J. Bot.*, Vol. 20, pp. 92–113, 1933.
48. GRIME, J. P., Shade-avoidance and shade-tolerance in flowering plants. In *Light as an Ecological Factor*, eds. R. BAINBRIDGE, G. C. EVANS, and O. RACKHAM, pp. 187–207, Oxford, 1966.
49. GRISEBACH, A. H. R., *Die vegetation der Erde nach ihrer klimatischen Anordnung*, Leipzig, 1872.
50. HAECKEL, E., Über entwicklungsgang und Augfabe der Zoologie, *Jenaischer Zeitschr. für Naturwiss*, Vol. 5, pp. 353–70, 1869.

51. HARE, F. K., The boreal conifer zone, *Geographical Studies*, Vol. 1, pp. 4–18, 1954.

52 HARE, F. J., *The Evapotranspiration Problem*, McGill University, Montreal, 1961.

53. HEILMAN, P. E., Relationship of availability of phosphorus and cations to forest succession and bog formation in interior Alaska, *Ecology*, Vol. 49, pp. 331–6, 1968.

54. HENSON, W. R., Chinook winds and red-belt injury to lodgepole pine in the Rocky Mountain Parks area of Canada, *For. Chron.*, Vol. 28, pp. 62–4, 1952.

55. HOFMANN, G., Die Thermodynamik der Tanbildung, *Ber. dtsch. Wetterd.*, 3, Nr. 18, 1955.

56. HORTON, R. E., Rainfall interception, *Mon. Wea. Rev.*, Vol. 47, pp. 603–23, 1919.

57. HORTON, R. E., Transpiration of forest trees, *Mon. Wea. Rev.*, Vol. 51, p. 569, 1923.

58. HUDSON, J. C., Some meteorological observations in Barbados, *Min. Agric. Lands Fish.*, *Barbados, Bull.*, No. 33, 1963.

59. HUGHES, A. P., The importance of light compared with other factors affecting plant growth. In *Light as an Ecological Factor*, eds. R. BAINBRIDGE, G. C. EVANS, and O. RACKHAM, pp. 121–47, Oxford, 1966.

60. ILJIN, W. J., Einfluss des Welkens auf die Atmung der Pflanzen, *Flora, Jena*, Vol. 116, pp. 379–403, 1923.

61. IVERSEN, J., *Viscum, Hedera* and *Ilex* as climatic indicators, *Geol. Fören Förh*, Vol. 66, pp. 463–83, 1944.

62. JARVIS, C. S., Floods, In *Hydrology*, ed. O. E. MEINZER, pp. 531–60, New York, 1942.

63. JONES, R. L., Dew as a factor in plant water balance, *N.Z. Dept. Sci. Ind. Res. Int. Ser.*, No. 12, pp. 66–72, 1956.

64. KINNE, O., Über den Einfliss des Salzegehaltes und der Temperatür auf Wachstum, Form und Vermehrung bei dem Hydroidpolypen *Cardylophora caspia* (Pallas), *Thecata, Clavidae. Zool. Hahrb.*, Vol. 66, pp. 565–638, 1956.

65. KITTREDGE, J. Jr., Natural vegetation as a factor in the losses and yields of water, *J. For.*, Vol. 35, p. 1011, 1937.

66. KITTREDGE, J. Jr., The magnitude and regional distribution of water losses influenced by vegetation, *J. For.*, Vol. 35, p. 775, 1938.

67. KITTREDGE, J. Jr., *Forest Influences*, New York, 1948.

68. KÖPPEN, W., *Die Klimate der Erde; Grundriss der Klimakunde*, Walter de Gruyter, Berlin, 1923.

69. LARSEN, E. B., Importance of master factors for the activity of Noctuids, *Saertr. Ent. Medd.*, Vol. 23, pp. 352–74, 1943.

70. LEE, C. H., Transpiration and total evaporation. In *Hydrology*, ed. O. E. MEINZER, Ch. 8, 1942.

71. LEMON, E., Energy and water balance of plant communities, In *Environmental Control of Plant Growth*, ed. L. T. EVANS, pp. 55–77, New York, 1963.

72. LEVITT, J., *The Hardiness of Plants*, New York, 1956.

73. LEVITT, J., Hardiness and the survival of extremes: A uniform system for measuring resistance and its two components. In *Environmental Control of Plant Growth*, ed. L. T. EVANS, pp. 351–651, New York, 1963.

74. LIEBIG, J., *Die organische Chemie in ihrer Anwendung vor Agricultur und Physiologie*, Braunschweig, 1840.

75. LIVINGSTON, B. E. and SHREVE, F., The distribution of vegetation in the United States, as related to climatic conditions, *Carnegie Inst.*, Pub. No. 284, Washington DC, 1921.

76. LOACH, K., Shade tolerance in tree seedlings. I. Leaf photosynthesis and respiration in plants raised under artificial shade, *New Phytol.*, Vol. 66, pp. 607–21, 1967.

77. MARSH, F. L., Water content and osmotic pressure of sun and shade leaves of certain woody prairie plants, *Bot. Gaz.*, Vol. 102, pp. 812–15, 1941.

78. MASON, H. L. and LANGENHEIM, J. H., Language analysis and the concept of environment, *Ecology*, Vol. 38, pp. 325–40, 1957.

79. MATHER, J. R., The measurement of potential evapotranspiration, *Publications in Climatology*, The Johns Hopkins University Laboratory of Climatology, Seabrook, New Jersey, Vol. 7, No. 1, 1954.

80. MEINZER, O. E., Occurrence, origin and discharge of ground water. In *Hydrology*, ed. O. E. MEINZER, pp. 385–443, New York, 1942.

81. MERRIAM, C. H., Results of a biological survey of the San Franscico Mountain Region and the desert of the Little Colorado, Arizona, *North American Fauna*, No. 3, pp. 1–396, 1890.

82. METEOROLOGICAL OFFICE, *Handbook of Meteorological Instruments, Part I*, HMSO, London, 1956.

83. MIDDLETON, H. E., The moisture equivalent in relation to the mechanical analyses of soils, *Soil Sci.*, Vol. 9, pp. 159–67, 1920.

84. MILTHORPE, F. J., The income and loss of water in arid and semiarid zones. In *Plant Water Relationships in Arid and Semi-arid Conditions*, pp. 9–36, UNESCO, Paris ,1960.

85. MONTEITH, J. L., Evaporation at night, *Neth. J. Agric. Sci.*, Vol. 4, pp. 34–8, 1956.

86. MONTEITH, J. L., The reflection of short-wave radiation by vegetation, *Quart. J. R. Met. Soc.*, Vol. 85, p. 386, 1959.

87. MONTEITH, J. L., Dew: facts and fallacies. In *The Water Relations of Plants*, eds. A. J. RUTTER and F. H. WHITEHEAD, pp. 37–56, London, 1963.

88. MOSS, A. E., Effects of wind-driven salt water, *J. For.*, Vol. 38, pp. 421–5, 1940.

89. MULLER, W. H. and MULLER, C. H., Association patterns involving desert plants that contain toxic products, *Am. J. Bot.*, Vol. 43, pp. 354–61, 1956.

90. MURRAY, D. B. and NICHOLS, R., Light, shade and growth in some tropical plants, In *Light as an Ecological Factor*, eds. R. BAINBRIDGE, G. C. EVANS, and O. RACKHAM, pp. 249–62, Oxford, 1966.

91. NEBIKER, W. A., Evapotranspiration studies at Knob Lake, June–

September 1956, *McGill University Sub-Arctic Research Papers*, No. 3, McGill University, Montreal. 1957

92. NITSCH, J., The mediation of climatic effects through endogenous regulating substances. In *Environmental Control of Plant Growth*, ed. L. T. Evans, pp. 175–92, New York, 1963.

93. OBERLANDER, G. T., Summer fog precipitation on the San Francisco peninsula, *Ecology*, Vol. 37, pp. 851–2, 1956.

94. ODUM, E. P., *Fundamentals of Ecology*, Saunders, Philadelphia, 2nd edn, 1959.

95. OLMSTEAD, C. E., Experiments on photoperiodism, dormancy and leaf age, and abscission in sugar maple, *Bot. Gaz.*, Vol. 112, pp. 365–93, 1951.

96. OOSTING, H. J., *The study of plant communities*, San Francisco, 2nd edn, 1956.

97. OOSTING, H. J. and BILLINGS, W. D., Factors affecting vegetational zonation on coastal dunes, *Ecology*, Vol. 23, pp. 131–42, 1942.

98. OOSTING, H. J. and KRAMER, P. J., Water and light in relations to pine reproduction, *Ecology*, Vol. 27, pp. 47–53, 1946.

99. PATTON, R. T., The factors controlling the distribution of trees in Victoria, *Roy. Soc. Victoria, Proceedings*, Vol. 42, pp. 154–210, 1930.

100. PEARL, R. Y., MATHEWS, R. H., SMITH, L. P., PENMAN, H. L., HOARE, E. R., and SKILLMAN, E. E., The calculation of irrigation need, *Min. Agric. Fish. Tech. Bull.*, Vol. 4, 1954.

101. PENMAN, H. L., Natural evaporation from open water, bare soil and grass, *Proc. R. Soc. Series A*, Vol. 193, pp. 120–45, 1948.

102. PENMAN, H. L., Weather and water in the growth of grass. In *The Growth of Leaves*, ed. F. L. MILTROPHE, London, 1956.

103. PENMAN, H. L., *Vegetation and Hydrology*, Comm. Agricultural Bureau, Farnham Royal, 1963.

104. REYNOLDS, E. R. C. and LEYTON, L., Measurement and significance of throughfall in forest stands. In *The Water Relations of Plants*, eds. A. J. RUTTER and F. H. WHITEHEAD, pp. 127–41, London, 1963.

105. RICE, L. A., Studies on deciduous forest animal populations during a two-year period with differences in rainfall, *Am. Mid. Nat.*, Vol. 135, pp. 153–71, 1946.

106. ROUSE, W. R. and WATTS, D., Two studies in Barbadian climatology, *Climatology Research Series No. 1*, McGill University, Montreal, 1966.

107. ROWAN, W., Relation of light to bird migration and developmental changes, *Nature*, Vol. 115, pp. 494–5, 1925.

108. ROWE, P. B., Some factors on the hydrology of the Sierra Nevada foothills, *Trans. Am. Geophys. Union*, Vol. 22, pp. 90–100, 1941.

109. RUNE, O., Plant life on serpentines and related rocks in the north of Sweden, *Acta Phytogeogr. Suec.*, Vol. 31, 1953.

110. SALISBURY, E. J., The oak-hornbeam woods of Hertfordshire, *J. Ecol.*, Vol. 4, pp. 83–117, 1916.

111. SALISBURY, F. B., Some chemical and biological investigations of materials derived from hydrothermally-altered rocks in Utah, *Soil Sci.*, Vol. 78, pp. 277–94, 1954.

112. SCHULMAN, E., Longevity under adversity of conifers, *Sci.*, Vol. 119, pp. 396–9, 1954.

113. SCHWABE, W. W., Morphogenetic responses to climate. In *Environmental Control of Plant Growth*, ed. L. T. EVANS, pp. 311–34, New York, 1963.

114. SEMMENS, E. S., Chemical effects of moonlight, *Nature*, Vol. 159, p. 613, 1947.

115. SEMPER, K., *Animal Life*, New York, 1881.

116. SERRA, L., The hydrological check on a catchment area, *Soc. Hydrotechnique de France, compte rendu des 3ème journèes de l'hydraulique*, 12–14 April, 1954, pp. 29–35, 1954.

117. SHELFORD, V. E., Physiological animal geography, *J. Morphol.*, Vol. 22, pp. 551–618, 1911.

118. SHELFORD, V. E., *Animal Communities in Temperate America*, Chicago, 1913.

119. SHELFORD, V. E., *The Ecology of North America*, Urbana, Illinois, 1963.

120. SHIRLEY, H. L., Reproduction of upland conifers in the Lake States as affected by root competition and light, *Am. Mid. Nat.*, Vol. 33, pp. 537–612, 1945.

121. SHREVE, F., The influence of low temperatures on the distribution of the giant cactus, *Pl. Wld*, Vol. 14, pp. 136–46, 1911.

122. SHULL, C. A., Measurement of the surface forces in soils, *Bot Gaz.*, Vol. 62, pp. 1–31, 1916.

123. SIBBONS, J. L. H., The climatic approach to potential evapotranspiration, *Adv. Sci.*, Vol. 13, pp. 354–6, 1956.

124. SLAYTER, R. O., Climatic control of plant water relations. In *Environmental Control of Plant Growth*, ed. L. T. EVANS, pp. 3–52, New York, 1963.

125. SMITH, G. W., The determination of soil moisture under a permanent grass cover, *J. Geophys. Res.*, Vol. 64, pp. 477–83, 1959.

126. STEAD, D. G., *The Rabbit in Australia*, Sydney, 1935.

127. STEELE B., Soil pH and base status as factors in the distribution of calcicoles, *J. Ecol.*, Vol. 43, pp. 120–32, 1955.

128. STEWARD, F. C., Effects of environment on metabolic patterns. In *Environmental Control of Plant Growth*, ed. L. T. EVANS, pp. 195–212, New York, 1963.

129. TANSLEY, A. G., *The British Islands and their Vegetation*, Cambridge U.P., 1953.

130. TANSLEY, A. G and ADAMSON, R. S., Studies of the vegetation of the English chalk. III. The chalk grasslands of the Hampshire–Sussex Border, *J. Ecol.*, Vol. 13, pp. 177–223, 1925.

131. THOMAS, M. D., Effect of ecological factors on photosynthesis, *Ann. Rev. Pl. Physiol.*, Vol. 6, pp. 135–6, 1955.

132. THORNTHWAITE, C. W., A contribution to the report of the Cttee. on transpiration and evaporation, 1943–44, *Trans. Am. Geophys. Union*, Vol. 25, pp. 686–93, 1944.

133. THORNTHWAITE, C. W., An approach towards a rational classification of climate, *Geog. Rev.*, Vol. 38, pp. 55–94, 1948.

134. THORNTHWAITE, C. W. and HARE, F. K., Climatic classification in forestry, *Unasylva*, FAO, Rome, Vol. 9, pp. 51–9, 1955.

135. THORNTHWAITE, C. W. and MATHER, J. R., The water-balance, *Publications in climatology*, Vol. 8, No. 1, The Johns Hopkins University Laboratory of Climatology, Seabrook, New Jersey, 1955.

136. TUBERVILLE, H. W. and HOUGH, A. F., Errors in age counts of suppressed trees, *J. For.*, Vol. 37, pp. 417–18, 1939.

137. VIEHMEYER, F. J., Evaporation from soils and transpiration, *Trans. Am. Geophys. Union*, Vol. 19, pp. 612–15, 1938.

138. VOGELMANN, H. W., SICCAMA, T., LEEDY, D., and OVITT, D. C., Precipitation from fog moisture in the Green Mountains of Vermont, *Ecology*, Vol. 49, pp. 1205–07, 1968.

139. WADSWORTH, R. M. ed., *The Measurement of Environmental Factors in Terrestrial Ecology*, Blackwell Scientific Publications, Oxford and Edinburgh, 1968.

140. WAISEL, V., Dew absorption by plants of arid zones, *Bull Res. Council Israel*, Vol. 6D, pp. 180–6, 1958.

141. WALTER, H., *Grundlagen der Pflanzenverbreitung*, Stuttgart, 1951.

142. WALTER, H., The water supply of desert plants. In *The Water Relations of Plants*, eds. A. J. RUTTER and F. H. WHITEHEAD, pp. 199–205, London, 1963.

143. WARD, R. C., Observations of potential evapotranspiration (PE) on the Thames flood plain, 1959–1960, *J. Hydrol.*, Vol. 1, pp. 183–94, 1963.

144. WARD, R. C., *Principles of Hydrology*, McGraw-Hill, Maidenhead, UK, 1967.

145. WARMING, E., *Oecology of Plants* (Trans. by P. GROOM and I. B. BALFOUR), Oxford, 1909.

146. WATT, A. S., On the origin and development of blow-outs, *J. Ecol.*, Vol. 25, pp. 91–112, 1937.

147. WATTS, D., Human occupance as a factor in the distribution of the California Digger Pine (*Pinus Sabiniana*), *Unpublished dissertation*, University of California at Berkeley, 1959.

148. WATTS, D., Man's influence on the vegetation of Barbados, 1627 to 1800, *University of Hull, Occasional Papers in Geography*, No. 4, Hull, UK, 1966.

149. WATTS, D., GALLOWAY, J. H., and GRENIER, A., Evapotranspiration studies at Knob Lake in the summers of 1957, 1958 and 1959, *McGill Sub-Arctic Research Papers*, No. 10, McGill University, Montreal, 1960.

150. WEAVER, J. E., The ecological relations of roots, *Carnegie Inst. of Washington*, Pub. No. 292, Washington DC, 1919.

151. WEAVER, J. E. and CLEMENTS, F. E., *Plant Ecology*, New York, 1938.

152. WENT, F. W., The concept of a phytotron. In *Environmental Control of Plants Growth*, ed. L. T. EVANS, pp. 1–4, New York, 1963.

153. WEST, W., Notes on the flora of Shetland, with some ecological observations , *J. Bot., Lond.*, Vol. 50, pp. 265–75, 297–306, 1912.

154. WHITEHEAD, F. H., Wind as a factor in plant growth. In *Control of the Plant Environment*, ed. J. P. HUDSON, London, 1957.

155. WHITEHEAD, F. H., The effects of exposure on growth and development. In *The Water Relations of Plants*, eds. A. J. RUTTER and F. H. WHITEHEAD, pp. 199–205, London, 1963.

156. WHITTAKER, R. H., The ecology of serpentine soils, *Ecology*, Vol. 35, pp. 258–88, 1954.

157. WHYTE, R. O., *Crop Production and Environment*, London, 1960.

158. WIJK, W. R. van and de VRIES, D. A., Evapotranspiration, *Neth. Journ. Agric. Sci.*, Vol. 2, pp. 105–18, 1954.

159. WILDE, S. A., Influence of forest cover on the state of the ground water table, *Soil Sci. Soc. Am., Proc.*, Vol. 17, pp. 65–7, 1953.

160. WILLIS, A. J. and JEFFERIES, R. C., Investigations on the water relations of sand-dune plants under natural conditions. In *The Water Relations of Plants*, eds. A. J. RUTTER and F. H. WHITEHEAD, pp. 168–89, London, 1963.

161. WILM, H. G., The influence of forest vegetation on water and soil, *Unasylva*, Vol. 11, pp. 160–4, 1957.

5

Population Limitations Within Ecosystems

5.1 Introduction

In all ecosystems, there may also be demographic elements of restraint which help to control the physiological and numerical growth of organisms. Some of these result from behavioural, social, or evolutionary factors, others from competition processes. All are examined within this chapter, mainly in the context of animal populations, since these are, perhaps, most closely influenced by them; however, it should be noted that they are known to affect the organizational structure of certain plant communities, too.

Despite the presence of so many animals on the earth's surface, it is still very difficult to enumerate and observe the behaviour of even one group of these in their natural environment, since most tend to remain under dense cover and stay away from habitats which are dominated by man. While writing about the rich and diverse fauna of the Rio Negro catchment of South America in the middle of the last century, Wallace[100] commented on the immense obstacles which had to be overcome before even limited population surveys could be made meaningful in the context of space and time. Indeed, it was only in the very late nineteenth century that the first successful attempts at taking faunal censuses were made by means of straining water through fine nets in order to count the plankton it contained. More detailed and sophisticated analyses of marine populations followed soon afterwards, stimulated in particular by Petersen's work in Danish coastal waters;[86] these culminated in a series of monographs from 1911 onwards, some of which were produced in collaboration with Jensen.[53] As in the elucidation of biospheric patterns of energy transfer (see pp. 27 to 28), the formidable complexity of land ecosystems was sufficient reason to delay their examination by faunal demographers until much later, for even though Shelford[90] gave the first comprehensive account of animal communities on land at almost the same time as Petersen began to publish his Danish material, this was not concerned with the counting of populations, but rather with an understanding of relationships between the component organisms on a non-quantitative basis. It was left to Lotka[67] and Elton[26, 27] in the 'twenties to put forward general theories of the community development of organisms, since when many more related papers have been presented, centred particularly around the theme which Elton[30] has suggested is the ultimate *raison d'etre* of all faunal demographic research: the discovery and measurement of the major dynamic interrelationships between all populations of organisms living within a given area over a certain period of time.

THE DEMOGRAPHY OF ORGANISMS

For the purposes of this chapter, *populations* of organisms may be said to refer to groups of the same species or variety which exist within definable limits of space and time. All populations possess *individual* (biological) and *collective* characteristics, both of which need to be considered as discrete parts of the whole. Isolated organisms may be studied within the general framework of their birth, growth, life, and death, while the distribution, density, and social habits of groups of organisms must also be examined in order to complete the demographic picture.

Even a brief glimpse at any natural ecosystem will confirm that many different types of organisms coexist or compete within it in a complex manner, and each of these is dependent upon a *habitat range* which is related to the availability of energy and food. Elton[30] has noted that the size of a habitat range tends to increase towards the higher trophic levels of an ecosystem, so that an inverse *pyramid of habitats* is often seen (see Fig. 5.1). Thus, herbivorous animals are usually confined to a very few species of plants, so that the distributions of both closely coincide. At secondary consumer levels, predators and parasites live off wider habitats; and at still higher trophic levels, as with shrews or insectivorous birds, the numbers of habitats

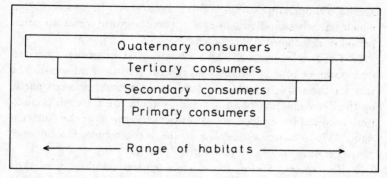

Fig. 5.1 A generalized pyramid of habitats

visited are numerous. Hawks and owls range very widely, but the least restricted consumer of all is, of course, man. The general effect of this is to create a situation in which relationships between consumer organisms and the environment are much less clearly seen among higher trophic levels than elsewhere; the only known exceptions to this rule are certain mammal and bird herbivores (e.g., rabbits, deer, and the wood pigeon *Columba palumbus*) which are not necessarily limited to narrow habitat ranges, due to wide variations in their seasonal food requirements and patterns of behaviour.

5.2 Methods of estimating population

The single most important preliminary to the study of populations is the devising of an accurate census. This is not always easy, for the many problems associated with the counting of motile organisms, and the large number of variables which may rapidly affect their rates of production and growth, ensure that most censuses will be only relatively precise, even within the context of very restricted time and space. Usually, one of the following techniques is chosen:

1. *The total count method*, in which all organisms within a given area are enumerated, regardless of age, sex, stage of development, or location. Practical considerations ensure that this technique is rarely employed, since the number of faunal components within any one ecosystem is frequently so great as to make the undoubted accuracy obtained not worth the time and effort needed to put it into effect. Where they are used, total counts may be equated with human population censuses taken among western societies. However,

199

a more common application of this is a *total count of a stage or class*, in which only those individuals of a particular group, or at a certain period of development, are noted.

2. *The sampling method*, which assumes that judgements derived from samples of a population may, with statistical safeguards, be true for the whole population. Successful sampling depends partly on the determination of an optimum sample size for each population, followed by sophisticated statistical treatment of the collected data. While *numerical sampling* is the most common technique, *biomass sampling* may also be effectively used.

3. *The marking method*, in which a known number of marked individuals within a population are turned loose, on the assumption that these will return later to their usual habitats where they can be recaptured. At this point, a count by means of the following simple equation, which notes the proportion of marked to unmarked organisms, gives sufficient information to calculate total population size:

$$\frac{Pm^1}{Pu^1} = \frac{Pm^2}{Pu^2} \tag{5.1}$$

where Pm^1 refers to the total number of marked animals released, Pu^1 to the totals of unmarked animals (i.e., the total population), and Pm^2 to the marked, and Pu^2 to the unmarked animals caught in the second census. These ideas were first proposed by Lincoln[66] in connection with the ringing of birds and also, theoretically, by Ford and Ford,[37] both in 1930. More recently, organisms have also been marked with cellulose paint or with radioactive materials. Dowdeswell, Fisher, and Ford[23] have found painting markers to be so effective on *Lepidoptera* that special processes have been adopted to calculate the population size of certain butterflies from the data so obtained (e.g., in Fisher and Ford,[32] Ford,[33] and Leslie, Chitty, and Chitty[65]). Paint has also been successfully employed on tsetse flies[52] and several birds,[8] but has not proved satisfactory among small mammal groups in which individuals tend to mix less freely with the general population.[14] Radioactive tracer techniques among insect and other assemblages have been developed especially by Kettlewell,[54] using sulphur-35 for species with a relatively long life history (e.g., *Lepidoptera*), and phosphorus-32 for individuals which are short-lived. Sulphur-35 has a half-life of 87·1 days[17] and phosphorus-32 one of 14·3 days.[55]

It is also possible to use several indirect means of estimating total populations, most of which measure some facet of their past activity. Thus, one may count the numbers of shed deer antlers, the pelts of fur-bearing animals, the frequency of bird calls, the density of rabbit and bird tracks, and so on. But these are notoriously unreliable techniques and are rarely employed today.

5.3 Birth, death, and growth rates

Once populations have been enumerated, they may be further analysed by a consideration of their rates of birth, growth, and death. These are usually far from constant, and variations in each can reflect not only differences in the reproductive and growth capabilities of particular organisms, but also their changing reactions to each other and to their environment.

Birth rates (or *rates of natality*) may be represented either as a *potential*, which is rarely realized in nature and relates to a theoretical maximum production of new organisms per unit time under ideal conditions, or as an *actual* value which is produced under specific restrictions of environment and social behaviour. It is possible to approximate potential birth rates in the laboratory or even, for very short periods, under field conditions; but one must be very careful when calculating them from large populations in natural ecosystems, for maximum rates of production, even for the same species, may vary widely as a response to differences in their total size and density (see also section 5.9).

Similar concepts may also be applied to the consideration of *rates of death*, or *mortality rates*, which are usually measured in terms of the numbers of deaths per unit time, *or* per unit time per individual. If there were no limiting factors, a theoretical minimum constant death rate, known as the *potential death rate*, would be reached for each population; this may also be referred to as the *physiological longevity capacity* of populations.[10] However, in nature, most organisms die sooner than they might as a result of environmental or social stress, so that the *actual death rate* varies considerably from the potential. Of course, extremely wide variations exist in the actual death rates of different population groups: to take only two examples, grasshoppers which live under the exposed conditions of the American Great Plains are characterized by extremely high mortality, while, in contrast, the death rates of honey bees living in the protected environment of hives with a well-developed social organization are low. Actual death rates which approach the potential have been noted by Pearl and Parker for *Drosophila melanogaster*,[82] by Noyes[77] and by Pearl and Doering[81] for the rotifer *Proales decipiens*, and by Wiesner and Sheard[102] for laboratory populations of albino rats.

Actual death rates may also be correlated fairly easily with the different age groups of specific populations, in which case they become of special interest to biogeographers, for then they help to give some indication of life expectancy and of potential causes of death among organisms at various stages in their life cycle. Such information may be presented most clearly through the use of *life tables* and *survival curves*, the former having been first used by Pearl and Parker in 1921,[82] and the latter at a slightly later date. One example of a life table is that presented for Dall mountain sheep, in Table 5.1. As these sheep graze on pastures in a reserve within

Table 5.1 *A life table for Dall mountain sheep (adapted from E. P. Odum*[78]*)*

Age interval (yrs)	Age as % deviation from mean lifespan	Number dying in age interval out of 1000 born	Number surviving at beginning of age interval 1000 born	Mortality rate/1000 alive at beginning of age interval	Expectation of life, or mean life-time remaining
0–0·5	−100	54	1000	54·0	7·06
0·5–1	−93·0	145	946	153·0	—
1–2	−85·9	12	801	15·0	7·7
2–3	−71·8	13	789	16·5	6·8
3–4	−57·7	12	776	15·5	5·9
4–5	−43·5	30	764	39·3	5·0
5–6	−29·5	46	734	62·6	4·2
6–7	−15·4	48	688	69·9	3·4
7–8	−1·1	69	640	108·0	2·6
8–9	+13·0	132	571	231·0	1·9
9–10	+27·0	187	439	426·0	1·3
10–11	+41·0	156	252	619·0	0·9
11–12	+55·0	90	96	937·0	0·6
12–13	+69·0	3	6	500·0	1·2
13–14	+84·0	3	3	1000	0·7

Data from E. S. Deevey Jr.[21] and A. Murie.[75]

Mount McKinley National Park, Alaska, the actual death rates refer to conditions in a natural environment, in which the main forms of death are by predation, especially from wolves. The table shows that their mean age of death is greater than seven years, and, in addition, that once the sheep have lived to beyond their first year, their chances of further survival are reasonably good.

Similar information may be gained from an examination of survival curves, several types of which may be seen in Fig. 5.2. In this, it is clear that while a linear relationship may exist between the variables of mortality and age those organisms which tend to die early in life (e.g., oysters) or relatively late (e.g., man) will naturally have very different curves. In extreme cases, in which the

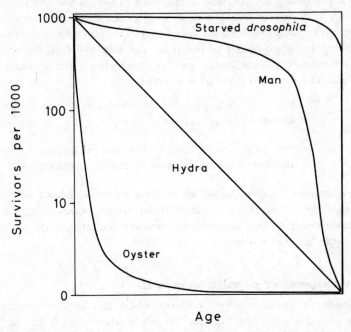

Fig. 5.2 Survival curves for selected organisms (from an original diagram in E. P. Odum,[78] using data from E. S. Deevey Jr[22])

actual death rates approach the potential, as with *Drosophila* populations under laboratory conditions,[80] all individuals will die at about the same age, thus producing a somewhat angular curve. While it is very easy to determine from these the age at which death is most likely to occur for particular organisms, it should be remembered that, even for populations of the same species, such curves may not always be constant, and indeed, they could change appreciably if environmental and social conditions are modified; thus, they often become unquestionably more concave in very densely crowded habitats.

Although many authorities believe that death rates are much more likely to be affected by changing environmental conditions than birth rates, and may therefore be more variable, the relative effect of the two on the overall *growth rate* of populations is hard to determine. Certainly, both need to be considered; high birth rates may be cancelled out by extremely severe death rates, possibly even leading to a population decline, and a low birth rate does not necessarily mean a declining population, as long as the death rate remains correspondingly low. Relationships between the two may be clarified by dividing the number of births by the number of deaths on a percentage basis, thus giving positive or negative rates of change, which may then be expressed as follows:

$$\frac{\Delta N}{\Delta t} = \quad \text{the } growth\ rate \text{, i.e., the rate of change in the number of organisms per unit time; or} \qquad (5.2a).$$

$$\frac{\Delta N}{N\Delta t} = \quad \text{the } specific\ growth\ rate \text{, or the rate of change in the number of organisms per unit time per organism. } (5.2b).$$

It is normally preferable for all of these to be calculated over a period of one year or more, although for certain populations (e.g., among insect species) a consideration of seasonal growth rates may at times be more realistic.

5.4 Patterns of population growth

Studies of population growth patterns, which date back to the early theoretical work of Malthus[71] in 1798, came of age when Verhulst,[98] in 1838, created the first mathematical representation of the nature of population increase as it occurs under the limiting environmental conditions of natural ecosystems. When plotted graphically, this is characterized by the familiar S-shaped growth curve of Fig. 5.3. At the time, Verhulst's ideas were not widely read and, partly because of this, failed to be fully appreciated until Pearl and Reed[83] rediscovered identical equations for growth patterns while working independently during the 1920s.

It was Verhulst who first suggested that all populations have a characteristic *growth form* which is determined by birth and death rates within a population and the dispersal of organisms from one unit area into another. If, for the moment, we discount the effect of dispersal, it is clear that the determination of an equation of growth

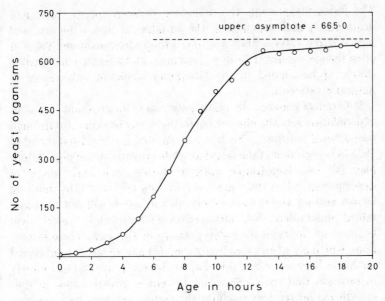

Fig. 5.3 The sigmoid growth curve, as represented by the growth of yeast in a culture. Environmental resistance will flatten the growth curve as it approaches the upper asymptote level (from an original diagram in R. Pearl, *The biology of population growth*, Alfred A. Knopf, Inc., New York, 1930)

which is true for all ecosystems is dependent on two assumptions: *first*, that mean rates of birth and death among specific organisms remain the same for similar age groups, when no limiting factors exist; and *second*, that populations with identical patterns of birth and death have a fixed age pattern, again with no limiting factors in the environment. So far, all the evidence supports these suppositions. Under these conditions, and over short periods of time, populations will tend to increase by means of the parameter r, in the following equations:

$$\frac{\Delta N}{\Delta t} = rN; \qquad \therefore \quad r = \frac{\dfrac{\Delta N}{\Delta t}}{N} \qquad (5.3)$$

For longer periods, the cumulative effect will be:

$$N_t = N_0 e^{rt}, \qquad (5.4)$$

where N_0 represents the population at the beginning of the period, N_t the population at time t, and e is the base of natural logarithms.

The index r represents the difference between specific birth and death rates, and depends on the variables of age, structure, and reproductive rates within any one group; its maximum value is often the *biotic potential* for any given area, and a figure which, while closely approximated in the laboratory, is found only rarely in natural ecosystems.

Differences between the biotic potential of an area and the totals of population actually observed in the field may be explained through *environmental resistance*. This is a concept first used by Chapman[13] in 1928 in recognition of the fact that while growth rates of populations may fall into logarithmic patterns during their early stages of development, when there are few crowding effects or other limiting factors and an ample food supply, this situation will not be maintained indefinitely, for, ultimately, environmental controls will begin to set limits on the *carrying capacity* of any area. These restrictions, which are of an exceedingly complex nature, are to be analysed further in later sections of this chapter: here, it is appropriate merely to mention that their effects are to reduce growth rates, so that growth curves tend to establish themselves between two common forms. One, which is J-shaped, showing a rapid increase in numbers at first along an exponential curve before stopping abruptly as environmental resistance becomes suddenly effective (see Fig. 5.4), may be expressed as follows:

$$\frac{\Delta N}{\Delta t} = rN, \text{ with a definite limit on } N. \tag{5.5}$$

However, this situation is probably less common in natural ecosystems than the S-shaped or sigmoid growth curve originally defined

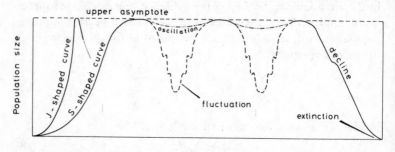

Fig. 5.4 Idealized population growth curves, from initial development to extinction (from an original diagram in W. C. Allee *et al*[2], following Gause[40])

by Verhulst[98] and Pearl and Reed.[83] In this, population growth takes place slowly at first, then increases to a maximum logarithmic phase before being reduced as environmental resistance builds up with increasing age; eventually, an equilibrium with the environment may be reached at the upper asymptote of the sigmoid curve, which is equivalent to the carrying capacity of the area. Such a curve is equivalent to:

$$\frac{\Delta N}{\Delta t} = rN\frac{(K - N)}{K} \tag{5.6}$$

in which K is a constant referring to the upper limit on population growth.

Once the size of a population has reached its upper limit, several possibilities for further development are open to it. It may maintain itself in equilibrium at the upper asymptote, a situation which is perhaps best manifested in populations of simple organisms yielding under controlled laboratory conditions (see Allee *et al.*,[2] who worked with yeast organisms). But it is more likely to show slight variations in numbers either in the form of minor oscillations or more major fluctuations. Eventually, the population may start to decline, and could even become extinct.

Several practical considerations make it difficult to study these postulated developments within the confines of natural ecosystems. For one thing, most populations are past the stage of rapid growth when they are censused, and, indeed, the initial phases of growth may often be clearly seen only on land which has been freshly invaded. Moreover, at the equilibrium stage, most natural populations are so sensitive in reaction to each other that it is hard to isolate them for study, and these reactions in themselves are subject to timelag effects of a complex nature which add to the difficulties of analysis. The theoretical implications of such timelags have been discussed by Wangersky and Cunningham.[101] Similarly, the initiation of a reduction in the size of a natural population is very difficult to discern, and usually it is only when numbers have become drastically reduced that their decline is examined in detail by research workers.

Nevertheless, with care, all of the theoretical features of population growth and decline may be observed in nature, for the continually-changing environment is always inducing modifications in the numbers and ranges of organisms which live within it. In freshly colonized areas, the very rapid numerical increases which are characteristic of the early stages of population growth may often be seen, as in the case of pheasants (*Phasianus colchicus*) on Protection Island in Washington State, USA, whose complement was raised by natural means from 8 to 1898 over a period of $5\frac{1}{2}$ years.[25] At times, particularly in bird populations, such increases may be accompanied by a remarkable extension of range. One of the best-known examples of this is that of the fulmar petrel (*Fulmarus glacialis*) which in Britain bred only on the island of St Kilda until shortly before the end of the nineteenth century. Since then, it has established itself on nesting sites in many parts of the western coasts of Scotland, Ireland, England, and Wales, and occasionally on eastern coasts, too.[36]

Once a population has reached the asymptote level, it is only rarely in natural ecosystems that it maintains perfect equilibrium, for usually one of several types of oscillation or fluctuation develop, which may follow *seasonal* or *annual* patterns. Seasonal changes, associated particularly with climate, may be found in most parts of the world. Many organisms in middle and high latitudes show immense fluctuations in numbers between winter and summer, with the greatest variability usually occurring in groups which have limited breeding seasons and short life cycles. Similar seasonal patterns of abundance and decline may also be seen in the tropics, as is the case for mosquitoes in the lowlands of eastern Colombia;[4] there, the frequency of occurrence is intricately related to the totals of rainfall and standing water and the availability of food (see Fig. 5.5). More complex seasonal irruptions, for which no known explanatory mechanism has been determined, may also be found in some marine environments; for example, the quantities of phytoplankton can suddenly grow to totals five or six times greater than normal and then decline again equally rapidly.

However, in general, seasonal changes are easier to explain than the annual fluctuations or oscillations which develop in many populations. These are difficult to categorize since they may be relatively minor oscillations, close to the asymptote, or much more major fluctuations; they may also occur in cycles or at irregular intervals. Moreover, while some may be closely associated with cyclical variations in food supply (e.g., from plant species), others appear to have no such connection. Perhaps, one may clarify the situation by considering first those extreme cases in which irregular fluctuations are known to be associated with particularly severe environmental factors. Thus, it has been observed that the total

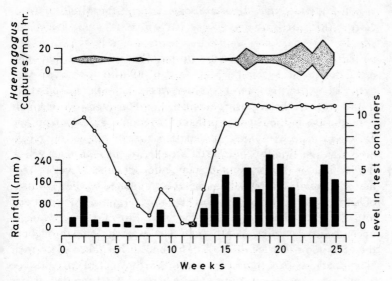

Fig. 5.5 Relationships between mosquito frequency, rainfall totals and standing water in eastern Colombia (from an original diagram in M. Bates[4])

number of herons in Britain decreases markedly after each hard winter, this having been particularly noticeable in recent years—in 1940–42, 1945, 1947, and 1963, although in each of these instances the population soon returned to its normal size once the winter had ended.[61] Or again, in Tasmania, fluctuations in the numbers of range sheep correlate well with variations in annual rainfall; in particular, drought years with poor grass development appear to be responsible for a rapid reduction in sheep numbers.[20] There are many other similar examples.

These major fluctuations are perhaps less common in mature populations close to the equilibrium level than the apparently controlled cyclical oscillations which are to be found within many ecosystems. The amplitude of these may be large or small, although all will tend to fall within well-defined limits. Thus, while the maximum numbers in bird populations rarely exceed twice the minimum, some insect populations living within coniferous plantations in Germany customarily oscillate between numbers whose maxima are 40 000 times in excess of the minima.[89] Even greater periodic oscillations may cause irruptions of organisms well away from their homelands, as in the case of the migratory locust (*Locusta migratoria*), which may devastate some areas of the Middle East or North Africa at least once every 40 years.[12] It is not known yet whether there is any connection between oscillation size and periodicity, even though studies to date have revealed that most oscillations occur within the framework of a 9 to 10, or a 3 to 4 year cycle. The latter has been best observed among animal populations of the northern hemisphere, of which the most renowned example on both sides of the Atlantic is that of the lemming (*Lemmus* spp. and *Dicrostonyx* spp.),[28] whose populations reach numerical peaks every 4 years. Usually, these peaks are followed by a normal cyclical decline but, if excessively high population densities develop, they may give way to a rapid fall in numbers stimulated by the migration of lemmings away from their normal habitats towards the coast and into the sea, where large numbers may drown. This seemingly uncontrolled lemming 'death rush' has repercussions throughout the ecosystems in which they live, for once it has taken place their chief predators (e.g., the snowy owl in North America) may also be forced to emigrate in order to search for new food supplies. Thus, Gross[44] and Shelford[91] have concluded that appearances of snowy owls in the continental USA are timed to correspond well with the major declines in lemming populations further north. Mice, voles, foxes, and many other northern rodents are also subject to 4 year cycles, while, in contrast, 10 year cycles have been noted for the lynx[78] and its main predator the snowshoe hare, although the full relationships between these two are by no means clear.

Generally, the regularity of these cycles is best seen among mammal and bird populations, and while similar phenomena are known to exist among populations of smaller organisms (*e.g.*, insects) their clarity in these instances is much more blurred. While an understanding of oscillations and fluctuations at the equilibrium level is obviously essential to a full interpretation of population dynamics, a satisfactory explanation of their controlling mechanisms has not yet been forthcoming. A review of the current major theories of control is presented in section 5.9.

The rather special cases of population decline and extinction, for which few records exist, will be examined in section 6.7. In natural ecosystems, they are induced mainly through the inability of certain species to adapt to changing environmental circumstances sufficiently quickly to ensure their continued survival.

5.5 The importance of age

Since it has already been established that birth and death rates may vary appreciably with age, it is important in the analysis of any population to consider *age structure*, or the percentage or number of organisms in any one age group. The most significant of these groups to biogeography may be broadly categorized as prereproductive, reproductive and postreproductive, the precise length of which will vary with different species. Thus, some species (e.g., modern man) have several years in which to produce their offspring, whereas, in contrast, the adult life of Ephemeridae, or members of the mayfly family, lasts for only a few days. Studies of age structure are particularly relevant in the determination of population vigour and stability, for while most groups tend to become stable if given sufficient time, those with a large proportion of reproductive adults will obviously have a greater predisposition towards stability than others. Lotka[67] has suggested that, in theory, once stability has been achieved, it is usually maintained, but, while this is generally true in most ecosystems, it is not always the case, and, indeed, if permanent changes in environment or social behaviour should develop, then new stable situations will undoubtedly be created out of the old.

Age structure may most clearly be seen through the compilation of age-sex pyramids, as in Fig. 5.6, in which the total number of individuals for a given population are subdivided according to age and sex. In these instances, the most stable age structure is to be found in Fig. 5.6a, which represents a situation in which birth and death rates are approximately equal, whereas Figs. 5.6b and 5.6c indicate unbalanced human populations, the one in Connemara, Ireland, and the other at Harlow, one of Britain's New Towns. That in Connemara is an extreme example of instability caused through the emigration of large numbers of adults in young age groups,[95] while some degree of imbalance at Harlow has resulted from the sudden immigration and settlement of large numbers of young adults with children. Although the use of age-sex pyramids may not be practicable for populations of all organisms, their immense significance where they can be used lies in their capacity to show at a glance the inherent stability or instability of a group; they may therefore be counted as important tools to be applied in the general elucidation of the vast number of variables which may influence the behaviour of populations in natural ecosystems.

211

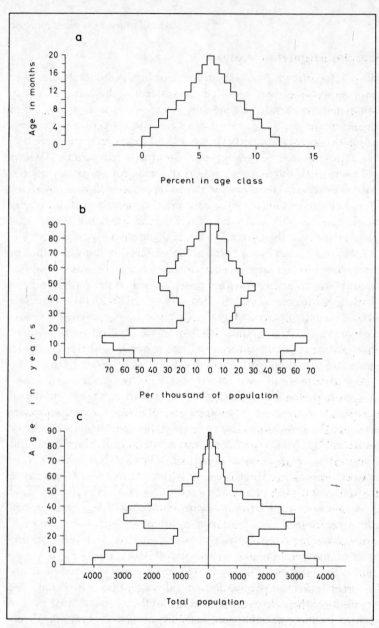

Fig. 5.6 Age-sex pyramids for (a) laboratory populations of the vole; and for human populations in (b) Connemara, Eire, and (c) Harlow New Town, England (from original diagrams in E. P. Odum,[78] using data from P. H. Leslie and R. M. Ransom;[64] and D. G. Symes[95])

5.6 Genetic factors

At this stage, it may be useful to deviate slightly from the major theme of the present chapter in order to consider some interrelationships between the genetic and evolutionary factors which affect populations, and others which have been noted thus far. Since Darwin's theory of natural selection was put forward,[19] there has been surprisingly little attention paid to the role which evolution can play in the numerical development of populations, and vice versa. This may result from two important considerations, the first being that few biogeographers or ecologists are also geneticists, and the second associated with the fact that there is only a limited amount of field material which is capable of being used jointly within these disciplines. Ford[36] has pointed out that, apart from domesticated animals and cultivated plants, a disproportionately large amount of research into these problems has been based on two species, *Drosophila melanogaster* and *Neurospora* sp.; yet it is scarcely possible to speak of the ecological relationships of the latter, and little is known of the larval ecology of the former.[34]

Genetic factors are of especial interest to the biogeographer in that marked upward fluctuations in the size of populations appear, at times, to trigger off a rapid evolution of new forms. Thus, for a fast-increasing population within a favourable environment, greater genetic variability than normal will occur, since some environmentally controlled aspects of selection will have been relaxed, and there may be more possibilities for mutation. Once environmental circumstances again become more restrictive, as when colder winters or a more arid climate prevails, those new genotypes* which are not suited will quickly be eliminated, so that the population will once more become genetically more uniform. Under these conditions, notable increases or decreases in genetic diversity usually coincide with the actual period of population change.

Fluctuations in population numbers are therefore of immense importance since they permit evolution to take place much more rapidly than would normally be the case. Newly created forms thus have the opportunity of interacting not only with each other, but also with the original genotypes in a favourable period, so that any

---

* In this discussion, the term *genotype* refers to the genetic differences, and the term *phenotype* to the differences in external appearance, which are to be seen among individuals of population groups (see also pp. 279–280).

especially advantageous variety will be well established when more limiting environmental conditions return. It follows also that, after rapid population growth, genotypes of populations may be somewhat different to those present at the beginning of the growth period. As Darwin suggested, the net effect of such a genotypic differentiation will be to increase the chances of survival of those types which best interact with each other for the good of the individual or the population as a whole.

In order to illustrate these points, one may refer to the example of the marsh fritillary butterfly (*Melitaea aurinia*), a colony of which was studied by Ford and Ford[37] over a period of 19 years; in addition, previous collectors had left good specimens of earlier local phenotypes which dated back a further 36 years. This colony was confined to lowland fields in Cumberland, England, and was effectively separated from others in the area by thick woods, heather, and agricultural land which the butterfly never penetrated. Their larvae fed most commonly on *Scabiosa pratensis*, but also used honeysuckle (*Lonicera periclymenum*) in adjacent hedges; they could also be reared on introduced snowberry (*Symphoricarpus rivularis*). Their young tended to hibernate in webs on the *Scabiosa*, while the imagines (from one generation per year) were on the wing only in the latter half of May, and the first half of June. Ford notes that these were slow-flying and rarely wandered more than 100 yards from their normal habitats.

Studies of this colony commenced in 1917, when only two to three individuals could be collected per day, but a large increase in population took place between 1920 and 1924, by which time vast numbers were present; from then until 1935 (when observations ceased), the butterflies remained at high-density levels. Before population growth began in 1920, a constant form had been present which allowed for very few variants, but during the increase in numbers of the succeeding four years, an extraordinary range of new varieties began to emerge, characterized by differences in colour patterns, form, and shape. Indeed, many were so deformed that they quickly became eliminated once stability had been resumed after 1924 (see Fig. 5.7). However, by this time, the most common form was markedly different from that which had been collected prior to 1920, thus indicating that the butterfly had made full use of the opportunity for evolution which the expanding population had presented.

214

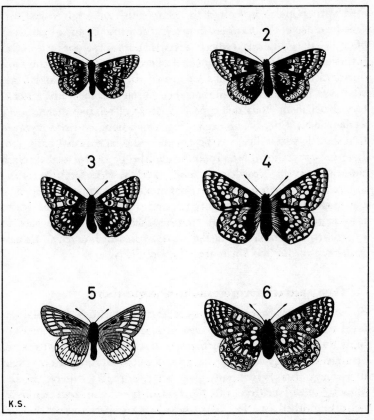

K.S.

Fig. 5.7 Phenotypic differentiation in the marsh fritillary butterfly, Cumberland. 1, male upperside, 1899; 2, female underside, 1899; 3, male upperside, 1935; 4, female upperside, 1933; 5, male underside, 1926; 6, female underside, 1926. 1–2 represent common forms before population growth occurred; 3–4 are common forms after the main growth period; and 5–6 are extreme phenotypes which were only recorded during the phase of maximum growth. See text for further explanation. (Based on photographs in *Butterflies*, E. B. Ford, New Naturalist Series, Collins, London, 1962)

It is not clear whether an example such as this is representative of similar situations throughout the world, for so few related studies have been completed. Indeed, Ford himself has noted that fluctuations in numbers may not always result in genotypic evolution, since other powerful methods of selection may be operating concurrently in an opposite direction to preserve some well-established phenotype. But there are some indications that evolution can be closely associated with population growth in many other natural ecosystems. One instance of this has been noted in the Mount San Jacinto area of southern California, where a particular karyotype* (the CH inversion) of *Drosophilia pseudo-obscura* becomes much more common during favourable conditions for growth between April and June, after which it decreases in importance. Birch[7] has shown by laboratory experiments that this type is capable of relative increase *only* under the conditions of an expanding population, so providing proof that genetic variability may be augmented on a seasonal basis, too. Inverse support for these proposals has also been provided through the discovery by Dowdeswell, Ford, and McWhirter[24] that some organisms use the technique of rapid numerical increase as a specific means of providing new genotypes, and so adjusting themselves to fast-changing environmental circumstances; such is the case in experimental populations of the meadow brown butterfly (*Maniola jurtina*) in the Tresco Farm area, Isles of Scilly.

5.7 The spatial arrangement of organisms

So far within this chapter, some general characteristics of populations have been discussed without considering the spatial arrangement of individual organisms within them. Variations in these patterns of arrangement may be of great importance to studies of ecosystem dynamics in that they usually give rise to, or reflect, different modes of social behaviour. In nature, four major types of arrangement may be distinguished (see Fig. 5.8), these being *random, uniform, bunched random* and *bunched uniform*.

Random arrangement is relatively rare, since it should only occur when all environmental factors affecting population growth are equally important. This is usually not the case for, as noted in chapter 4, most populations are controlled by one (or a few) major limiting factors, and this situation does not pave the way for randomness. Indeed, it is more common, though still quite unusual, to find uniformly-spaced individuals, particularly where physical factors have stimulated intense competition, as in certain sage brush (*Artemesia tridentata*) plant populations of desert and semidesert areas in the southwestern USA, where amounts of available water are very

---

* The term *karyotype* refers to the differences in chromosomal structure, according to number, size and shape, which are to be seen among individuals of population groups.

restricted. Uniform distribution may also result from active antagonism between organisms.

The most frequently-observed patterns of spatial arrangement are those in which a bunching of individuals is present. Cole[16] has suggested that some form of bunching is to be expected in many plant

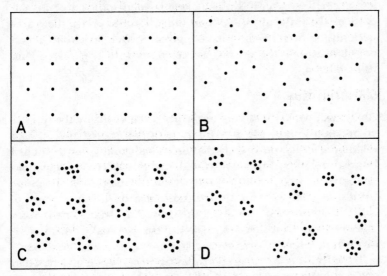

Fig. 5.8 Patterns of spatial arrangement among organisms. A, uniform; B, random; C, bunched uniform; D, bunched random (modified from an original diagram in E. P. Odum[78])

and animal populations, since both seed and asexually reproduced plants will tend to yield their greatest number of offspring around the parent, and most animals show a predisposition not only towards rearing their young in bunched groups, but also towards establishing aggregations within their own populations, either through sexual attraction or by social instinct. The latter is certainly the case for many mammals, birds and insects, and may well hold true for other groups of organisms, too. Bunching may also result from the differential availability of particular environmental components (e.g., water, nutrients, etc.) which are essential to life. Furthermore, it can help certain organisms to modify their environment for their own good, so reducing the potential dangers of predation, climate, and disease. Thus, bees in a hive may generate sufficient heat among themselves to exist in cold weather which would undoubtedly kill them off in isolation;[78] indeed, many of the highly developed social organizations of insects, such as *Isoptera* termites, ants, and bees, may only be maintained through bunching on a massive scale. These features have led to the idea that, for some species at least, the chances of survival may be considerably reduced where abnormally low as well as excessively high population densities are present, particularly in the absence of bunching. In other words, there may not only be overcrowding in any given environment, but under-crowding as well: this notion has been termed *Allee's principle*,[2] after its founder.

5.8 Migration

Analyses of populations may be further complicated by the transfer of organisms into and out from particular ecosystems through migration, for it is only rarely that individual groups may be studied in areal isolation. Migration can alter the balance of populations not only on large continents, but also in the most isolated oceanic islands, a point which is to be taken much further in chapter 6. Little is known even now about the general or detailed patterns of movement which it involves, though these are usually broadly sub-divided into those of *emigration, inmigration*, and *seasonal migration*.

The effectiveness of migration as an agent of change in population size and structure is dependent on several factors, among the most important of which is the relative immobility of most organisms, for, whether movement is active or passive (e.g., in the case of winged seeds), individuals show a tendency to remain within short distances of their areas of origin wherever possible. Indeed Wolfenbarger[103] has concluded that it is usual for migration patterns to show a logar-ithmic relationship between the number of organisms present and the distance travelled from their source area, at least for non-human populations. Even so, many plants (e.g., spores and seeds) can be transported huge distances by the wind or through other means; and more motile organisms often journey hundreds or thousands of miles when they are required to do so. These general considerations are made much more complex by the fact that most migration patterns, especially those which are seasonal in nature, are highly selective, so that their effects are not felt equally everywhere; in addition, the movement of even highly motile organisms can be restrained to some extent by environmental barriers, such as major oceans or high mountain ranges.

If a population is large and stable, then it is probable that migration has little overall effect on it, for inmigration may approximate outmigration, and even where this is not the case, the excessive development of one or the other often results in equivalent, balancing changes in birth or death rates, so maintaining an equilibrium situation. However, if a population is inherently instable, or even has a tendency towards instability, then migration may result in rapid changes of population structure. Thus, in Connemara (see Fig. 5.6), the continued emigration of large numbers of young adults to Britain and the USA has helped to accentuate a rapidly developing trend towards population decline, since those people who remain are in the older age groups and so may not be expected to maintain the vigour of the population in the area. The effects of seasonal migration patterns are more difficult to determine, and probably vary widely within different ecosystems and among different populations.

5.9 Density-dependent and density-independent controls

All aspects of demography in natural ecosystems are regulated to some extent by biogeographical factors which restrict the total growth and development of populations to an upper asymptote level. These factors may be divided into those which are *density-dependent* and others which are *density-independent*, two terms which relate to the fact that while some environmental controls of population appear to be constant regardless of population density (the latter case), others are modified as population density changes. In nature, distinctions between the two are not always easy to determine, since one control may operate in both ways at the same time, but for the present it will be convenient to study the effects of each in isolation from the other.

Density-independent controls usually derive from conditions of the physico-chemical milieu within which the population has become established. Thus, for terrestrial ecosystems, they may be associated with characteristics of temperature, wind and storms, precipitation, atmospheric humidity, energy income, and the availability of food, while in aquatic environments they are often related more specifically to the physical and chemical qualities of the water, its movement, and the light penetration within it. One good example of a density-independent reaction is that recorded in coastal waters of the Texas Gulf during January 1940, following a severe drop in local air temperature to $-9\,^{\circ}\mathrm{C}$ as cold northerly winds penetrated southwards from the Great Plains and Canada.[45] The result was a death rate greatly in excess of normal for all fish species over a wide area during the following three months. Flounders were particularly vulnerable, with a decline in numbers of over 90 per cent.

Similar effects may be induced by the introduction of pollutants into aquatic environments, as when pulp-mill waste caused the physiological dwarfing of individuals in oyster populations of the York Rifer, Virginia;[39] these were then unable to store normal levels of glycogen in their bodies, so that vigour was reduced, and this, in turn, led to a major decrease in population size. It has already been noted (see pp. 208 to 209) that many bird populations are severely affected by adverse weather conditions in a density-independent way, and entomologists have reported identical effects among some insect species.

In contrast, density-dependent controls are associated predominantly with biotic interactions between organisms. Rates of birth, death, and migration are known to change as population density is modified, as also does the speed of post-embryonic development among insects. Moreover, the individual growth of organisms, their rate of oxygen consumption and their resistance to noxious agents, the aggregating activity of bacteria and protozoa, and even sex determination and the morphological alteration of certain anatomical forms may all be subject to change through density differences. In addition, it has been inferred previously that growth forms of populations may be best developed when some degree of crowding is present (see Allee *et al.*[2]), while density influences may also act indirectly on populations through, for example, reducing the relative amount of food per organism as the size of the population increases, so limiting its ultimate development.

Both density-dependent and density-independent factors also need to be considered as possible causal mechanisms for the oscillations and fluctuations within mature populations at the equilibrium stage, which were described previously. Indeed, one of the major current controversies in faunal biogeography is centred around the question as to which might be the most effective means of accounting for these.

In this connection, three predominant theories of population dynamics need to be reviewed. The first, developed by Nicholson in 1933,[76] assumes that most populations are normally balanced and stable, so that any fluctuations are relatively restricted and will be created by density-dependent media. In challenging this view of density-dependent control, the second theory, introduced by Andrewartha and Birch in 1954,[3] suggests that since some populations are known to have become extinct, many will be in a state of potential imbalance. From this, they infer that most fluctuations will be irregular and, perhaps, can be correlated particularly with minor climatic vacillations. The third theory, propounded by Wynne-Edwards,[104] puts forward the notion that, while density-dependent factors are important, the availability of food is the ultimate controlling element in population growth. It will be useful to examine all these theories in more detail.

Perhaps because the idea of density-dependence was first put forward in purely mathematical terms (Lotka[67] and Volterra[99]), it did not become widely known until Nicholson[76] argued that the steady oscillations of numbers in any population over a period of years could only be explained by some factor which encouraged growth after a decline or vice versa (or else the population would either increase indefinitely or become extinct), and that such a factor could only be density-dependent. It is known that such oscillations occur commonly among mammal populations and, since the total numbers of these are not permanently changing, it follows that birth and death rates must be kept in balance in such a way. Similar effects have also been recorded within bird and reptile populations, in which stability is maintained by the development of an increased death rate, should the birth rate be augmented; this is particularly true for blue tit (*Parus caerulus*),[92] starling,[62, 88] and wall lizard (*Lacerta sicula*)[58] groups. It is also highly probable that identical controls will be found among certain insect species. However, while agreeing with the general idea of density-dependence, Lack[59] has recently argued that some modification to this theory is necessary in order to take into account the considerable differences in the availability of food which may be present for populations in natural ecosystems, and, if we accept this, then it follows that at least two types of density-dependent regulation may exist, the one varying quite quickly where populations are limited directly by food shortage, and the other showing delayed and indirect variations within a predator-prey relationship.

These views were originally challenged by Andrewartha and Birch[3] on the grounds that, while the idea of density-dependent

221

control formed an attractive theory, no instance of this had at that time been observed in natural ecosystems. Instead, they suggested that limitations of numbers in animal populations may derive from three important density-independent influences: first, a shortage of material resources, such as food or nesting material; second the relative degree of accessibility of such resources in relation to an individual's capacity for finding them; and third, the availability of time which, if restricted, might affect a population's ability to develop. In all instances, they also claimed that local extinctions of populations could occur anywhere, should environmental limitations be excessive, to be followed by a possible later reinvasion by the same species from outside. It is clear that all of these limiting factors could be associated with restrictive and severe climatic conditions, which the authors held to be the ultimate influence on population size. Since these ideas were propounded, they have received criticism from several sources, particularly from Lack,[59] who suggests that they are weakened by their failure to consider the role of competition for food among animals, which he thinks is a much greater cause of food shortage than any created by adverse climatic conditions. Moreover, both Lack[61] and Solomon[94] (among others) have stated that Andrewartha and Birch failed to provide convincing proof that density-dependence does not operate, even under the conditions which they describe; in other words, all fluctuations may still be centred around a theoretical population level which is determined by density-dependent factors—and many examples are now known which suggest that this is indeed the case.

The third possibility, cited by Wynne-Edwards,[104] is based on an observation by Lack that a dispersion of individuals to extremely favourable sites is common among many bird populations, in order to obtain a maximum amount of food for their young. While Lack's original proposition was that this took place only among those pairs of birds which bred first, Wynne-Edwards concluded that it might form a significant general control in population growth, in that there are a limited number of such sites available for them. He also inferred that group-balancing (or *homeostatic*) mechanisms would emerge among populations through natural selection, so that while populations always maintained themselves close to an optimum level which the food supply could theoretically support, they would never rise above this. This notion has also been criticized by Lack on the grounds that it is not demonstrable in nature and, furthermore,

222

the stability produced from homeostacy could equally well result from the density-dependent factors already described. Indeed, Lack's most recent suggestion[61] is that stability such as this could be best obtained from a combination of, first, the evolution through natural selection of reproductive rates which are as large as the environment and the organism's capacities will allow, and second, the establishment of density-dependent mechanisms within mature populations, which ensure that death rates equal birth rates so that equilibrium is maintained.

Some of the differences of opinion voiced within these arguments may be partially explained by referring to the varying terms of reference of their proponents. For instance, Andrewartha and Birch have been working mainly with insect populations, which have only a seasonal growth pattern, while Lack, Nicholson, and Wynne-Edwards are all associated with investigations of larger animals, especially birds. One may only conclude at present that many more case studies of animal populations need to be completed before the relative merits of each of these several theories can be further clarified.

COMPETITION BETWEEN ORGANISMS

It is evident from the preceding section that one of the factors which may encourage or inhibit population growth is that of competition. Despite the doubts expressed by Andrewartha and Birch, it is probable that, except in populations of very low densities, competition between organisms for the same commodity (food, building material, etc.) is to be found in every ecosystem. This may be between plants, as in the struggle for light in tropical rain forests (see pp. 129 to 130), or for water in many desert and semidesert areas; between animals, this being a much more complex phenomenon; or even between plants and animals. Such competition can be between individuals of the same species (*intraspecific competition*), or at a more intricate level between members of different species (*interspecific competition*).

5.10 Intraspecific competition: territoriality

Among populations whose controls are known to be density-dependent, intraspecific competition is often extremely important. It may occur simply as direct competition between large numbers of organisms, following which the weakest are eliminated at the expense of the strongest; or it may develop as a much more precise

223

struggle for space, in which the tendency is for all individuals save one or two to be driven out of a *home territory* or *range*. Of these two possibilities, the former usually results from a need to conserve adequate amounts of food or other resources within the framework of an optimum population size, while the latter frequently arises from a desire for isolation, particularly during the breeding season. This is best observed among birds, but is also common among many other animals which have a complicated behavioural routine

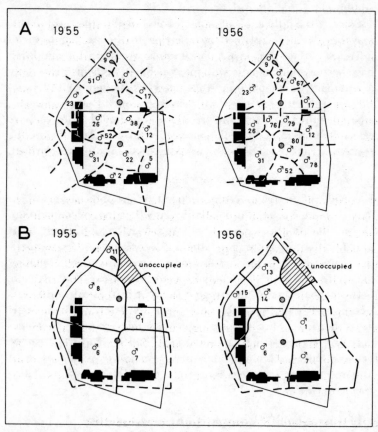

Fig. 5.9 Territories of male blackbirds (*Turdus merula*) and male song thrushes (*Turdus philomelos*) in the Oxford Botanic Garden at the start of the breeding seasons in 1955 and 1956. A, blackbirds; and B, song thrushes (from original diagrams in D. W. Snow, *A study of blackbirds*, Allen & Unwin, Ltd., London, 1958; and P. W. Davies and D. W. Snow, Territory and food of the song thrush, *British Birds*, Vol. 58, pp. 161–75, 1965)

associated with reproduction and the care of offspring. Technically, it is referred to as the practice of *territoriality*.

It was Howard in 1920[49] who first intimated that the definition of territory by resident bird populations in the fields, hedgerows, and heaths of western Europe was an essential prerequisite to pairing, egg-laying, and the successful rearing of offspring. The process of acquiring territory appears to begin well before the breeding season, when the occupier claims his right of ownership through powerful displays, or song, which serve as a warning to other individuals to keep away. Often, males who do not succeed in establishing territory will not breed, while those with their own territory will have a high probability of doing so. Once the nesting process has begun, females will join in the defence of territory, and should either male or female be killed, another will quickly move in to take its place, thus indicating some possible control of population numbers. Frequently, individuals will keep the same territory over a period of several years, assuming that they remain healthy and vigorous. Two patterns of observed territoriality may be seen in Figs. 5.9 and 5.10, the former

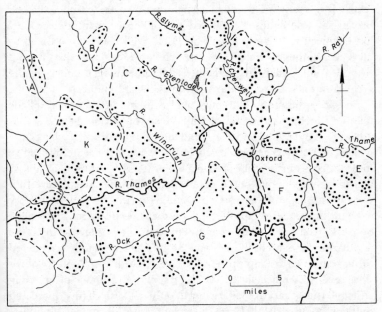

Fig. 5.10 Winter roosts and associated summer feeding areas (A–K, bounded by dotted lines) of rooks (*Corvus frugilegus*) in the Oxford region, 1931–32. Dots are breeding rookeries (from an original diagram in C. S. Elton,[30] after W. B. Alexander[1])

225

on a relatively small scale, and the latter covering a much broader stretch of terrain.

However, it is not yet certain whether the effects of this practice directly aid the regulation of population numbers, for as Roe and Simpson have pointed out,[87] the established patterns of territoriality show considerable variation among different groups of organisms. Even for bird populations, Lack[61] has shown that two critical questions still need to be answered: first, does such behaviour actually limit the number of pairs (or individuals) in an area, as opposed to merely having the effect of spacing them out, with the total numbers determined by some other factor; and, second, if it does, are such limits to population growth also related to food availability? Wynne-Edwards[104] has answered both these queries in the affirmative partly on the grounds that the more highly favoured areas within any environment will always be occupied first, to be followed later by settlement in less amenable localities. These proposals, also accepted by Gibb,[41] are based to some extent on results obtained from studies of great tit (*Parus major*) groups in Holland,[57] individuals of which form territories much earlier in the year within food-rich broad-leaf woodlands as opposed to coniferous plantations. They have, however, been criticized by Lack, who claims that the sampling techniques on which they were based were of poor quality, and who also notes that, in England,[60] fluctuations in numbers of the great tit appear to be similar in both types of habitat.

Probably the most commonly held contemporary concept of the manner in which territoriality may affect population density is that introduced by Huxley in 1934,[51] and later supported by Tinbergen,[96] which compares territories to rubber discs, in which greater resistance to pressure occurs as that pressure increases. Accordingly, at higher density levels, fluctuations in population totals will take place within very fine limits, and there may be many years in which these are very close to the upper asymptote. But this view has also been challenged by Lack, who asserts that exceptional years may occur in which the population far exceeds in numbers that which is normally considered to be the upper asymptote level. Thus, in Marley Wood, Oxfordshire, even though before 1961 the maximum number of breeding pairs of the great tit had reached only 50, this figure increased suddenly to 86 in 1961. If the theory of territorial control of population numbers holds true, then effective territorial pressure within Marley Wood could only have been exerted in this year, and even then there is no logical reason for assuming that the limits of pressure had been reached. However, some control may still have existed through the modification of territory size by individuals, according to the amount of available food which was present. In this case, it would be expected that an increase in bird numbers in Marley Wood could be equated with a corresponding augmentation of caterpillar populations which are their main food. But this also does not apply,[60, 84] for in the 'abnormal' year of 1961 there were slightly fewer caterpillars available than in preceding years.

In the light of his experience, Lack[61] has suggested that density-dependent behaviour, at least for the great tit, may not be correlated with territorial behaviour or with patterns of predation and disease. Nor does he consider the availability of food to be an important factor in the determination of numbers, at least in the breeding season. Instead, he intimates that better relationships might exist between food supply and total population *in winter*, although this is difficult to prove. However, for certain other species (e.g., the tawny owl, *Strix aluco*), there is little doubt that territorial behaviour does appear to influence and limit the total population, even though it may also permit a slow growth in it; moreover, in many cases, once the owner of a territory dies he is quickly replaced, and some form of population control can be inferred from this fact alone. But this may be very indirect, and both its nature and functional mechanisms, and its degree of effectiveness, are still far from being fully understood.

5.11 Interspecific competition: the ecological niche

In addition to intraspecific competition, that between individuals and populations of different species (*interspecific competition*) is to be found in all ecosystems. This is usually most intense when occurring between allied forms within the same genus, when it might even become critical to the survival of a species. Perhaps the best modern definition of this very complex phenomenon is that given by Miller in 1967,[72] following Clements and Shelford,[15] which describes it as being the active demand by two or more species at the same trophic level for a common resource, or other material need, which is actually or potentially limiting; it is within the framework of this statement that the remainder of this section has been written.*

* Other definitions of interspecific competition have been propounded in the past by Crombie in 1947,[18] Solomon in 1949,[93] Birch in 1953,[6] and Milne in 1961.[73]

In order to clarify some of the problems attending any study of interspecific competition, it will be useful first to reconsider the mathematical representations of population growth devised by Lotka,[68] Volterra,[99] and Gause;[40] these state that each individual addition to a given population (N) will reduce the growth capacity of that population by a constant increment, so that sigmoid growth curves develop, in which N approaches an upper asymptote K, this being the equivalent of the carrying capacity of the ecosystem. If two species with carrying capacities K_1 and K_2 compete, and assuming that any individual has an identical effect on the growth potential of both populations, the situation may be expressed as follows:

$$\frac{dN_1}{dt} = r_1 N_1 \frac{K_1 - N_1 - \alpha N_2}{K_1} \qquad (5.7)$$

$$\frac{dN_2}{dt} = r_2 N_2 \frac{K_2 - N_2 - \beta N_1}{K_2} \qquad (5.8)$$

when N_1 and N_2 represent the numbers of each species, r_1 and r_2 refer to their rates of increase, and K_1 and K_2 have been determined through growing each species separately in the environment. The inhibitory effect of N_1 on N_2 is represented by β/K_2, and the reciprocal effect of N_2 on N_1 by αK_1. Thus, the outcome of competition will depend on the inequalities $\alpha > K_1/K_2$ and $\beta > K_2/K_1$.

As shown by Volterra and Lotka, these equations suggest that, given the same resources in an ecosystem, similar species cannot coexist with each other, a supposition which Gause[40] has also inferred from laboratory experiments (Fig. 5.11). In nature, this presupposes that species will be ecologically identical in terms of the *niches* which they inhabit, an assumption which has been challenged by Hardin[47] who sees no evidence for this. In this connection, Gilbert and his associates[42] have also pointed out that there has been a tendency in the study of field ecosystems to notice only those situations in which certain species have been excluded from particular niche habitats through competition, and which therefore support Gause's theories, whereas, in contrast, if species do coexist, it is often stated that they occupy different niches. Certainly, enough doubt exists to justify caution in the broad application of mathematical formulae and experimental laboratory results such as these to the more complex patterns of competition in natural ecosystems.

Taking this argument one stage further, it is also essential to be able to understand why there should be such a diverse number of species in any one ecosystem. A study of even a small habitat, such as a pond, will provide ample proof that a very great range of species indeed may live in it, yet there is no apparent reason why this should be so. It is true that classical food chains and food webs

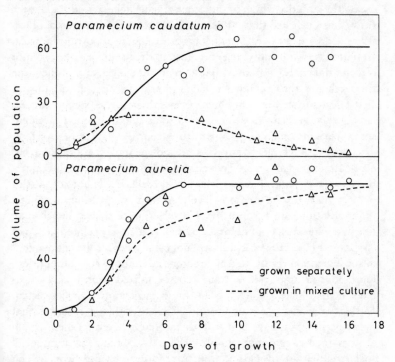

Fig. 5.11 Competition between closely related species of protozoa with identical niches. When separate, *Paramecium caudatum* and *P. aurelia* follow normal sigmoid growth curves in controlled populations with a constant food supply; when together, similar populations of *P. caudatum* decline rapidly (modified from an original diagram by A. E. Emerson in W. C. Allee *et al*, *Principles of Animal Ecology*, W. B. Saunders, and Co., Philadephia, 1942[2])

(chapter 2) leave room for the development of a variety of forms at different trophic levels, but this, in itself, cannot explain the very large numbers which are customarily present, many of which are often closely related to each other. Nor, as Mitchell has pointed out,[74] is it clear why several species should share one trophic level, since it is technically possible for only one to be adapted to take the whole of the energy flow through that level.

Part of the answer may be found in the highly organized functional nature of most ecosystems, in which all organisms are partially dependent on each other, so maintaining a certain degree of stability among themselves; in fact, those with the greatest diversity are usually the most stable. Does the very diversity of organisms, then, create the stability which might otherwise be lacking, and if so, how? Bray[11] has suggested that it does, in that part of the function of diversity is to create as many paths of energy exchange as possible between trophic levels and individual organisms, thus reducing the amount of potential energy left unavailable in any one level, and at the same time reducing the dependence of energy transfer on a few organisms.

It follows from the above that species diversity, especially among closely related forms, is associated in some measure with functional specialization, which may be used to encourage the transfer of food or energy (see Klopfer[56]); and it may aid the movement of chemical elements too (chapter 3). Indeed, all natural communities contain both specialists and non-specialists, and Miller[72] has noted that this is true at every taxonomic level. Accordingly, through genetic differentiation, most groups have a selection of species which may live in a wide variety of habitats, as well as a closely related group of specialist forms. However, it is still not clear whether the endowment of specialization results in any inherent advantages for competition, for should an organism become too specialized it may lead to inefficiencies of performance, so that time and energy may be wasted. In this connection, Macarthur and Macarthur[69] have recorded that many populations have homeostatic mechanisms within them which are designed to maintain stability through reducing the number of species which are overspecialized. Any future general theory which seeks to explain competition processes must take this into consideration.

Another facet of interspecific competition which needs to be examined carefully is the concept of the niche, for the original use of this term, as presented by Elton[27] and Grinnell,[43] appears now to be relatively unsophisticated in the light of present knowledge. While Elton defined niches in functional terms, referring to their role within a community and noting that similar niches existed in every ecosystem regardless of whether they were occupied by identical species, Grinnell was more particularly concerned with their precise location, in representing them as the ultimate distribution units for each species. Grinnell also assumed that no two species could occupy the same niche. Of these two concepts, that of Elton is unquestionably the broader and most useful, partly because Grinnell's cannot be used to define conditions of competition unless it is greatly modified. Elton also provides for the possibility of *vacant niches* which may be occupied eventually by invading species. These, and other niche concepts, have been discusssed in greater detail by Miller.[72]

The importance of the niche concept to biogeography is that it serves to define precisely the conditions in which competition and coexistence between species might occur. Thus, the interactions which take place between two individuals or populations in a given space and time usually fall within a *fundamental niche*, which an individual may occupy in the absence of competition with other species, or a *realized niche* which is the actual niche to be occupied when competition is in progress. These relationships have been analysed further by Hutchinson[50] using set theory in which an infinite number of variables is represented within a restricted number of dimensions. Taking the case of fundamental niches, Hutchinson proposed that two major types of intersection could exist, as shown in Euler diagrams for two linearly-presented environmental variables x and y (see Fig. 5.12). In these diagrams, it is assumed that all

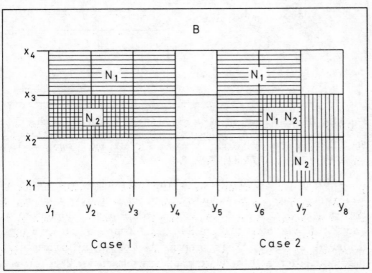

Fig. 5.12 Relationships between fundamental niches (N_1 and N_2) defined by two variables (x and y) in a biotope (b) (from an original diagram in R. S. Miller[72])

points in the fundamental niche imply equal probability of survival, with zero probability for the same species outside the niche; the model is also associated with a single instant in time. $\mathcal{N}_1$ refers to an *n*-dimensional hypervolume equivalent to the fundamental niche of species S_1, and $\mathcal{N}_2$ will form a similar hypervolume for species S_2. Moreover, if *B* is a volume of space in which the biotope of species S_1, S_2, etc., exists, the biotope is complete relative to S_1, since all the

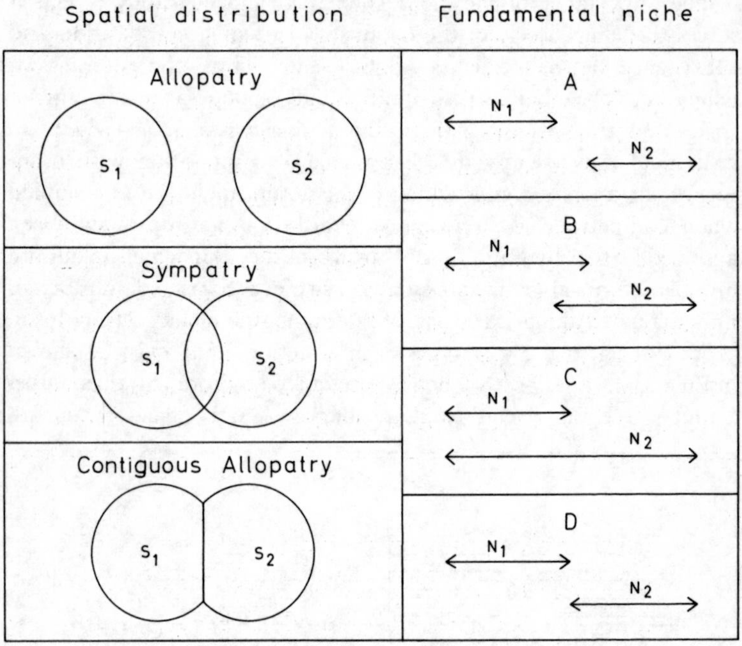

Fig. 5.13 Distribution patterns of two species (S_1 and S_2) and their possible niche relationships ($\mathcal{N}_1$ and $\mathcal{N}_2$). A, non-intersection; B, intersection equivalent to case 2, Fig. 5.12; C, intersection equivalent to case 1, Fig. 5.12; and D, contiguous non-intersection (from an original diagram in R. S. Miller[72])

points in $\mathcal{N}_1$ are represented in *B*. The two fundamental niches $\mathcal{N}_1$ and $\mathcal{N}_2$ may have no points in common, being entirely separate; or they may *intersect*. If so, $\mathcal{N}_1 . \mathcal{N}_2$ is the *subset* of points common to $\mathcal{N}_1$ and $\mathcal{N}_2$. In Fig. 5.12, two possible cases of intersection have been delimited: in case 1, $\mathcal{N}_2$ is a proper subset of $\mathcal{N}_1$, the former being inside the latter, and therefore smaller; while in case 2, $\mathcal{N}_1 . \mathcal{N}_2$ is a proper subset of both $\mathcal{N}_1$ and $\mathcal{N}_2$. In case 1, two possibilities exist for the survival of species S_1: first, if S_1 is dominant, all the elements of

$N_1.N_2$ will be eliminated, thus causing in time the removal of S_2. This implies that coexistence is impossible in an *included niche* $(N_1.N_2)$ if S_1 is a superior competitor. The alternative situation, in which S_2 is superior, is one in which both survive, with S_1 being excluded from the intersection subset, but still remaining in other parts of N_1.

It is important to realize the full implications of these theories, particularly since included niches are relatively common in natural ecosystems, for they help to explain not only the survival of some species whose tolerance range is relatively narrow, but also the maintenance of species diversity in some areas.

It has already been noted that the precise patterns of competition in many natural ecosystems may perhaps best be seen where new species invade restricted habitats. In more stable situations, competition appears to be a continuing and constant phenomenon, stimulated either through intersection of *sympatric* groups (see Fig. 5.13), with the possible displacement of one group, or through contact along a well-defined line in *contiguously allopatric* populations. The latter is an especially interesting case which may only be explained through the existence of severe and almost equal competition along the line of contact, except on occasions when the line corresponds to an important environmental discontinuity, as with a major change in soil or rock type. But this is not to say that competition is ubiquitous in nature, for some groups may be *allopatric* in distribution and so will not meet each other; and also, as Elton and Miller[31] have pointed out, there are very many niches in which empirical evidence for the existence of competition is almost totally lacking.

More frequently in natural ecosystems, the simple patterns of competition outlined above are blurred by the complexity of inter-specific and environmental relationships. Those of contiguous allopatry are perhaps easiest to find, if only because the contact line between the two population groups is so well marked. Thus, in the case of two species of triclads (*Planaria montenegrina* and *P. gonocephala*) which inhabit mountain streams in Britain,[5] the former is present only in waters with a temperature of above 13° to 14°C, and the latter in waters colder than this. This particular example is thought to have been brought about by the competitive exclusion of each species from the areas to which they were least suited. Other relationships are often much more complex, as with slightly different species

233

of the ichneumon wasp (*Megarhyssa* spp.),[97] which at first sight appear to coexist in sympatry with each other to the extent of using identical foods, nesting places, and oviposition sites. But there is no sign of active competition between them, and this apparent paradox may only be elucidated by studying their larval distributions, in which an allopatric situation may be observed; the adult populations merely maintain this pattern, so that they are technically not sympatric.

While a full interpretation of the complex relationships between organisms at an interspecific level is not yet forthcoming, it is certain that a considerable degree of interspecific competition exists in most natural ecosystems, particularly among closely related forms. While this may often give rise to some restriction on population size, it may also stimulate the evolutionary processes of natural selection so that species diversity is maintained or increased. A further consideration of long-term effect of these several possibilities may be found in sections 6.6 and 6.7.

5.12 Positive and negative competition

It may also be useful to consider briefly the processes of competition in the light of their ability to produce *positive* or *negative* reactions within populations for, while competition is usually restrictive for at least one of the groups, there may be occasions in which both benefit. It used to be thought that negative competition was found only in the initial phases of ecosystem development but, in view of recent research summarized in section 5.11 (particularly in connection with reactions between organisms in included niches, or with contiguous allopatric distributions), it seems more likely now that this may occur at all stages of growth. Odum[78] has also included the cases of *parasitism* and *predation* among negative reactions, even though both of these are exceedingly complex situations, and their exact role in competition processes is not always clear. Certainly, both predation and parasitism may take place without active competition between individuals within the same niche, and on these occasions it may be more true to say that such organisms are merely dependent on, rather than competing with, each other. However, both reactions may help to control population size at particular niche levels within ecosystems.

Positive reactions may be subdivided into those of *commensalism*, in which one population dominates; *protocooperation*, when groups are almost equally benefited; and *mutualism*, in which populations have become completely dependent on each other. All are important to ecosystem growth, for, as Darwin noted, cooperation in nature is as important as competition for the well-being of most organisms. Commensalism may occur among motile populations which exist in conjunction with sessile organisms, as when seaside crabs have smaller organisms living on their shells. Examples of protocooperation may often be seen when certain plants and animals are partially dependent on each other, as in the case of the termite (*Atta* sp.) and the sand-box tree (*Curatella americana*) in interior savannas of Guyana. Here, the tree cannot become established before termites have broken through an extremely hard hardpan layer which lies immediately below the soil surface, and which the tree roots themselves cannot penetrate; while, in turn, the termites live in part off substances secreted from the tree. This close association may to some extent be responsible for the curiously local distribution of both tree and termite in many of these areas. The case of mutualism may also be widely observed, and one of the best examples is to be found in a lichen, which in itself is a combination of algae and fungi or, in other words, of autotrophic and heterotrophic organisms.

5.13 Competition and succession

Although the actual processes of competition remain identical throughout, certain competetive trends between organisms and populations will tend to vary as ecosystems develop and become more complex. In general, little is known about the details of these relationships, but four aspects of change appear to be particularly significant.

First, those animals and plants which are important to ecosystems during their pioneer stages of development will often give way to a large number of different species when the ecosystem becomes 'older' (see Fig. 5.14). This change will not always be clear-cut, for some species may persist longer than others if they have particularly wide tolerance ranges, and, in addition, some modification to their degree of tolerance to changing circumstances may result from fluctuations in their numbers and diversity. Usually, an inverse relationship exists between species diversity and the tolerance range which each possesses. *Second*, the weight of biomass (i.e., the standing crop) will tend to increase as the ecosystem matures, while food and

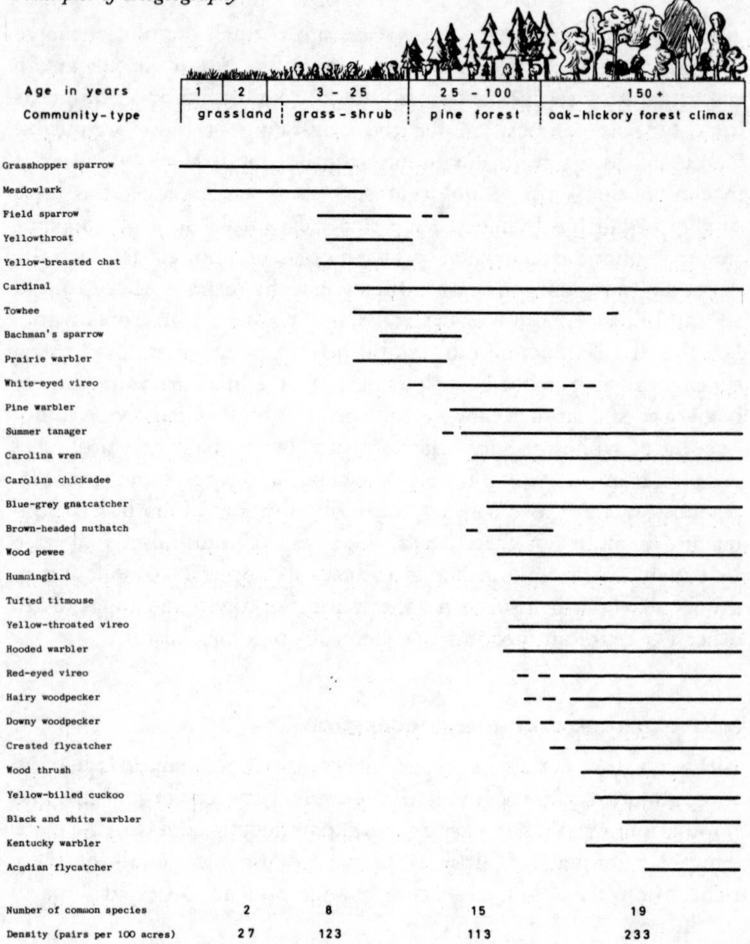

Age in years	1	2	3 - 25	25 - 100	150 +
Community-type	grassland		grass-shrub	pine forest	oak-hickory forest climax

Grasshopper sparrow

Meadowlark

Field sparrow

Yellowthroat

Yellow-breasted chat

Cardinal

Towhee

Bachman's sparrow

Prairie warbler

White-eyed vireo

Pine warbler

Summer tanager

Carolina wren

Carolina chickadee

Blue-grey gnatcatcher

Brown-headed nuthatch

Wood pewee

Hummingbird

Tufted titmouse

Yellow-throated vireo

Hooded warbler

Red-eyed vireo

Hairy woodpecker

Downy woodpecker

Crested flycatcher

Wood thrush

Yellow-billed cuckoo

Black and white warbler

Kentucky warbler

Acadian flycatcher

Number of common species	2		8	15	19
Density (pairs per 100 acres)	27		123	113	233

Fig. 5.14 Schematic diagram of secondary succession on abandoned farm land in the US Piedmont region. (Modified from an original diagram in E. P. Odum, *Ecology*, © Holt, Rinehart & Winston Book Co., N.Y., 1963, using data from D. W. Johnston & E. P. Odum, *Ecology*, Vol. 37, pp. 50-62, 1956)

energy webs and biogeochemical cycles become more complex. In addition, and partly resulting from this, changes in the organic structure of the ecosystem begin to occur, giving rise to (*third*) a renewed diversity of species and a greatly increased potential for competition among the available niches, even though new niches may be continually created by the developing biomass and the

236

increased stratification of plant associations. This new diversity is particularly noticeable among heterotrophs and microorganisms; it may be found within a wide range of species and genera, but at times could be restricted to a few. In either case, the net effect is to further the degree of complexity, and the efficiency of food/energy transfer throughout the ecosystem. *Finally*, a decrease in net production may eventually take place, coupled with an increase in the total respiration of the ecosystem as populations approach asymptote levels. Some of the more theoretical aspects of these changes, particularly as they affect plant communities, are to be further evaluated in section 6.4.

REFERENCES

1. ALEXANDER, W. B., The rook population of the upper Thames region, *J. Anim. Ecol.*, Vol. 2, pp. 24–35, 1933.
2. ALLEE, W. C., EMERSON, A. E., PARK, O., PARK, T., and SCHMIDT, K. P., *Principles of Animal Ecology*, Saunders, Philadelphia, 1949.
3. ANDREWARTHA, H. G. and BIRCH, L. C., *The Distribution and Abundance of Animals*, Chicago, 1954.
4. BATES, M., Observations on climatic and seasonal distributions of mosquitoes in eastern Colombia, *J. Anim. Ecol.*, Vol. 14, pp. 17–25, 1945.
5. BEAUCHAMP, R. S. A. and ULLYOTT, P., Competetive relationships between certain species of freshwater triclads, *J. Ecol.*, Vol. 20, pp. 200–08, 1932.
6. BIRCH, L. C., The experimental background to the study of the distribution and abundance of insects. I. The influence of temperature, moisture and food on the innate capacity of increase of three grain beetles, *Ecology*, Vol. 34. pp. 698–711, 1953.
7. BIRCH, L. C., Selection in *Drosophila pseudo-obscura* in relation to crowding, *Evolution*, Vol. 9, pp. 389–99, 1955.
8. BLACKWELL, J. A. and DOWDESWELL, W. H., Local movement in the blue tit, *British Birds*, Vol. 44, pp. 397–403, 1951.
9. BODENHEIMER, F. S., Population problems of social insects, *Biol. Rev.*, Vol. 12, pp. 393–430, 1937.
10. BODENHEIMER, F. S., *Problems of Animal Ecology*, Oxford U.P., 1938.
11. BRAY, J. R., Notes toward an ecological theory, *Ecology*, Vol. 39, pp. 770–6, 1958.
12. CARPENTER, J. R., Insect outbreaks in Europe, *J. Anim. Ecol.*, Vol. 9, pp. 108–47, 1940.
13. CHAPMAN, R. N., The quantitative analysis of environmental factors, *Ecology*, Vol. 9, pp. 111–22, 1928.
14. CHITTY, H., *Report of the Bureau of Animal Populations*, Oxford, 1938.
15. CLEMENTS, F. E. and SHELFORD, V. E., *Bioecology*, New York, 1939.

16. COLE, L. C., A theory for analysing contiguously distributed populations, *Ecology*, Vol. 27, pp. 329–41, 1946.

17. COOK, L. M. and KETTLEWELL, H. B. D., Radioactive labelling of lepidopterous larvae, *Nature*, Vol. 187, pp. 301–02, 1960.

18. CROMBIE, A. C., Interspecific competition, *J. Anim. Ecol.*, Vol. 16, pp. 44–73, 1947.

19. DARWIN, C., *On the origin of species by means of natural selection, or the preservation of favoured races in the struggle for life*, London, 1859.

20. DAVIDSON, J., On the growth of the sheep population in Tasmania, *Tr. Roy. Soc. S. Aust.*, Vol. 62, pp. 342–6, 1938.

21. DEEVEY, E. S., Jr., Life tables for natural populations of animals, *Quart. Rev. Biol.*, Vol. 22, pp. 283–314, 1947.

22. DEEVEY, E. S., Jr., The probablity of death, *Scient. Am.*, Vol. 182, pp. 58–60, 1950.

23. DOWDESWELL, W. H., FISHER, R. A., and FORD, E. B., The quantitative study of populations in the *Lepidoptera*, *Ann. Eugen.*, *London*, Vol. 10, pp. 123–36, 1940.

24. DOWDESWELL, W. H., FORD, E. B., and McWHIRTER, K. G., Further studies on the evolution of *Maniola jurtina* in the Isles of Scilly, *Heredity*, Vol. 14, pp. 333–64.

25. EINARSEN, A. S., Some factors affecting ring-necked pheasant population density, *Murrelet.*, Vol. 26, pp. 39–44, 1945.

26. ELTON, C. S., The dispersal of insects to Spitsbergen, *Trans. Ent. Soc.*, *London*, Vol. 73, pp. 289–99, 1925.

27. ELTON, C. S., *Animal Ecology*, London, 1927.

28. ELTON, C. S., *Voles, Mice and Lemmings: Problems in Population Dynamics*, Oxford UP, 1942.

29. ELTON, C. S., *The Ecology of Invasions by Animals and Plants*, London, 1958.

30. ELTON, C. S., *The Pattern of Animal Communities*, Methuen, London, 1966.

31. ELTON, C. S. and MILLER, R. S., The ecological survey of animal communities, with a practical system of classifying habitats by structural characters, *J. Ecol.*, Vol. 42, pp. 460–96, 1954.

32. FISHER, R. A. and FORD, E. B., The spread of a gene in natural conditions in a colony of the moth *Panaxia dominula, L.*, *Heredity*, Vol. 1, pp. 143–74, 1947.

33. FORD, E. B., The experimental study of evolution, *Rep. Aust. Assoc. Adv. Sci.*, Vol. 28, pp. 143–54, 1953.

34. FORD, E. B., Evolution in progress. In *Evolution after Darwin*, ed. SOL TAX, Vol. 1, pp. 181–96, University of Chicago Press, 1960.

35. FORD, E. B., *Butterflies*, New Naturalist Series, London, 1962.

36. FORD, E. B., *Ecological Genetics*, London, 1964.

37. FORD, H. D. and FORD, E. B., Fluctuation in numbers and its influence on variation in *Melitaea aurinia*, *Trans. Roy. Ent. Soc.*, *London*, Vol. 78, pp. 345–51, 1930.

38. GALTSOFF, P. S., Ecological changes affecting the productivity of oyster grounds, *Trans. N. Am. Wildl. Conf.*, Vol. 21, pp. 408–19, 1956.

39. GALTSOFF, P. S., CHIPMAN, W. A., HASLER, A. D., and ENGLE, J. B., A preliminary report on the cause of the decline of the oyster industry

of York River, Va., and the effects of pulp-mill pollution on oysters, *Investigational Report* No. 37, US Department of Commerce, 1938.

40. GAUSE, G. F., *La théorie mathématique de la lutte pour la vie*, Paris, 1935.

41. GIBB, J., The importance of territory and food supply in the natural control of a population of birds, *Sci. Rev.*, Vol. 20, pp. 20–1, 1962.

42. GILBERT, O., REYNOLDSON, T. B., and HOBART, J., Gause's hypothesis: an examination, *J. Anim. Ecol.*, Vol. 21, pp. 310–12, 1952.

43. GRINNELL, J., Geography and evolution, *Ecology*, Vol. 5, pp. 225–9, 1924.

44. GROSS, A. O., Cyclic invasions of the snowy owl, and the migration of 1945–1946, *Auk*, Vol. 64, pp. 584–601, 1947.

45. GUNTER, G., Death of fishes due to cold on the Texas coast, January 1940, *Ecology*, Vol. 22, pp. 203–08, 1941.

46. HALDANE, J. B. G., Animal populations and their regulation, *New Biol.*, Vol. 15, pp. 9–24, 1953.

47. HARDIN, G., The competitive exclusion principle, *Science*, Vol. 131, pp. 1292–7, 1960.

48. HARDY, A. C., Zoology outside the laboratory, *Rep. Brit. Ass.*, No. 6, pp. 213–23, 1949.

49. HOWARD, L. E., *Territory in Bird Life*, New York, 1920.

50. HUTCHINSON, G. E., Concluding remarks, *Cold Spring Harb. Symp. Quant. Biol.*, Vol. 22, pp. 415–27, 1957.

51. HUXLEY, J. S., A natural experiment on the territorial instinct, *British Birds*, Vol. 27, pp. 270–27, 1934.

52. JACKSON, C. H. N., The use of the recovery index in estimating the true density of tsetse flies, *Trans. R. Ent. Soc.*, London, Vol. 84, pp. 530–2, 1936.

53. JENSEN, P. B., Studies concerning the organic matter of the sea bottom, *Rep. Danish Biol. Sta.*, Vol. 22, pp. 1–39, 1915.

54. KETTLEWELL, H. B. D., The use of radioactive tracers in the study of insect populations (*Lepidoptera*), *Nature*, Vol. 170, p. 584, 1952.

55. KETTLEWELL, H. B. D., Labelling locusts with radioactive isotopes, *Nature*, Vol. 175, pp. 821–2, 1955.

56. KLOPFER, P. M., *Behavioural Aspects of Ecology*, Englewood Cliffs, New Jersey, 1962.

57. KLUYVER, H. N. and TINBERGER, L., Territory and the regulation of density in titmice, *Archs. neerl. zool.*, Vol. 10, pp. 265–89, 1953.

58. KRAMER, G., Veränderung von Nachkommenziffer und Nachkommengrösse sowie der Altersverteilung von Inseleidechsen, *Z. Naturf.*, Vol. 1, pp. 700–10, 1946.

59. LACK, D., *The Natural Regulation of Animal Numbers*, Oxford, 1954.

60. LACK, D., A quantitative breeding study of British tits, *Ardea*, Vol. 43, pp. 91–124, 1958.

61. LACK, D., *Population Studies of Birds*, Oxford, 1966.

62. LACK, D. and SCHIFFERLI, A., Die Lebensdauer des Stares, *Orn. Beob.*, Vol. 45, pp. 107–14, 1948.

63. LEOPOLD, A., SOWLS, L. K., and SPENCER, D. L., A survey of over-populated deer ranges in the United States, *J. Wildl. Mgmt*, Vol. 11, pp. 162–77, 1947.

64. LESLIE, P. H. and RANSOM, R. M., The mortality, fertility and rate of natural increase of the vole (*Microtus agrestis*), as observed in the laboratory, *J. Anim. Ecol.*, Vol. 9, pp. 27–52, 1940.
65. LESLIE, P. H., CHITTY, D., and CHITTY, H., An estimation of population parameters from data obtained by means of the capture-recapture method, *Biometrika*, Vol. 40, pp. 137–69, 1953.
66. LINCOLN, F. C., Calculating waterfowl abundance on the basis of banding returns, *Circ. US Department of Agriculture*, No. 118, 1930.
67. LOTKA, A. J., *Elements of Physical Biology*, Baltimore, 1925.
68. LOTKA, A. J., The growth of mixed populations, two species competing for a common food supply, *J. Wash. Acad. Sci.*, Vol. 22, pp. 461–9, 1932.
69. MACARTHUR, R. H. and MACARTHUR, J. W., On bird species diversity, *Ecology*, Vol. 42, pp. 594–8, 1961.
70. MACLULICH, D. A., Fluctuations in the numbers of the varying hare (*Lepus americanus*), *Univ. Toronto Studies, Biol. Ser.*, No. 43, 1937.
71. MALTHUS, T. R., *An Essay on the Principle of Population as it Affects the Future Improvement of Society*, London, 1798.
72. MILLER, R. S., Pattern and process in competition. In *Advances in Ecological Research*, ed. J. B. CRAGG, Vol. 4, Academic Press, N.Y., 1967.
73. MILNE, A., Definition of competition among animals, *Symp. Exp. Biol.*, No. 15, pp. 40–61, 1961.
74. MITCHELL, R., A study of sympatry in the water mite genus *Arrenurius*, *Ecology*, Vol. 45, pp. 546–58, 1964.
75. MURIE, A., Dall sheep, In Wolves of Mount McKinley, *National Parks Service (U.S.A.) Fauna*, No. 5, Washington DC, 1941.
76. NICHOLSON, A. J., The balance of animal populations, *J. Anim. Ecol.*, Vol. 2, pp. 132–178, 1933.
77. NOYES, B., Experimental studies on the life-history of a rotifer reproducing parthenogenetically (*Proales decipiens*), *J. Exp. Zool.*, Vol. 35, pp. 255–56, 1922.
78. ODUM, E. P., *Fundamentals of Ecology*, Saunders, Philadelphia, 2nd edn., 1959.
79. ORIANS, G. H., Natural selection and ecological theory, *Am. Nat.*, Vol. 96, pp. 257–63, 1962.
80. PEARL, R., *The Biology of Population Growth*, New York, 1930.
81. PEARL, R. and DOERING, C. R., A comparison of the mortality of certain lower organisms with that of man, *Science*, Vol. 57, pp. 1923.
82. PEARL, R. and PARKER, S. L., Experimental studies on the duration of life: introductory discussion of the duration of life in *Drosophila*, *Am. Nat.*, Vol. 55, pp. 481–509, 1921.
83. PEARL, R. and REED, L. J., On the growth of the population of the United States since 1790, and its mathematical representation, *Proc. Nat. Acad. Sci.*, Vol. 6, pp. 275–88, 1920.
84. PERRINS, C. M., Population fluctuations and clutch size in the great tit, *Parus major*, *J. Anim. Ecol.*, Vol. 34, pp. 601–47, 1965.
85. PETERSEN, C. G. J., The sea bottom and its production of fish food. A survey of the work done in connection with valuation of the Danish Waters from 1883–1917, *Rep. Danish Biol. Sta.*, Vol. 25, pp. 1–62.

86. RILEY, G. A., Biological oceanography. In *Survey of Biological Progress*, Vol. 2, pp. 79–104, 1952.

87. ROE, A. and SIMPSON, G. G. eds., *Territoriality: A Review of Concepts and Problems. Behaviour and Evolution*, Newhaven, Connecticut, 1958.

88. SCHEIDER, W., Die Lebensdauer und Brutgrösse beim mitteldeutschen Star., *Proc. Int. Orn. Cong.*, Vol. 11, pp. 516–21, 1955.

89. SCHWERDTFEGER, F., *Grundriss der Forstpathologie*, Berlin, 1950.

90. SHELFORD, V. E., *Animal Communities in Temperate America as Illustrated in the Chicago Region: A Study in Animal Ecology*, Chicago, 1931.

91. SHELFORD, V. E., The abundance of the collared lemming in the Churchill area, 1929–1940, *Ecology*, Vol. 24, pp. 472–84, 1943.

92. SNOW, D. W., The annual mortality of the blue tit in different parts of its range, *British Birds*, Vol. 49, pp. 174–7, 1956.

93. SOLOMON, M. E., The natural control of animal populations, *J. Anim. Ecol.*, Vol., 18 pp. 1–35, 1949.

94. SOLOMON, M. E., Dynamics of insect populations, *A. Rev. Ent.*, Vol. 2, pp. 121–42, 1957.

95. SYMES, D. G., The population resources of the Connemara Gaeltacht, *University of Hull, Department of Geography, Miscellaneous Series* No. 1, Hull, UK, 1965.

96. TINBERGEN, N., The functions of territory, *Bird Study*, Vol. 4, pp. 14–27, 1957.

97. TOWNES, H. and TOWNES, M., Ichneumon-flies of America north of Mexico, 2. Subfamilies *Ephialtinae, Xoridinae, Acaenitinae, US Nat. Mus. Bull.*, No. 216, pp. 1–676, 1960.

98. VERHULST, P. F., Notice sur la loi que la population suit dans son accroissement, *Correspondence mathématique et physique*, Vol. 10, pp. 113–21, 1838.

99. VOLTERRA, V., Varriazioni e fluttuzaoni del numero d'individui in specie animali conviventi, *Atti Acad. naz. Lincei Memorie*, Vol. 2, pp. 31–113, 1926.

100. WALLACE, A. R., *A Narrative of Travels on the Amazon and Rio Negro . . .*, London, 1853.

101. WANGERSKY, P. J. and CUNNINGHAM, W. J., Time lag in population models, *Cold Spring Harb. Symp. Quant. Biol.*, Vol. 22, pp. 329–38, 1957.

102. WIESNER, B. P. and SHEARD, N. M., The duration of life in an albino rat population, *Proc. R. Soc., Edinburgh*, Vol. 55, pp. 22–37, 1934.

103. WOLFENBARGER, D. O., Dispersion of small organisms, *Am. Mid. Nat.*, Vol. 35, pp. 1–152, 1946.

104. WYNNE-EDWARDS, V. C., *Animal Dispersion in Relation to Social Behaviour*, Edinburgh, 1962.

6

The Time Factor: Dynamic Aspects of Ecosystems

6.1 Introduction

It is one of the premises of this book that the functional mechanisms which assist the growth and development of ecosystems allow for and, indeed, to a large extent depend upon the ability of individual plants and animals to live in equilibrium with each other and within their environment. Where this is not the case, some breakdown in the efficiency of energy transfer and biogeochemical cycling may ensue, leading to restrictions in the patterns of life of particular communities. In order to emphasize these considerations, an intrinsically static situation has so far been assumed in the discussion, in which all component organisms of ecosystems have in some measure become adjusted to life in a relatively stable physical environment, although it is true that some indications as to the consequences of change induced by man have been given from time to time. However, the fact is that such stability is rarely found in nature, and this is particularly true of the last few million years, during which climatic fluctuations associated with the Pleistocene glacial periods have helped to modify the physical milieu continually, often at a very rapid rate. Mountain-building and vulcanism, variations in sea level, the acceleration and deceleration of rates of erosion and sedimentation, and, possibly, even such long-term factors as continental drift, have all contributed towards other processes of change, as have also biotic interactions initiated by the spread of organisms to new territory. Accordingly, all ecosystems must be regarded not only as complexes

242

which give rise to ever more intricate relationships between plants and animals, and the pathways of energy and chemical element movement between them, but also as entities perpetually seeking adaptation to environments which are inevitably being qualified in the course of time.

Within such a setting, the aim of this chapter is to search for an understanding of some of the more significant responses to external change found within the living organisms of any ecosystem. These may be broadly categorized into two groups. *First*, there are those which are ecological in the sense that they reflect the more direct physical reactions between organisms and their environment; and, *second*, there are those resulting from evolutionary trends which can affect their ability to survive or compete successfully. As inferred by Charles Darwin[37] both are inextricably interwoven, although, paradoxically, the study of each has remained markedly disparate until very recently; now, however, the two related disciplines of ecology and genetics are becoming increasingly interlinked as the realization grows that there is a considerable overlap of interest between them.

ECOLOGICAL ASPECTS OF CHANGE

The concept of ecological change derives from ancient origins. Over 2000 years ago, Plato recorded the negative effects of forest removal on the environment of Greece, suggesting that this led to soil erosion, the lowering of water tables as runoff increased, and problems of regeneration among several native plant species. However, it was not until the sixteenth and seventeenth centuries, when forest came to be cleared from western Europe in ever greater quantities to contribute towards the need for timber in house construction, merchant and military shipping, charcoal production, and home fuel supplies, that the awareness of such consequences spread more widely among the general population, particularly in the minds of gamekeepers and foresters whose concern it was to maintain a balance between rates of growth and harvest among trees and birds in the semi-natural woodland and swamp ecosystems with which they were entrusted. The basic concepts of plant-animal interrelationships which emanated from these erstwhile conservationists proved to be sound and invaluable in the subsequent, more precise formulation of ecological principles.

Given uniform conditions of environment and access, it is probable that, in time, many species of plants, animals, and microorganisms would become widely scattered over the face of the earth and in the oceans and atmosphere, except in cases where, for various reasons, rates of dispersal are extremely slow, or the ability to survive and compete is restricted. But even a brief glance at present distributional patterns is sufficient to show that these idealized conditions are not met, in that most organisms do not extend throughout their potential range. Indeed, the vast majority are now seeking to expand the area which they occupy in competition with others through complex dispersal and colonization procedures. At the same time, all communities are striving to reach a state of equilibrium with their environment, but, as has already been suggested, their environment is continually changing. The result is that world ecosystems today show an extreme fluidity, some effects of which are to be described and analysed below.

6.2 World patterns of distribution among organisms

The biosphere is made up of well in excess of two million organisms, over two-thirds of which are animals and the remainder plants. An examination of any distributional map will serve to emphasize the known fact that the ranges of these vary very widely, for some are continuous and others discontinuous, and within this broad framework there may be narrow or wide niche patterns. Certain groups may be *cosmopolitan* and found in every part of the world, as is the case for a few rats and bats, grasses and ragworts (*Senecio* spp.), but most have not managed to extend themselves so far, and for a preliminary explanation of this one must consider the age of their evolution, their opportunities for dispersal, and the nature of the changing environments of the recent past.

It is now generally agreed among geologists and palaeoclimatologists that the earth has experienced several major phases of relatively rapid environmental change. That which affects us most directly can be traced back to Cretaceous times (Table 6.1), when land masses were more uniform in elevation and climate less varied than today. Under these conditions, any newly evolving organisms would tend to spread unchecked throughout continents, and to many islands as well, as long as they were sufficiently well adapted to compete successfully with existing forms. One manifestation of this may be seen in the establishment of contemporary luxuriant subtropical/ warm-temperate forests with very similar component species over

244

Table 6.1   Recent divisions of geological time

Geological subdivision		Age before present day (million years) 0
QUATERNARY (Pleistocene)		3
TERTIARY	Pliocene	7
	Upper and middle Miocene	18
	Lower Miocene	26
	Oligocene	38
	Eocene	54
	Palaeocene	65
CRETACEOUS		

areas as far apart as Spitsbergen, Greenland, and the American sub-tropics. But towards the end of the Cretaceous period and in the early Tertiary, important changes in world environments began to initiate the breakdown of this ancient pattern. At first, the transgression of ocean onto land helped to reduce the extent of continental land masses and sever the links between them, so that isthmuses, for example in the Bering Straits, the vicinity of Panama, and between New Guinea and Australia, came to be covered by sea water. At the same time, a few earth tremors associated with mountain-building and vulcanism were felt, and these were later to grow, until they reached their maximum intensity in the Miocene period, when most of the present high mountain ranges of the world were thrust upwards; in some instances (particularly in western North America), these upthrusts created new atmospheric wave structures and climatic effects as the high-speed air currents of the upper troposphere were reached by their summits. Superimposed upon this was an additional gradual cooling of world climates during the Tertiary period, from which devolved the more severe fluctuations of the Pleistocene ice ages (section 6.5). The long-term result of all these events was to give rise to a massive worldwide diversification of plant and animal forms in three major ways: the disintegration of the old

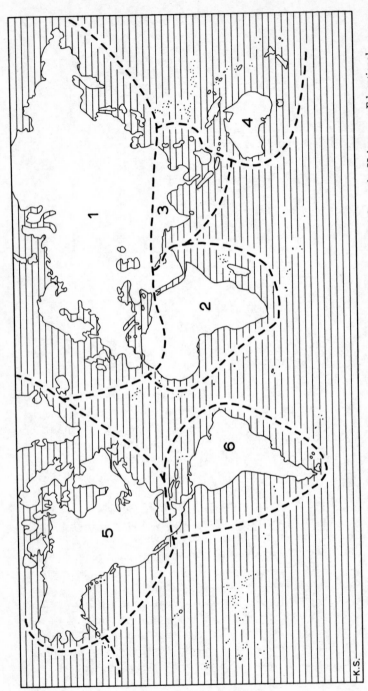

Fig. 6.1 Wallace's Realms (modified from an original diagram in W. George, *Animal Geography*, Heinemann Educational, London, 1962[59]) 1. Palaearctic; 2. Ethiopian; 3. Oriental; 4. Australasian; 5. Nearctic; 6. Neotropical

K.S.

cosmopolitan groups, the evolution of new regional and local populations to meet the changing conditions of environment, and the subsequent spread of these into a wider range than that of their area of origin.[46] In other words the present multiplicity of biological life forms can be traced in no small measure to these tremendous habitat upheavals of post-Cretaceous times.

In the late nineteenth century, the apparent distributional chaos of such organic diversity first began to be sorted out when Sclater[124] reached the conclusion that, despite immense variations in their ranges, all bird populations could be grouped into six broad zoo-geographical divisions. These were later redefined by Wallace[142] as an exercise in the delimitation of more complex regions, based on the range extent of all the then known animals, both vertebrate and invertebrate. To this task, he brought an immense amount of personal knowledge and experience gained in explorations of the Amazon and Rio Negro river basins of Brazil between 1848 and 1852, and of the Malay archipelago between 1854 and 1862, as well as a huge collection of animal forms. *Wallace's Realms*, as they subsequently came to be termed, are outlined in Fig. 6.1: they consist of the Neotropical and Nearctic regions, covering the whole of the New World; the Palaearctic of Europe and much of northern and central Asia; the Oriental of India, much of South-east Asia, Formosa, the Philippines, and the Malaysian islands west of the Celebes and Lombok; the Australasian, of islands east of Borneo, Java, and Bali, including Australia, New Zealand, and many islands in the Pacific; and the Ethiopian, incorporating Africa south of the Sahara, and Madagascar. At times, the line of division between particular realms is very sharp, as in the case of that separating Australasian and Oriental fauna in the East Indies which is thought to have arisen along a narrow zone of intense crustal disturbance dating from Mesozoic times; more often, though, they are much less clearly distinguishable. While these major areal divisions form a useful basis for further discussion, it should be remembered that as our knowledge of range patterns increases they may become substantially modified, both as to the details of their boundaries, and also with regard to the formulation of important new subregions which may be characterized by relatively rapid evolution and which may, to some extent, replace the old (e.g., within the south-west Pacific, as shown in Fig. 6.2).

Since, being less motile, plants are more at the mercy of their environment than animals, it is to be expected that zoogeographical regions are inherently less precisely delimited than botanical regions, and in many cases this is undoubtedly true. Taken as a whole, plant distributions are often exceedingly intricate, both on a macro- and a micro-scale; despite this, however, some generalizations may be made. First, there are very few cosmopolitan plants, and most of

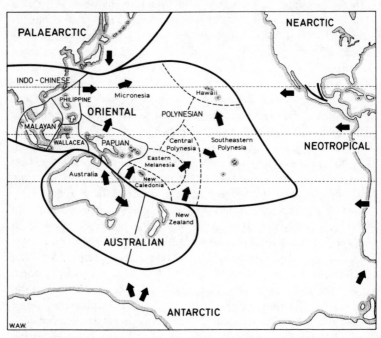

Fig. 6.2 Faunal regions, and directions of faunal movement in the south-west Pacific (from an original diagram in R. L. Usinger[139])

those which are widely established can be subdivided into subgroups of arctic-alpine, temperate, and pan-tropical species. About 100 *arctic-alpines* exist, among which grasses, sedges, and perennial herbs or dwarf shrubs (e.g., moss campion *Silene acaulis*, and the willow *Salix herbacaea*) are most common. *Temperate* plants which cover extensive areas, located predominantly in the northern hemisphere, are often characterized by advanced seeding and dispersal techniques, with a considerable degree of adaptability to different climates; many are now classified as weeds, such as dandelion (*Taraxacum agg.*), common plantain (*Plantago lanceolata*), and wild

248

Table 6.2   Floristic provinces of the world (from an original diagram in
Pierre Dansereau, Biogeography: An Ecological Perspective,[32]
copyright © 1957, Ronald Press, New York, after R. Good[63])

Kingdom	Subkingdom	Province
Boreal		1. Arctic and subarctic 2. Euro-Siberian 3. Sino-Japanese 4. Western and Central Asiatic 5. Mediterranean 6. Macaronesian-transition 7. Atlantic North American 8. Pacific North American
Palaeotropical	African	9. North African-Indian desert 10. Sudanese Park steppe 11. North-east African highland steppe 12. West African rain forest 13. East African steppe 14. South African-transition 15. Madagascan 16. Ascension and St Helenan
	Indo-Malaysian	17. Indian 18. Continental South-east Asian 19. Malaysian
	Polynesian	20. Hawaiian 21. New Caledonian 22. Melanesian and Micronesian 23. Polynesian
Neotropical		24. Caribbean 25. Venezuelan-Guyanan 26. Amazon 27. South Brazilian 28. Andean 29. Pampas 30. Juan Fernandez
South African		31. Cape
Australian		32. North and east Australian 33. South-west Australian 34. Central Australian
Antarctic		35. New Zealand 36. Patagonian 37. Southern temperate oceanic islands

oats (*Avena fatua*). *Pan-tropical* species are also frequently weeds (grasses, Leguminoseae, Euphorbiaceae), or at least occur in man-disturbed areas; of these, the coconut (*Cocos nucifera*) has been disseminated most widely, although grasses, such as *Cynodon dactylon*, may be found in areas as far apart as the West Indies and Florida, Hawaii, and southern Africa. In contrast, it is rare to find pan-tropical species in undisturbed tropical rain forest, for most plants here fall into American, African, or South-east Asian categories and do not grow naturally outside of their respective subregions.

Many more, and indeed the vast majority of plants, have ranges which are much more restricted than those described above. These plants are deemed to be *endemic* in character, frequently reflecting the evolution of new forms within geographically-circumscribed areas in post-Cretaceous times (see section 6.7); they may be further subdivided into *broad endemic* or *narrow endemic* groups. Broad endemics usually grow within a single floristic province which has a distinctive plant assemblage found nowhere else. On a world-wide scale, 37 of these have been delimited by Good, some of which are very large (e.g., the Euro-Siberian), and others very small (Hawaii) (see Table 6.2 and Fig. 6.3). In contrast, narrow endemics are much more restricted in extent, often being associated with particular local climates or edaphic situations, such as the presence of serpentine rock (section 4.9). Sometimes they may be comprised of new, aggressive and rapidly spreading species, or, at other times, of old relict forms, such as the California redwood (*Sequoia sempervirens*), which are now confined to a much more limited area than heretofore. Finally, some plant groups are unusual in that their ranges are *discontinuous* on a continental scale, being separated by gaps of several hundreds of miles. Sometimes, this may be explained fairly easily, as in the cases of the tulip trees (*Liriodendron* spp.) and the sweet gum (*Liquidambar* spp.) which have maintained themselves to the present day in eastern Asia and eastern North America from a former hemispheric distribution in early Tertiary times, but which have not survived in the intervening region of western Europe due to the lack of warm southern refuge areas there to which they could retreat during the phases of Pleistocene ice advance. More puzzling discontinuities between northern and southern hemispheres may also exist, and these are much more difficult to interpret: thus, the creosote bush (*Larrea divaricata*) occurs naturally in the more arid parts of both California and Chile.

250

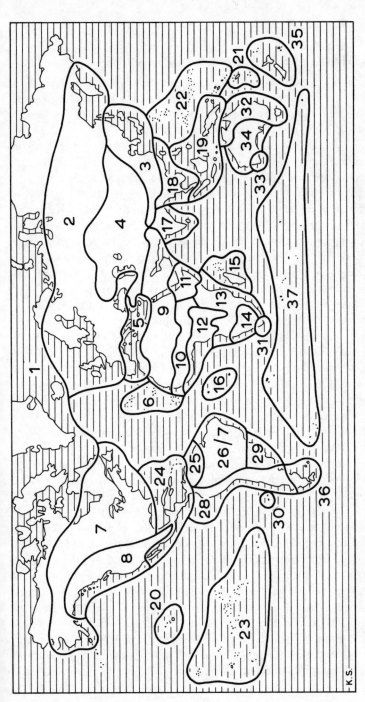

Fig. 6.3 Floristic regions of the world (modified from an original diagram in Pierre Dansereau, *Biogeography; an ecological perspective*,[32] copyright © 1957, The Ronald Press, Co., New York, after R. Good[63])

K.S.

6.3 Modes of dispersal of organisms

Currently, most forms of plants and animals are seeking to extend the areas which they occupy in a process which Elton[46] has termed *randomization* or, in other words, an attempt by all species to occupy as great a quotient of the earth's surface as they can. If this continues, a greater uniformity of species may be expected in time as their *ecological range* (that which they actually occupy) broadens towards their *tolerance range*, a term which represents the complete milieu of environmental circumstances in which they can survive and reproduce. Both these concepts need careful application, for as organisms undergo continual adaptation and evolution in response to changing habitats, their tolerance and ecological ranges may be altered differentially. At present, the small number of cosmopolitan species is witness to the fact that very few plants and animals inhabit their full tolerance range, and one may assume that many are now migrating in response to their inclination to reach this goal. Indeed, it is probably true to say that the distributional patterns of most are changing more quickly today than ever before. For the biogeographer, it is important to consider the implications of this to ecosystem organization, and to evaluate the relative effectiveness of dispersal among different organisms and in dissimilar regions. In order to accomplish this, he must take into account the three basic propositions indicated below.

1. It is a commonplace to say that the ability of organisms to disperse from their home areas may vary appreciably not only through the consequences of differential competition, but also because of their morphological and physiological variety. In the case of the larger animals, George[59] has pointed out that this may be due to factors of size, psychology, the relative rates of birth and death, or simply the speed of locomotion. Elsewhere, the indirect effects of environment may prevent range extension, as in the case of the giant panda, whose limited occurrence seems to be due to the restricted distribution of certain types of vegetation on which it feeds. Even where all other conditions are suitable, favourable areas for colonization may not, at times, be reached if they are isolated by barriers such as high mountains, oceans, or incompatible climates or vegetation groups. Thus, many mammals may be unable to cross mountains or extensive areas of fresh or salt water, and the same is often true for amphibia. Probably the most effective barrier to animal

colonization is the presence of a large ocean, and the results of this may be seen in the impoverished fauna of many oceanic islands which lie far from the nearest continental mainland. Thus, Gressitt and Yoshimoto[66] have recorded that terrestrial mammals (except for the semi-ubiquitous bat and rat) are almost totally lacking in many Pacific islands, along with amphibia and freshwater fishes. Molluscs, such as land snails, may be common, though the reasons for this are not known,[85] and, while seabirds are often distributed widely, the same is not true for the landbird fauna. Among reptiles, skinks and geckos are frequently observed, though snakes are usually confined to islands close to the mainland; turtles can, however, travel across many hundreds of miles of ocean water. By far the most dominant animals on oceanic islands are insects, and these may have been dispersed from continents either on rafts or logs, on the feet or feathers of birds, by man, or probably, in most cases, through the medium of tropical storms which can carry them considerable distances, often against the direction of normally-prevailing winds. The paucity of animal species on oceanic islands means that many of their ecological niches are empty, a situation which may give rise, at times, to the explosive spread of new immigrant populations if they should happen to find one which is agreeable to them. Zimmerman[155] has suggested that this has occurred fairly frequently in Hawaii, with evolutionary consequences which may be profound (see pp. 289 to 290).

2. It is inherent within the nature of plant species that, being largely immotile, they have evolved specific and well-developed aids to dispersal which may be extremely efficient. Some plants are especially well endowed in this respect, as in the case of the common reed (*Phragmites communis*), which may spread through the windblow of its tufted and plumed seeds, the growth of long runners, and by rhizomes capable of floating long distances on water. But these instances are relatively unusual, and most will have only one agency of dissemination. As far as their reproductive capacities in general are concerned, plants err on the pessimistic side, particularly among sexually-reproducing species, which tend to bear large numbers of seeds, though relatively few of these survive to beyond the seedling stage, the remainder having been eliminated through grazing, competition, or other processes.

Much of our detailed knowledge of the mechanisms of plant dispersal was first gleaned by Guppy[67] and Ridley,[121] and this was later supplemented and reexamined by Dansereau and Lems[33] and

Lems.[87] Guppy was especially concerned with the effectiveness of ocean currents as a means of seed carriage, and after an immense collation of facts and personal observation, Ridley outlined 10 ways in which plants could scatter their progeny; these are indicated below in a slightly modified form. First, some seeds can be transported very great distances by wind, as in the case of the temperate-latitude dandelion (*Taraxacum agg.*) and rose-bay willow herb (*Chamaenerion angustifolium*); usually, these are modified for windblow, either through the formation of plumes, as in the two instances above, wings (e.g., the sycamore, *Acer pseudoplatanus*), long hairs (some cottons), fine spores (mushrooms, ferns), dust seeds (orchids), and so on. Even in strong winds, the heavier seeds tend to fall within the immediate vicinity of the parent plant, except when convectional air currents are rising strongly, or under tornado, hurricane, and blizzard conditions, but the lighter spores and dust seeds may, along with many insects, be carried many thousands of miles away. Formerly, it used to be thought that large oceans created a major barrier to wind-dispersal such as this, but sufficient recent evidence has come to light in the Pacific area (see Fosberg,[56] Thorne,[137] Zimmerman[155]) to contradict this notion, and now, in Zimmerman's words, one has to accept the existence of a 'veritable aerial plankton drifting over the oceans', often in the high-altitude jet streams, from which substantial colonization of some plants, even on isolated islands, may ultimately take place.

Second, other plants may be disseminated by water, among which are those whose leaf stalks act as floats (e.g., the tropical water hyacinth, *Eichornia crassipes*), those which spread by means of water-tolerant rhizomes, and those with seeds that can withstand long periods of floating at sea (e.g., the mangrove *Rhizophora mangle*). At times, rapidly-flowing rivers or rain wash may also carry seeds for very long distances, especially in the fringe areas of mountain regions. Rafting of seeds can occur in certain oceans, as in the case of the rotting vegetation which drifts out to sea from the mouths of the Orinoco and Amazon rivers, or through iceberg rafts under arctic and subarctic conditions, though these are usually not a major factor in dispersal.

Fruits or seeds are often distributed widely, either advertently or inadvertently, by (third) mammals, (fourth) birds, (fifth) reptiles and lower animals, and (sixth) by man; and in these instances they may move externally (*epizoically*) or internally (*endozoically*). Many internally carried seeds often have an attractive colour, scent, or taste, while epizoic transportation may be encouraged by specific adaptations which add to the ease of seed uptake by roaming animals or birds. The latter were considered so important by Ridley that he accorded them separate categories in his classification of dispersal mechanisms, so that simple adhesion formed a seventh agency, adhesion through hooks (e.g., the burdock, *Arctium* spp.) an eighth, and adhesion due to viscid modification a ninth. Ridley's tenth means of dispersal was through jactitation, or the mechanical throwing of some seeds considerable distances from a parent tree. Thus, the Para rubber tree (*Hevea brasiliensis*) and the South American sand box tree (*Hura crepitans*) can both eject seeds up to 40 feet away from their source. One may also mention an additional factor in dissemination, that of gravity, which in itself can be a significant agency of range expansion in mountain areas, where heavy seeds or cones, such as those of the California Digger pine (*Pinus sabiniana*), may roll down slopes for considerable distances.[145] Most other plants in world ecosystems are dispersed fairly slowly by vegetative means, through the spread of underground stems, rhizomes or runners.*

3. A *third* proposition is that the possession of effective systems of reproduction and range extension among plants and animals does not necessarily mean that new areas *will* be colonized, for as Setchell[125] has indicated, establishment not dissemination is the key process of dispersal, and success here is dependent upon many factors. Chances of establishment are normally increased in inverse proportion to the distances travelled by the seeds or offspring,[35] or to put this in Clementsian terms, effective colonization is normally local in scale, close to the parent community. However, given time, even slowly spreading species may invade large areas, and, as has been shown, the most isolated of oceanic islands may eventually receive large numbers of plants, insects, and other organisms, either through the action of wind, ocean currents, birds, and other animals, or more recently, through man. In general, however, observations have con-

---

* Dansereau and Lems,[33] and Lems[87] have divided plants into the following 10 categories, based on their morphological adaptations for dispersal: *sarcochores*, with diaspores and fleshy; *desmochores*, with diaspores sticky or barbed; *sporochores*, with diaspores light and minute; *pogonochores*, with diaspores plumed; *pterochores*, with diaspores winged; *cyclochores*, with diaspores held on a wind-blown spherical frame; *ballochores*, with jactitation mechanisms; *auxochores*, with diaspores deposited by the parent plant; *sclerochores*, with diaspores without any visible adaptation; and *barochores*, with heavy diaspores.

firmed that the establishment of colonizing species is increasingly less likely to take place as the distances from their source areas grow, a consequence noted in the Pacific by Wallace[143] and Mayr[101] for bird populations, Ekman[45] for marine shelf fauna, and generally by Darlington;[36] it may also depend, to a large extent, on aspects of competition noted in chapter 5, and especially on the degree of competition. Thus, several authors have recorded that closed forest communities rarely receive invading plants, since competition for niches may be severe, so forcing colonizers to move elsewhere, or to attempt settlement in a niche which they do not normally inhabit. Thus, the white dead nettle (*Lamium album*), a woodland alien whose origin lies in the Caucasus Mountains, has been able to found new colonies in Britain only on open sites such as roadsides and wasteland;[46] and, indeed, such sites everywhere, whether natural or man-induced, often form the most suitable venues for colonizing plants, in view of their continually altering niche structure, the importance of which point is to be considered in greater detail later in this chapter (pp. 288 to 289). Partly because of this, most successful invaders are also pioneers, whose ultimate distribution in new territory may be severely restricted if for any reason the extent of sites favourable to them is subsequently limited. This may be illustrated by considering the history in Britain of the sea plantain (*Plantago maritima*) and the snail *Catinella arenaria*, both of which were widely located over large areas of open land immediately following the retreat of Pleistocene ice sheets, but which are now confined to very circumscribed districts close to the sea (in the case of the snail, to Braunton Burrows, a sanddune area of North Devon) as more mature ecosystems have developed. Effective dispersal onto open sites, however, is not a simple matter, and may be confused chronologically and areally by sudden explosions of populations which were formerly stable, and through rapid fluctuations of species numbers. In England, the former situation may be seen in the instance of Oxford ragwort (*Senecio squalidus*), a Sicilian native which escaped from the Oxford botanical garden in 1794 to spread slowly along railway lines to London, and then eventually undergo a very rapid extension in numbers and range after the Second World War;[83] and the latter by the Canadian water weed *Elodea canadensis* which, following its introduction to the River Trent in the eighteen-forties, propagated itself so quickly in the subsequent 20 years as to interfere with fishing and even the flow of water, since when it has declined both in numbers and apparent aggression.[43] It is possible that both these, and many other similar instances, involve the development of mechanisms of pre- and postadaptation to life in new environments. Differential competitive abilities may often also be seen, as, for instance, when invaders from floristically rich continental areas always have an advantage over those from oceanic islands. Other environmental or biotic restraints to colonization frequently occur, as with the spread of dog's mercury (*Mercurialis perennis*) in Ireland, which is prevented when ants associated symbiotically with it are not available to transport seeds to new habitats.[11]

Whether invasion is local or distant, continuous or intermittent, from one direction or ecotonic (from two or more major communities), dispersing organisms will always tend to organize themselves into species aggregations, which then automatically increase their chances of survival in a new area. This may take place on a very large scale, as when big sage brush (*Artemesia tridentata*), mesquite (*Prosopis juliflora*) and one-seed juniper (*Juniperus monosperma*) begin to move into disturbed areas of the American south-west,[21, 71] or on a much smaller basis, as when *Spartina townsendii* colonizes small areas of mud flats in Britain. Such is the pioneer stage of dispersal, or, to use Dansereau's term,[32] the period of *ecesis*; it is followed by the competition described above, with later adjustment to the biotic community, and only then may a species be said to be established in new territory.

6.4 Climax and polyclimax succession

At the same time that species are seeking to extend their ranges, all ecosystems experience further modification as they adjust towards new equilibria within changing environments. Once a balance has been reached within a given habitat, the ecosystem is said to be mature, and the component communities are described as being in a *climax* state, beyond which, in theory, they can develop no further. As suggested in chapter 4, it is likely that the climatic parameters of energy income, and heat and water availability, offer the most severe potential general limitations to the growth of organisms and ecosystems, so that most climax communities are taken to be *climatic* in nature, although edaphic and other climaxes may also arise.

It was Clements[25, 26] who first put forward the notion that, for any given environment, vegetation communities would change in structure and composition from a very simple form on bare rock and soil, to much more intricate multilayer combinations as they approached maturity. We have already seen in Fig. 5.14 (p. 236) that these principles may be applied within certain animal communities as well. This successional development is termed a *sere*, or a *prisere* (primary sere) if it evolves in habitats which have not previously supported organic life forms; secondary seres, tertiary seres, and so on, may also be defined. The speed at which these evolve can vary substantially with climatic conditions, so that in tundra localities it may be extraordinarily slow, and in the energy-rich wet tropics very rapid. One of the best known and most detailed analyses of seral development is that on Krakatoa, an island of volcanic origin lying in the Sunda Strait between Java and Sumatra, which exploded violently in 1883, covering and killing all life forms with a layer of ash up to 100 feet in thickness. Yet, only 50 years later, a rich and maturing tropical rain forest with over 250 component plant species and many epiphytes had reestablished itself; and, in addition, over 720 insect species had arrived, along with a few reptiles,[137] though there were still no amphibia or mammals, except for bats and rats. The stages in this recolonization process were as follows. After the cooling of ash deposits, some littoral vegetation spread quickly from neighbouring islands, probably carried by ocean currents, birds, and wind, and by 1886, spores of blue-green algae and some ferns had established themselves in a *pioneer cover* inland. *Secondary communities* of dense grass had formed in the interior by 1897, together with numerous ferns and a few shrubs with seeds that were easily wind-blown. By 1906, a stratified maritime woodland fringed the coast, while grasses maintained their dominance inland until 1919, after which a *tertiary community* of forest took over there, though this was characterized by an unusual paucity of species when compared to similar assemblages in adjacent islands. Since then, further adjustments to the community structure have occurred as new tree species continue to establish themselves, but now relative stability appears to have been achieved so that the vegetation may once again be close to a climax state. In this, as in most other instances involving vascular plants, the seral sequence has followed the classic pattern in which ever taller plant forms succeed each other as the community proceeds towards maturity: usually, the tallest species at each stage is termed the *dominant* and demands the greatest amounts of incoming solar energy. From this, it may be implied that while the relatively open and simple plant assemblages in the pioneer stages of development are characterized by heliophytes (and often xerophytes), the more closed and complex situations later in the sere usually have a greater number of mesophytes and sciophytes, except for the mature dominants themselves. As noted previously (section 5.13), seral changes also trigger other reactions in the environment, and affect substantially the patterns of energy and chemical element transfer, soil and microclimatic conditions, factors of competition, and so on.

Originally, it was argued by Clements that for any major regional climatic zone only one type of vegetation climax could be expected, even though this might be derived from a wide range of pioneer sites (e.g., dunes, swamp, bare rock, turf, and so on). Although these broad climax *formations* might differ in the details of their local species composition, they would nevertheless be generally so similar as to be clearly related to each other; thus, oak (*Quercus* spp.) and beech (*Fagus sylvatica*) communities in western Europe obviously have different dominants, but would still be part of the same regional summer-green climax formation. However, at any point on the sere the succession towards this state could be halted, either temporarily or permanently, by *arresting factors* of a topographic, edaphic, pedalogic or biotic nature, and these might then give rise to *subclimaxes* in which plants were held in an apparently stable situation by non-climatic controls; however, the normal pattern of seral development could again be resumed once the arresting factor was removed, in which case it would be referred to as a *subsere* (see Fig. 6.4). When man's activities have proved to be an arresting factor, any subsequent seral development is deemed to be a *plagiosere;* and if his effects are permanent or semi-permanent, a man-controlled *plagioclimax* formation may be created, as in the case of many British heather moorlands (section 3.9). Indeed, much of the world's vegetation today is plagioclimax, in the Clementsian sense of the word. Consequently, great care should be exercised in the application of the word 'climax' to any vegetation communities at the present time, and, in any case, it should only be applied to those which retain their essential stability longer than the lifespan of their component organisms. The delimitation of climax formations may also raise difficulties on a local scale, in that less and more favourable environmental circumstances than the norm may give rise respectively to stable *preclimax* or *postclimax* communities, as when water

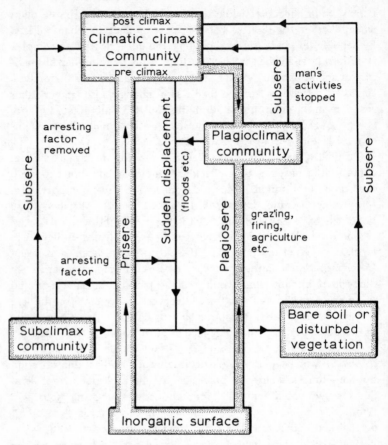

Fig. 6.4 The patterns of seral development (modified from an original diagram in S. R. Eyre, *Vegetation and soils*, Edward Arnold, (Publishers) Ltd. London, 1963[81])

availability in marginally-arid areas is reduced or improved by the presence of freely-draining or water-retentive soil and rock. *Disclimaxes* may also be present, when the introduction of a rapidly spreading new species causes an apparent climax, which, however, is inherently unstable.

It may be deduced from the above that the application of climax theory to particular ecosystems might cause certain problems. The greatest difficulty seems to lie with the assumption that only one climax formation (or a *monoclimax*) is found in specific regions, and,

260

indeed, this idea has received a good deal of criticism from the 'twenties onwards, when Tansley in the UK,[134] Nichols in North America,[112] Domin,[42] and several Soviet scientists, expressed their scepticism, and Gleason[60] went so far as to say that monoclimax as a theory was unworkable in nature. Gradually, an alternative concept of *polyclimax* was formulated, in which it was recognized that plant formations (and also to some extent their animal associates) might be equally influenced by factors other than climate, so that edaphic, pedalogic, topographic, halophytic, pyric (fire), and other climaxes might be delimited, given the right environmental circumstances; however, as Tansley[135] has pointed out, polyclimax does not necessarily rule out the idea of climate as a controlling medium in many instances. Relationships between the two theories are confused, and, as Cain[18] noted over 30 years ago, are not clarified by the existence of substantial conflicts in terminology; moreover, Whittaker[148] has suggested that the notion of polyclimax, while increasingly favoured as a general explanation of successional patterns in vegetation, has shown a recent trend towards divergence among many pathways. Nevertheless, most authorities agree that monoclimax theory falls down on the supposition that the majority of plant communities can be seen ultimately to merge into wider formations, which are distinctive units controlled by climate. In view of this divergence of opinion, it may be useful to reconsider the somewhat more sophisticated idea that all environments have a *biotic potential* of growth forms which can exist within it, towards which all ecosystems and their component communities will develop and within which all organisms will have genetic limits related to their tolerance of that environment, so giving rise usually to a 'mosaic of communities whose distribution is determined by a corresponding mosaic of habitats'.[136] Another suggestion is that the biotic potential may be seen as a varying population pattern of different species related to each other, which may be correlated with an equivalent pattern of changing environmental gradients. Both these two hypotheses, while biologically more convincing as an explanation of succession, have proved difficult to use in analysis so far; both still do not preclude the possibility that, granted timelag effects, climate may still control, to a very large extent, the mature pattern of plant and animal distribution in particular associations; and both are still based essentially on qualitative observations, so that they may be considerably modified as more quantitative studies are completed.

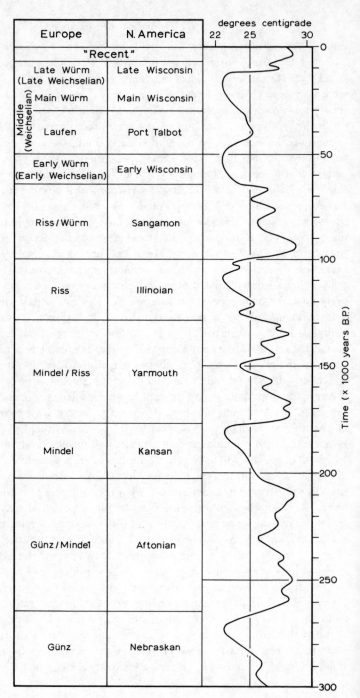

6.5 Climatic change and equilibrium within ecosystems

It may be appropriate to close this brief ecological examination of
fluidity in world ecosystems by turning to one of the most significant
contributary influences, that of climatic change, particularly as it
has affected western Europe. As stated earlier in this chapter, temperature and moisture conditions over the world as a whole have been
fairly uniform for much of the earth's history. But on occasions, in the
Eo-Cambrian (ca. 350 million years ago), the Permo-Carboniferous
(ca. 150 million years ago), and, more recently, the late Tertiary-
Pleistocene periods (Table 6.1), a general cooling has occurred,
superimposed upon which have been wider cyclical fluctuations in
temperature, particularly in mid- and high latitudes. These have
led, in turn, to the periodic growth of extensive ice caps in arctic and
subarctic regions, and the consequent displacement of other temperature and moisture zones towards the Equator. Such glacial phases,
and the intervening interglacials and interstadials* (so named because the ice caps partially or wholly melted from time to time), are
extremely complicated phenomena the origins of which are still a
matter of some dispute. Whether one envisages with Simpson[126]
that they were caused by variations in the output of solar energy, or
following interference with the zonal circulation patterns of the
atmosphere after the uplift of mountain ranges in the Tertiary, or
by other means, it is probably safest to say that, at present, we
simply do not have enough information to speculate with any
degree of certainty what the mechanisms which caused them
might be.

The most recent glacial period, with which this chapter is most
directly concerned, is commonly referred to as the *Pleistocene*, a term
used initially by Lyell in 1839[91] to describe recent geological sediments, and later modified by Forbes in 1846[55] to make its beginnings
correspond to the time when the post-Tertiary ice sheets first
formed. By then, the cooling which had been a feature of early

* It is now customary, at least in western Europe, to use the term *interglacial* to
describe a temperature period in which mean annual temperatures equalled or
exceeded those of the Flandrian climatic optimum (Table 6.3), whereas the term
interstadial indicates a somewhat cooler temperate period.

Fig. 6.5 General temperature curve of Pleistocene temperatures in the Caribbean,
with correlations to the continental glacial and interglacial periods (from C.
Emiliani,[48] *Journal of Geology*, Vol. 63, 1955, published by the University of Chicago
Press. Copyright by the University of Chicago. Used with permission)

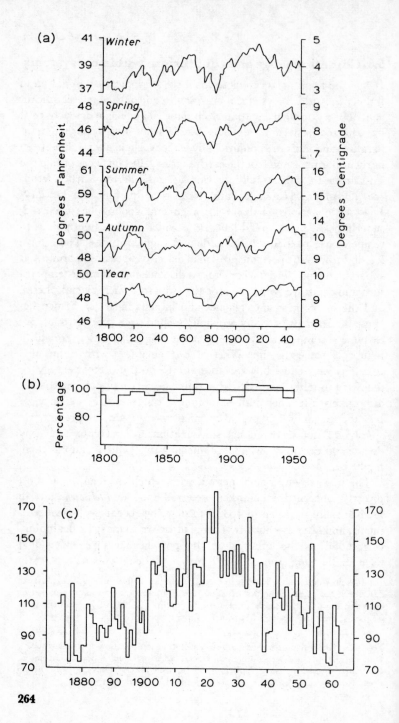

Tertiary times was already under way, and the subsequent major fluctuations in temperature about to commence. Over mid- and high latitudes as a whole, four important glacial phases can be detected in the Pleistocene period (as noted in Fig. 6.5), along with many more minor cold spells. Evidence from the tropics is less certain, though there now appears to be general agreement, based both on the location of relict species and on changes in the microfossil content of core samples taken from deep ocean sediments mainly in the Caribbean, that temperatures at the height of the glaciations were depressed here, too. The precise drop in mean annual temperatures at the glacial maxima remains a matter for conjecture, though from recent evidence it has been estimated by West[147] to lie somewhere between 5° and 8°C less than present-day values at sea level in middle and high latitudes, and by Emiliani[49] at between 2° and 3°C at sea level in the tropics. In the general amelioration of temperatures which has followed the retreat of ice from the last glaciation to high latitudes (where, of course, it still persists today in Greenland, the Canadian Arctic islands, and Antarctica), further minor temperature fluctuations have been experienced, as indicated in Fig. 6.6 and Table 6.3; and although present climatic trends are difficult to discern, it is now thought that temperatures, at least in north-west Europe, are beginning to fall again rather rapidly. Associated with these fluctuations are equivalent changes in the patterns of wind speed and direction, and rainfall.

Table 6.3   Recent climatic change in Britain (after R. G. West[147])

Dates	Climate
5000–3000 BC	*Flandrian climatic optimum.* Annual mean sea level temperatures ca. 2°C above present
AD 500	Climatic deterioration of the *subboreal*
AD 1000–1300	Slightly warmer conditions of the *lesser climatic optimum*
AD 1550–1850	'*Little ice age*'
AD 1850–1940	Increase in warmth
1940–present	Decrease in warmth

Fig. 6.6   Recent climatic changes in Britain: (a) average temperatures in central England (°C), shown by 10-year running means for each season, and for the year; (b) rainfall in England and Wales, plotted as decade averages which are shown as percentages of the averages 1900–39; (c) number of days classified as a 'westerly' type for each year since 1873 (from an original diagram in R. G. West, *Pleistocene geology and biology*, Longman Group Ltd., London, England, and John Wiley and Sons, New York[147] quoting G. Manley[96] and H. Lamb[86])

265

Evidence for these changes may be collated in several ways. Over large areas, glacial or periglacial deposits, permafrost, features of glacial erosion, lake-damming and overflow channels, and the presence of misfit streams in extremely broad valley floors, all point to the fact that many of our mid-latitudinal landscapes were formed, or moulded to some extent, either underneath ice sheets or in tundra conditions, where the outflow of streams from glaciers was considerable. On a different plane, marine erosion terraces which now lie well above sea level indicate the former existence of higher sea levels during interglacial phases. The equatorwards compression of climatic zones as glaciation occurred may be detected from the remnants of extensive old lake beds in areas which are now dry and support only desert vegetation, as in Utah, the Sahara, and parts of Central Asia. Major differences in the distribution of organisms can be inferred through the analysis of sediment cores taken from land, freshwater lakes, or oceans, which contain faunal or floral remains, or by means of the study of pollen grains (*palynology*) from suitable sites. The dating of all such cores and sediments, first attempted in detail for varve clays* which are particularly finely delimited in Scandinavia,[58,122] has recently been immensely encouraged by the introduction of radiocarbon techniques, involving especially carbon-14 molecules, whose half-life of 5568 years[81,89] may be used accurately on fossil wood, shells, and similar biologically derived materials deposited at up to 30 000 to 40 000 years before present. Beyond this, C-14 processes are satisfactory only if isotopic enrichment has taken place, in which case they are valid for up to 60 000 years. Older remains than this can sometimes be placed chronologically by utilizing isotopes of potassium and argon,[50] a practice which has been successful in clarifying Pleistocene events of between 1·5 and 3·5 million years ago in Africa, Europe, and America, though the chances of error here are much greater. Rates of isotope decay of U-235 and U-238, and relationships between Th-230/U-235 and Pa-231/Th-230, have also proved useful for dating,[15,151] though these are in a much more experimental stage of development. More recent patterns of change may also be observed at times through the application of dendrochronological techniques, in which variations in the annual increment of woody growth, as expressed in

---

* The term *varve* was originally applied by de Geer[58] to describe 'periodically laminated sediments in which the deposition for every single year can be discriminated'.

tree rings, have been successfully correlated with minor climatic fluctuations, notably in the case of pine and spruce in subarctic northern regions, and for *Pinus aristata* and *Pinus flexilis* in many of the semiarid areas of the south-western USA. These have been described in full elsewhere, as in Glock.[61]

In many regions, the effects of these climatic fluctuations upon the distribution of plant and animal communities have now been determined with a considerable degree of accuracy. In Scandinavia, Andersson[2] has recorded fossil specimens of hazel (*Corylus avellana*) and *Trapa natans* of Flandrian climatic optimum age at locations far to the north of existing polar limits for both species, as shown in Fig. 6.7. A much broader, long-term view indicates that the early effects of ice advance in Europe had a major general effect on the subtropical, warm-temperate forests established there in late Cretaceous and early Tertiary times (see also p. 250). Even during the general Pliocene cooling, remnants of this forest still persevered along the southern margins of the North Sea, where fossil remains of hickory (*Carya*), redwood (*Sequoia*), and hemlock (*Tsuga*) have been recovered. However, with further Pleistocene chilling, this forest came to be trapped between ice masses advancing from the north and those moving down alpine slopes; eventually, when it could no longer survive in or migrate from the changed conditions, it was obliterated. Accordingly, the first overriding effect of glaciation in western Europe was to leave an ecological void, which, after subsequent ice retreat in the first interglacial, came to be filled by a new, rapidly evolving flora of a distinctly different nature to the old, characterized by a greater number of herbaceous and cold-resistant species similar to those of the present day. The later phases of glaciation and deglaciation served both to emphasize and reaccentuate these important changes.

Further details of this immense modification and readjustment of plant and animal communities in the Pleistocene and post-Pleistocene periods may, perhaps, be best revealed by studies of events in relatively restricted areas. Much information may be gleaned from an analysis of post-Weichselian times (from ca. 9000 years ago) in the British Isles, during which ice retreated from a maximum advance which had left only approximately one-third of the southern part of the land area free from its cover (Fig. 6.8). The ensuing patterns of plant and animal recolonization in these islands can largely be explained by a consideration of three influential factors: *first*, the

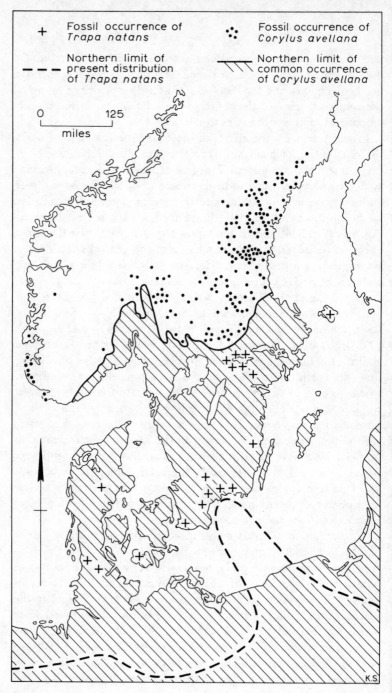

+	Fossil occurrence of *Trapa natans*
- - -	Northern limit of present distribution of *Trapa natans*
⋰	Fossil occurrence of *Corylus avellana*
⁄⁄⁄	Northern limit of common occurrence of *Corylus avellana*

0   125
miles

K.S

268

nature, distribution, and availability of refuge areas in which species might have survived the cold of glacial phases; *second*, the relative speed and effectiveness of species dispersal from these once the ice melted; and *third*, the existence of potential barriers which might serve to restrict this dispersal.

The location of refuges from which species could recolonize the British Isles has by no means been fully determined, even for post-Weichselian times. It was Forbes[55] who, over a century ago, first put forward the notion that the present faunal and floral elements of Britain could be arranged into five categories, as delimited both by their area of origin (often their refuge) and their date of entry into the country: these comprised a recent, *arctic-alpine* group; a postglacial, currently widely distributed *Germanic* group; pre-Weichselian *continental* and *Atlantic* groups, which, he suggested, had maintained themselves in favoured southern sites during glacial phases, and are now centred respectively in south-east and south-west England; and a *Lusitanian* element, which today is predominantly confined to the south-western extremities of Ireland and England, and which includes such warmth-loving species as the Mediterranean-based strawberry tree (*Arbutus unedo*).* The considerable discussion which has followed Forbes's proposition, particularly during the present century, has, if anything, served neither to confirm nor deny the validity of his view that a wide range of species remained in Britain during glacial phases. Thus, Clement Reid[119] suggested that, at glacial maxima, all temperate species in England must have been killed, while arctic-alpines survived only in the extreme south of England. Later, along with several Scandinavian authorities, Wilmott[149] contested this, in arguing that small nunataks (ice-free areas) within the glaciated zone might have supported a limited arctic-alpine flora through to post-Weichselian times, a thesis which has been used to explain the presence today of the occasional occurrence of very restricted arctic-alpine assemblages outside their normal areas of distribution, such as that on Teeside (Fig. 6.8), which includes mountain avens (*Dryas octopetala*), alpine saxifrage

* At present, the British distribution of *Arbutus unedo* is restricted to south-western Ireland.

Fig. 6.7 The present and past occurrence of *Corylus avellana* and *Trapa natans* in Scandinavia. Fossil records date from the Flandrian period (from an original diagram in G. Andersson[2])

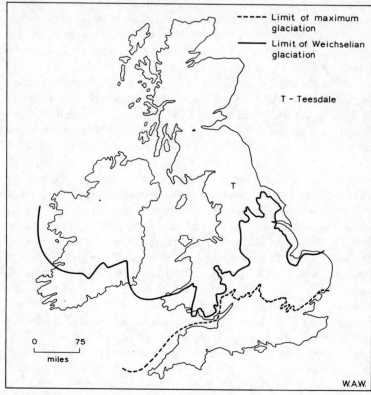

Fig. 6.8 Limits of glaciation in the British Isles

(*Saxifraga nivalis*), and spring gentian (*Gentiana verna*).[115] However, it is still not known whether any species or communities were able to survive in this way, since their existence might also be explained through their being forced upslope from a former much wider area of distribution, either because of climatic amelioration during the Flandrian period to which they could not adjust, or due to the increased competition from other rapidly spreading plants which was characteristic of that time.[116] The question as to how many species could remain in areas of mainland Britain south of the Weichselian ice front is also unresolved, though here West[147] has recently amassed an impressive array of evidence to suggest that a surprisingly diverse flora of open ground plants was, in fact, able to sustain itself under periglacial conditions, and in this there were four main

270

elements: *first*, a northern or montane group, including the previously mentioned mountain aven, the dwarf willow *Salix herbacea*, purple saxifrage (*Saxifraga oppositifolia*), and alpine meadow rue (*Thalictrum alpinum*); *second*, species commonly restricted to seashore sites today, such as sea pink (*Armeria maritima*), sea plantain (*Plantago maritima*), and herbaceous seablite (*Suaeda maritima*); *third*, many weeds, such as knot grass (*Polygonum aviculare agg.*) and creeping buttercup (*Ranunculus repens*); and *fourth*, plants of a more general southern distribution, as for example celery-leaved crowfoot (*Ranunculus scleratus*) and members of the duckweed family (*Lemna* spp.). The existence of these different plant groups may perhaps be accounted for by supposing a wide range of local climates within the broadly overriding tundra setting of the time, some of which may well have been quite warm for at least part of the year. Much less is known about the origins and history of the Lusitanian element of the British flora; however, it has been postulated that temperatures in the extreme west of Ireland may have remained high enough, even at the glacial maxima, to support some frost-tolerant trees or shrubs which may accordingly be pre- or interglacial in origin (e.g., *Erica mackiana*), although others which are markedly frost-intolerant (*Arbutus unedo*) almost certainly did not colonize the area until Flandrian times.

The patterns of relative dispersability in open areas left by the retreating ice are also complex, and not simplified by our inability to distinguish more than a few species among the genera identified by post-glacial pollen analysis. However, some broad indications as to the nature of such colonization in post-Weichselian times have recently been given by Godwin,[62] Matthews[100] and West,[147] and it is possible that these may be applicable to interglacial phases as well. Godwin has estimated that approximately one-half of the present British flora existed south of the ice sheets at the end of the Weichselian, and this consisted mainly of weeds and ruderals which quickly gained a competitive advantage on land left bare by ice melt. Later, as temperatures increased during the Flandrian, trees and other warmth-loving species gradually extended their range from Europe into Britain at the expense of the Weichselian flora, which came to be restricted to ever more limited sites such as sea shores or mountains, where the invading forest species in particular found severe environmental limitations to their growth. As shown in Table 6.4 and Fig. 6.9, often the first trees to colonize were *Pinus*, *Betula*, and *Corylus*,

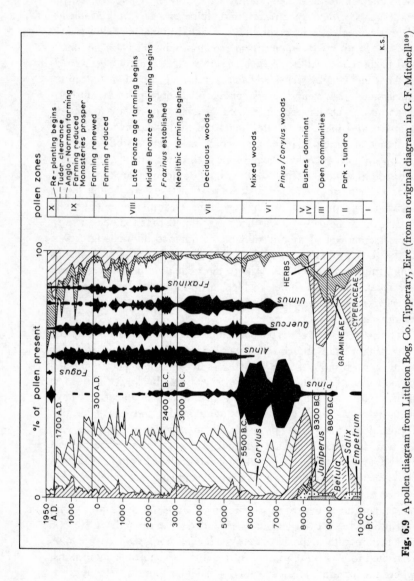

Fig. 6.9 A pollen diagram from Littleton Bog, Co. Tipperary, Eire (from an original diagram in G. F. Mitchell[109])

Table 6.4   Sequences of the late-Weichselian/Flandrian vegetational recolonization in England and Wales (modified from R. G. West[147])

Stage	Subdivision	Date	Pollen zone	General characteristics of vegetation (dominants)
			VIII—modern	Afforestation
	Sub-Atlantic	AD 1000 AD/BC	VIII	*Alnus—Quercus—Betula —(Fagus-Carpinus)*.
		1000 BC 2000 3000	VIIb	*Alnus—Quercus—Tilia* Deforestation. *Ulmus* decline.
	Atlantic	4000 5000	VIIa	*Alnus—Quercus—Ulmus —Tilia*
	Boreal	6000 7000	VI *Pinus-corylus*	c. *Quercus—Ulmus—Tilia* b. *Quercus—Ulmus* a. *Ulmus—Corylus*
			V	*Corylus—Betula—Pinus*
	Pre-Boreal	8000	IV	*Betula—Pinus*
		9000 10 000	III II	Park tundra *Betula* with park tundra
			I	Park tundra

The left side of the table is labeled: **Flandrian** (spanning from Pre-Boreal up through Sub-Atlantic) and **Late Weichselian** (spanning zones I–III).

followed shortly afterwards by elm (*Ulmus*) and oak (*Quercus*). At times, alder (*Alnus*) and willow (*Salix*) were also among the first arrivals. Towards the climatic optimum, a deciduous forest of alder, oak, elm, and lime (*Tilia*), with holly (*Ilex*) and ivy (*Hedera*), had established itself in lowland Britain, while pine had retreated upslope and to the north, reaching its largest areal extent at this time in Scotland. The subsequent climatic deterioration at first led to a decline in the extent of deciduous woodland, and later to podsolization, and the development of wet mire and peat habitats dominated by *Sphagnum* moss in some hilly areas. It was then, too, that many of the late Weichselian species became extinct, being no longer able to adapt to the increasingly wetter conditions; thus, dwarf birch (*Betula nana*) failed

273

to survive in Ireland, although it has continued to maintain itself to the present day in Scotland, where there were many more rugged and better-drained sites to which it could retreat. An additional contemporary development was the first anthropically induced phase of forest removal, initiated on hilly land by small-grain farmers of the Mesolithic and Neolithic periods (see also pp. 348 to 349). But, despite this generalized information, our knowledge is still sadly lacking when the more precise details of these changes are sought. West[147] has noted that immigration rates of most species and genera are often unknown, though he has been able to determine significant differences in the patterns of *Corylus* colonization, whose nuts have been found in deposits of all temperate stages within the Pleistocene period. In this instance, it is clear that in earlier interglacials hazel expanded at a later date with respect to the establishment of oak woods than in Flandrian times, in which it preceded the oakwoods as a separate dominant. The reasons for this differential behaviour are unknown, though they may be associated with corresponding variations in the distance travelled by the species from its refuge, or reflect changing degrees of competition and co-existence with other colonizing species. If similar features are found among other populations, they will obviously give rise to major regional incongruities in the post-Weichselian patterns of plant invasion within Britain as a whole. It is to be noted that the history of contemporary animal recolonization is even less well known, and a great deal of work needs to be completed before its perplexities are fully resolved.

Some areal differentiation in the patterns of plant and animal recolonization may also result from the existence of environmental barriers which can restrict the free movement of species from one district to another. In the British Isles, these are overwhelmingly associated with the severance of Britain from continental Europe, and of Ireland from Britain, as Flandrian sea levels rose eustatically. The result has been that many continental species have not been able to extend their ranges into Britain and Ireland in post-Weichselian times, so helping to preserve a general impoverishment of fauna and flora in both these islands, but especially in Ireland. Thus, only 21 land mammals are present in Ireland today, as opposed to 41 in Britain and 167 on mainland Europe; and there are only 1260 species of flowering plants, as compared to 2200 in Britain. Snakes are completely absent in Ireland, and, to name a few examples only, trees such as lime (*Tilia cordata*), beech (*Fagus sylvatica*), and hornbeam (*Carpinus betulus*), with herbs such as herb paris (*Paris quadri-folia*), have all apparently not succeeded in bridging the water gap, to establish themselves across the Irish Sea.

The intricacy of these interactions between organisms in changing environments is further accentuated when it is considered that they may be affected by events well outside their immediate sphere of influence. Thus, a recent rise in mean annual temperatures has resulted in a large increase in the summer population of oyster catchers (*Haemetopus ostralegus*) in the Faroes and Iceland; however, since there are no suitable local wintering sites for these birds, they are forced to migrate south in the autumn to Morecambe Bay, in north-west England, where they remain until the spring. Here, they have created an intense ecological imbalance in inshore waters by severely overgrazing cockle populations in their search for food—a disturbance which was unlikely to have occurred if the ecosystem had remained completely closed, and so restricted to those species present on a year-round basis.

EVOLUTIONARY ASPECTS OF CHANGE

As organisms are dispersed outside their areas of origin, or as the physical and demographic environments in which they live are changed, they need to adjust themselves to new conditions which may be very different to any they have experienced before; this is achieved through the twin processes of adaptation and evolution, both of which contribute substantially towards the stimulation of change within world ecosystems. Since Charles Darwin's detailed pioneer study of the Galapagos island finches, the general principle has been accepted that most species show a tendency to vary in their morphology and physiology, so that some forms are better suited to their environment than others, and so are much more likely to survive. In other words, the environment acts as a selective screen which eliminates all those individuals which are not adapted to it, and if the screen conditions change the types of organisms which pass through it may also be modified, giving rise to a *natural selection* of species which show a *fitness to survive*. It is inherent in this concept that the species variation which is so important to adaptation processes is the result of genetic traits passed on from generation to generation, the mechanisms of which were first outlined by Mendel's[108] research into the laws of inheritance in pea populations. More recently, attempts have been made to clarify the relationships between genetics and ecology, as they affect community change within ecosystems, by examining the following basic propositions in the synthesizing science of *genecology*:[74, 75,138] *first*, that wide-ranging species may show considerable variation in their morphology and physiology; *second*, that much of this variation can be correlated with habitat differences; and, *third*, that much of the variation is due to processes of natural selection operating within the framework of the genetic material available to the species as a whole. It is these propositions which are to be examined further in the remainder of this chapter.

6.6 Evolution as a reaction to changing environments: general considerations

That evolution is a continuing feature within nature is by now well established. Geological history has shown that, from the emergence of the very first forms of life, periodic outbursts of new varieties and species have occurred, the main stream of which has given rise to the *evolutionary tree* of organic development from which the current multiplicity of life forms has originated. New techniques of analysis, such as that indicated in Fig. 6.10, are now beginning to uncover many of the complex interrelationships of this 'tree'. Much of the development which has taken place therein has occurred as a response to changing environments or modifications in the patterns of life, as when early animals and plants moved from aquatic to land ecosystems. The rate at which it was achieved is difficult to determine, though, given favourable circumstances, it may have been very rapid; certainly at the present time, bearing in mind the general instability of post-Pleistocene climates and ecosystems, and the severity of landscape changes wrought by man, the production of new forms may be very fast indeed.

In order to fully appreciate the effects of environment on evolution, it is first necessary to examine briefly the controlling mechanisms of selection and the present system of species grouping. Dating back to Linnaeus, the latter is based on the delimitation of life forms according to their most significant anatomical characteristics, and the assumption of a hierarchy of *rank names* from which smaller and smaller groups of related organisms can be differentiated. At the apex of the hierarchy is the *empire*, a now-disused term representing the universe, and beneath this, three *kingdoms* (animal, vege-

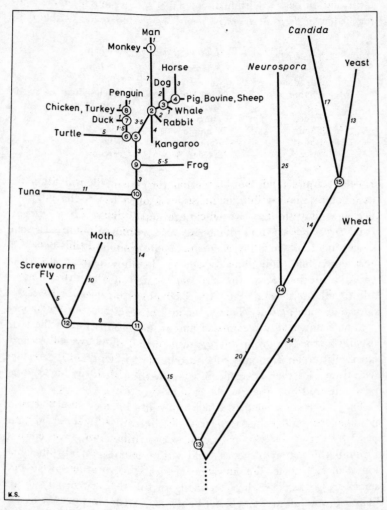

Fig. 6.10 A phylogenetic tree showing the derivation of presentday organisms, constructed on the basis of a computer analysis of homologous proteins of cytochrome C, a complex substance found in similar versions in different species. The computer programmes determine the sequences of the unknown ancestral proteins shown at the nodes of the tree (numbered circles), and the number of mutations that must have taken place along the way (numbers on branches). (From M. O. Dayhoff, *Computer analysis of protein evolution*.[38] Copyright © July 1969 by Scientific American, Inc. All rights reserved)

table, and mineral), with further smaller subdivisions of, in descending order, *phyla, classes, orders, family, genus*, and *species* (Table 6.5).

Table 6.5   The Linnaean grouping of organisms

	Man	Daisy
Kingdom	Animal	Plant
Phylum	Chordate	Tracheophyte
Class	Mammal	Angiosperm
Order	Primate	Asteral
Family	Hominid	Composite
Genus	Homo	Bellis
Species	Sapiens	Perennis

It will be appreciated that, for normal purposes of identification, the use of all divisions within the hierarchy is unwieldy, so that, in practice, it is only the last two which are noted (daisy- *Bellis perennis*; man- *Homo sapiens*), thus giving rise to the common binomial system of classification which has been employed throughout this book. On occasions where individuals of one species show a great degree of variation in their form, the term *variety* may also be utilized as an additional subdivision (e.g., the Barbados sour grass: *Andropogon intermedius* var. *acidulus*). Today, the idea of species as outlined above is almost universally recognized among taxonomists, and the term usually applied to groups of organisms which show overall resemblance, distinction from any other group, and a degree of persistence with time. However, the concept is not without its drawbacks, some of which are to be discussed later in this section.

The inheritance factors which transmit species characteristics from generation to generation were fully identified by T. H. Morgan in 1910.[111] It was he who first realized that inheritance was effected through the action of genes, held within strings of chromosomes found in cell nuclei. For any one species, the number of chromosomes and, therefore, the gene-fixing properties are nearly always identical, and all are capable of being carried from parents to later generations. In sexually-reproducing organisms, identical sets of chromosomes are received from each parent in separate cells termed *gametes*, which then fuse to form a *zygote* possessing twice the normal number of chromosomes (in genetic terms this is a *diploid* cell, with $2n$ chromosomes). In order to retain the standard features of the species, the zygote must divide at a later date in a process termed *meiosis*, in which the double chromosomes are split equally, so creating two identical daughter cells, each with a *haploid* (n) chromosome number. Usually, meiosis occurs just before reproduction, though this is not always the case. Slightly different mechanisms are required for those organisms which do not reproduce sexually (i.e., the vegetative reproducers in plants), in which two identical daughter cells are produced from an original through a more direct form of cell division termed *mitosis*. For sexually-reproducing organisms, it is a matter of chance which of the split chromosomes are passed on to the next generation following meiosis. This situation may be examined

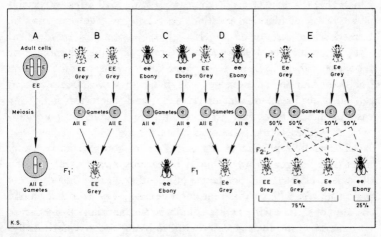

Fig. 6.11 Genotypic mixing in *Drosophila*. (Modified from P. B. Weisz, *Elements of Zoology*,[146] Copyright McGraw-Hill, Inc. New York, 1968. Used with permission.) A, Meiosis, in which gene pairs in adult cells become reduced to single genes in the gametes; B, C, and D, indicate the results of mating by EE × EE, ee × ee and EE × ee genotypes (P) in the filial population (F₁); and case E indicates the results of mating heterozygous Ee phenotypes (F₁) in the filial (F₂) populations. See text for further explanation

further by considering the case in Fig. 6.11 of two *Drosophila* individuals, each possessing a pair of genes producing grey coloration, on pairs of chromosomes in each cell; these may be said to have a grey *phenotype* (or appearance), resulting from a particular *genotype* (EE) which produces that colour. Upon the production of gametes, meiosis occurs, and then a mating of EE × EE genotypes may follow, which, since they are genetically identical for this characteristic, will produce EE offspring of a grey colour. Similarly, if ebony-coloured *Drosophila* with a recessive ee gene structure should mate, only ebony ee offspring will be produced, However, complications arise when the grey EE flies are mated with black ee specimens,

279

in which case, Ee offspring are realized; these will also be grey in colour since E, which is the dominant gene, can still exert its influence, while the recessive e cannot. From matings of the Ee generation, it is clear through the laws of statistics that only 50 per cent of offspring will have Ee genotypes, the remainder being 25 per cent EE and grey, and 25 per cent ee and black, always assuming a random and freely mingling population; in later generations, the patterns of mixing will become even more complete. Genes such as these, which affect the same trait (in this case, colour) in different ways, and which occur at identical locations on a chromosome pair, may be termed *allelic genes* (or *alleles*). If both alleles of a pair are identical (EE, ee), they are *homozygous*, and could be either dominant or recessive; where mixed, *heterozygous* alleles (Ee) consist of one dominant and one recessive gene. It should, however, be noted that not all genes are dominant or recessive, and where this is not the case, allelic mixtures may exert very different influences on organisms, according to the traits held by each gene.

All sexually-reproducing organisms do, of course, receive many more than one gene from each of their parents, and the transmission of each is effected independently of all others, providing the gene pairs concerned lie on different chromosome pairs. In other words, the laws of chance will operate equally in the inheritance of separate chromosomes and their gene contents, so giving rise to a statistically-controlled variation in the transfer of traits from generation to generation. However, any given chromosome pair may itself contain large numbers of genes, ranging *in toto* from a few hundred to a few thousand, and the inheritance pattern of these is somewhat more complex. It may be taken as largely unalterable that *linked* genes within the same paired chromosome are transmitted and inherited together, so that one may expect arrangements of genes in filial populations to be similar to those presented in Fig. 6.12. Occasionally, however, during meiosis, some paired chromosomes may twist around each other and then become broken, after which they can reform in reverse order, a phenomenon known as *crossing-over*. The subsequent rearrangement of genes which results from this will then be passed on to following generations, so giving rise to further genetic variation. This also implies that genes on a single chromosome must be arranged linearly, for only then could crossing-over take place: in practice, the frequency of crossing-over has been found to depend on the distance between each gene on a given chromosome.

These basic, Mendelian features of gene transfer in sexually-reproducing species account for much of the genotypic and phenotypic variation found in ecosystems, especially where the component organisms have complex gene structures, in which case the possibilities for gene recombination through chance or by crossing-over are

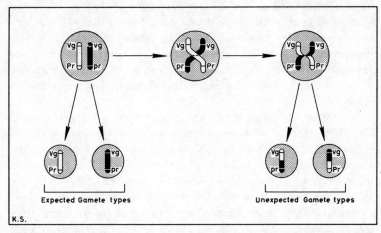

Fig. 6.12 Crossing-over in chromosomes (from an original diagram in P. B. Weisz, *Elements of Zoology*.[146] Copyright McGraw-Hill, Book Co. Inc. New York, 1968. Used with permission)

very great. But they are not the only source of variety, since *mutations* of particular genes can also occur through exposure to harmful radiation (X-rays, cosmic rays, see chapter 3), radioactivity from certain rock formations, or exposure to man-made chemicals, some of which may also be radioactive. Mutations are distributed randomly, at a normal rate of between 1 : 100 000 and 1 : 1 000 000 genes produced, and they are usually recessive, so that their effects are often supressed in ensuing populations; however, they may still be felt when recessive mutations become homozygous, or on the relatively rare occasions when dominant mutations are formed.

All these factors of inheritance form the basis of evolutionary trends which are perhaps best seen in large populations, when the gene characteristics of each individual organism contribute to a comprehensive *gene pool* which is then reorganized and reshuffled within that population many times, assuming that interbreeding takes place freely. This may lead to the development of many new phenotypes which might or might not survive depending on their fitness to do so, as determined by the selective screen of the environment; but if the processes of selection operate in their favour, new permanent phenotypes may be established from them as their traits are passed on to subsequent generations. In instances such as these, natural selection is usually effected through patterns of *differential reproduction*, by which those members of a population best suited to an environment tend to produce more offspring than others which are less capable of surviving. Thus, the process of evolution in any population involves two separate and distinctive steps: *first*, the appearance of a variety of phenotypes, and *second*, the determination of the fitness of these for any particular habitat through reproductive controls.

It may now be convenient to give some further consideration to the effects, extent, and expression of phenotypic variation as viewed within the framework of species classification. The fact is that the single term *species* may be interpreted in a variety of ways. At times, it has been given a morphological basis (a *morpho-species*) with the following essential characteristics: it has the lowest taxonomic rank, as delimited by its morphological features; it is monotypic, with very few variations; and it is static, with no references to change in space and time.[17] On a worldwide scale, this is not an entirely satisfactory view of the species concept, in respect of the phenotypic variation which one might expect to find among all species which occupy a number of very different habitats. Thus, the great tit (*Parus major*), whose range extends from western Europe to China, displays significant changes in coloration from place to place, as indicated in Fig. 6.13; and the same is true for European rabbits (*Oryctolagus cuniculus*) in Tasmania,[9] which have produced unusual black forms at high altitudes and a striking degree of polymorphism in blood group structure. Gates[57] has indicated that changes in the rates of photosynthesis and respiration among plant groups which move to new environments are also effective variations in phenotype, along with more obvious morphological responses. Indeed, most species which are widely distributed may be described as *polytypic*, having many distinguishable subtypes within their group, each of which may be sympatric or allopatric in their areal relationships with others (section 5.11). It was the recognition of these facts which gave rise to a more biologically-oriented definition of species, in which they are described as groups of 'actually or potentially inter-breeding populations which are reproductively isolated from other groups'.[103]

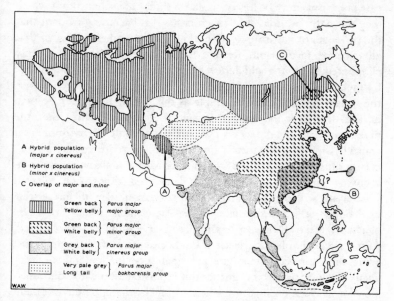

Fig. 6.13 Variations in the great tit (*Parus major*) (from an original diagram in A. J. Cain, *Animal species and their evolution*, Hutchinson, London, 1963)

In other words, species only become species when they can no longer breed with other similar or non-similar forms; they will then differ not only genetically, but also behaviourally, morphologically, physiologically, and ecologically from others. Even this definition may fall down on the infrequent occasions when two such species hybridize when they meet; under these conditions, the species concerned are often incorporated into a *superspecies* group, or they may themselves be categorized as *semispecies*, which could be inherently unstable. Good examples of this may be seen in the case of English oaks (common oak, *Quercus robur* and sessile oak, *Q. petraea*), which are known to interbreed freely under favourable circumstances.

From evidence which has been gathered in many parts of the world (see Darlington,[34] Dobzhansky,[40] Mather,[98] and Stebbins[130,131]), it is now clear that the variations found in polytypic species are frequently created as much through adjustment to a changing environment, or through the extension of their range to new sites, as through gene recombination *per se* or mutation. They may also be triggered off by the inflow of new competitors, in which

283

case organisms may be forced to seek a different niche to that which they normally occupy. In all such cases, it is becoming evident that the processes of natural selection affect first the phenotypes, so that all extreme forms unfit for particular environments are quickly eliminated, following which new genetic recombinations will emerge by means of processes which are not fully understood.[141] However, the situation is not quite as simple as this, for there appears to be a natural feedback mechanism in selection which ensures that the making of new genotypes is only partly the result of environmental control; in effect, this means that sexually reproducing species will offer a range of possibilities for evolution within a changing environment or in particular ecological niches. At times, one tolerant phenotype will be produced, which survives under all foreseeable environmental circumstances, though it is more usually the case that a few relatively non-specialized, or a large number of specialized, forms are available for selection. On occasions, organisms may exist with an unaltered phenotype and genotype in 'lenient' environments for long periods of time, and it is possible that the evolution of new forms often occurs only when a quickly changing environment has reached a non-lenient stage in which species are forced to modify their phenotypes equally rapidly in order to maintain themselves (see also section 5.6). Processes of natural selection may also be intense in exacting environments (e.g.,deserts[144]) and, as Mayr[104,105] has suggested, in the marginal parts of a species' range, where survival is often more difficult than elsewhere. In general, however, the evolution of new forms takes place most effectively when associated with active ecological change, coupled with some instability, conditions which are often displayed when (for example) unsettled niches are being filled by invaders, bare ground is freed from an ice cover, or ecosystems are severely disturbed by man.

It has been proposed by Wilson[150] that the colonization of such areas is often genetically a very difficult process in which very few species can succeed. Those which do are sometimes termed *fugitive species*,[78] adapting to the occupation of a wide range of transient habitats, and frequently exhibiting superior powers of dispersal and reproduction. At times, they may need to *preadapt* to the conditions of a neighbouring site before they can move onto it, as in the cases of the Oxford ragwort (pp. 256 to 257) and the 20 or so imported insects which have become widely established in the USA.[39] It follows that many plant and animal colonizers have a highly flexible genetic system in which, for example, the rates of gene cross-over can be readily adjusted; and many also possess a mating system involving a predominance of self-fertilization, enabling them to produce filial generations by themselves when separated by long distances from their parent assemblages. Allard[1] has shown that the latter is true for several of the world's most successful non-agricultural colonizers, namely pigweed (*Chenopodium album*), wild oats (*Avena fatua*), chickweed (*Stellaria media*), yellow sorrel (*Oxalis corniculata*), prickly lettuce (*Lactuca serriola*), storksbill (*Erodium cicutarium*), wall barley (*Hordeum murinum*), purslane (*Portulaca oleracea*), curled dock (*Rumex crispus*), gallant soldiers (*Galinsoga* spp.), wild radish (*Raphanus sativus*), and English plantain (*Plantago lanceolata*), many of which are, of course, classified as weeds. Other species, such as Monterey pine (*Pinus radiata*), show genetic plasticity in being able to cope satisfactorily with both sedentary and colonizing phases of growth.[8] Slightly different mechanisms may be found among animals, particularly mammals, in which, while adaptation is common, genetic variation may not occur.[76] In all these instances, the rate of selection among colonizing populations is usually at its most intensive during the log phase of growth, when numerical increase may be explosive (see pp. 252 to 253 with respect to Hawaii); then, as numbers even out in a density-dependent way, a much slower, finer pattern of selection takes place.[88]

The processes of evolution which give rise to new forms may therefore be seen most clearly in any ecologically disturbed area; indeed, in such locations, they may be absolutely essential to the continued existence of a species. Where selection is not effective, organisms may be forced into a different niche or into a new environment, or they may adopt behavioural traits which enable them to share the resources of more than one environment. This seems to be the case for several seasonally migrating American landbirds, whose beaks are neither sufficiently large nor variable to ensure survival in year-round competition with resident birds for the available food resources of any one ecosystem.[31] Where none of these reactions are possible, the species may become extinct. Usually, the selection of different forms continues in this way until a new species is evolved, at which point the size of the original gene pool is automatically reduced. However, the casual mechanisms and potential results of such an event are extremely intricate, and are given greater consideration in the next section.

6.7 Phyletic evolution and speciation

In the case of a uniform environment with universally ranging organisms, there is no apparent reason why species should not maintain themselves indefinitely, for little phenotypic variation will occur except that caused by gene recombination, and the chance that new mutations may be passed as dominants or as homozygous forms through subsequent generations is at a minimum. Under these conditions, populations are in genetic equilibrium and the rate of evolution is stable; the gene pool is only reshuffled, and not drastically rearranged.

However, as already noted, this situation does not apply in natural ecosystems where the environment is neither uniform nor static, and where most species are seeking to extend their ecological range, modifying their phenotypes and genotypes as they colonize new territory. Moreover, at least among higher animals, mating usually does not occur on a random basis, so that *selection pressure* on some genes may result, thus altering the genetic emphasis of a population. Group sizes may also vary appreciably, so that, in small populations, the chance selection of certain gene combinations may become established as the norm, thus putting them outside the general rules of random selection—a feature known as *genetic drift*. The consequences of these traits are examined more fully below.

Dobzhansky[40] has suggested that the genetic and environmental influences which give rise to the development of new forms may also, in time, result in the emergence of new species by means of *phyletic evolution* or the gradual production of a new species stock. Mayr[106] has agreed with this supposition in stating that processes such as this appear to be effective over the world as a whole, though the rate at which they take place may vary for particular groups. In all cases, phyletic evolution follows from the *adaptive radiation* of new phenotypes and genotypes, from which certain forms become much more suited to different or changing environments than others, and it is these which ultimately comprise the basis of new species. As a general rule, it is probable that, as van Steenis[140] has tentatively proposed, phyletic evolution is more common among animal groups than plant populations, since the patterns of evolution among the latter seem generally to be much more rapid and erratic (see *speciation*, later); however, it may still have been responsible for the emergence of such broadly based plant formations as deciduous hardwoods which first appeared in the lower mid-latitudes of the northern hemisphere in mid-Cretaceous times, where they were associated with broad-leaf evergreens of tropical affinity in a warm temperate climate. Axelrod[4] has intimated that the hardwoods might represent a contemporary adaptation to the moderate drought which was then beginning to develop in the cooler parts of the year, to give rise to a climate very similar to that existing today in Taiwan, southern Mexico, and Yunan (all areas which still support deciduous woody species, such as *Acer*, *Celtis*, *Prunus*, *Quercus*, *Rosa*, and *Salix*, each of which has evergreen subtropical rain forest relatives).

However, it is perhaps more usual to find that the pattern of species evolution in plants and many animals follows a much more rapid course than this, particularly when small groups of individuals become isolated from their parental gene pool as they are dispersed from their source region. Then, barriers of a geographical, physiological, or behavioural nature may form between the original and the separated group. Usually, geographical separation precedes the development of physiological and behavioural incompatibilities, but this is not always the case, and often the relationships between all three are extremely complex. Simple isolation may be seen in many areas, as in the case of oceanic islands, or when mountains have been raised sufficiently high to form 'islands' of cold climates close to their summits (Fig. 4.15, p. 173). More intricate barriers are known from many ecosystems, as in those occupied by flocks of certain tropical freshwater fish, which live in a state of micro-isolation induced by physical, chemical, and biotic controls, with the latter accentuated by a plethora of differential 'courtship' displays and dissimilar spawning grounds.[90] Separation among birds is achieved if precise mating procedures are not followed, or in frog and mammal groups when calls or smells are not recognized; in plants, it may also derive from differences in the time of effective pollination or, as for red and white campion (*Silene dioica* and *S. Alba*),[5] when pollination is performed only by precisely delimited insect groups.

Whatever the cause of isolation, the effect is to reduce immediately the random nature of gene exchange, and, if this continues for a long enough period of time, two separate and distinct gene pool stocks may emerge, giving rise to new species which can no longer reproduce with each other. This process is termed *speciation*. The mechanisms which induce it have been demonstrated in experiments involving *Drosophila* populations, from which it has become clear that

287

there is a greater chance of it being effective if a large number of small population groups exist—an effect termed the *Sewall Wright principle*, after its originator.[152] Later, Wright modified this view somewhat in proposing that the best conditions for rapid species formation may be found in small, *partially*-isolated populations, in which continuous evolution can occur even if the environment remains constant;[153] this supposition was later substantiated experimentally by Dobzhansky and Pavlovsky.[41] If, in contrast, the isolation is severe, then the gene pool from which new species are evolved may be very limited (technically it need only be that of two parents in the *founder population*), and the rate of evolution may accordingly be much slower, though such groups may still be very successful once they have overcome this 'bottleneck' stage of establishment in new territory.[104] Surprisingly little is known about the time factor in speciation, though it is thought that evolution may take place much more quickly in areas with a large number of niches or where niches are continually disturbed. Some plant groups may certainly preserve a close genetic identity even after several million years apart, as is the case with creosote bush (*Larrea divaricata*) populations of North and South America which first were separated in late Tertiary times.[10, 4] But for most, a period of between 10 000 to 1 million years seems to be sufficient for new species to evolve in this way;[69] thus, Britain has experienced no speciation following its separation from the European continent approximately 10 000 years ago, while Pacific islands with a more ancient history have many examples of it.

If, as suggested above, the rate of speciation increases as more niches become available, then one would expect that tropical associations, with their multi-niche structure, could support a large number of new plant and animal groups, and though some evidence exists to support this contention, the matter is by no means yet resolved. Certainly, there are many more species in the tropics than elsewhere: thus, 222 bird species have been recorded in Alaska, as compared to 286 in California, 477 in Nicaragua, and, as one progresses towards the Equator, 667 in Panama and over 1300 in both Colombia and Venezuela.[104] Most other faunal populations in the tropics are equally diverse, and the complexity of floral associations is well known; moreover, since competition is intense, no one area becomes over-enriched with numbers from any particular group, and most species become closely packed and rather spottily distributed.

Federov[52] has argued that these features are a manifestation of the occurrence of frequent speciation in the humid tropics, as developed in small population groups, all of which show a certain degree of isolation in their discontinuity. He also suggests that the process of self-fertilization, which is the rule in many tropical rain forest species, encourages niche isolation and adds to the number of new forms produced. On the other hand, Ashton,[3] who has worked with Dipterocarps in South-east Asia, has argued against this view in suggesting that the presence of such an immense number of species in the humid tropics dates from a past explosion of forms, when angiosperms invaded a 'vacant' humid environment then characterized by gymnospermous forest, and that diversification since has become an increasingly rare phenomenon, as selection for mutual avoidance in the geological and seasonal climatic stability of these areas has given rise to increasingly narrow specializations, ever smaller niche sizes, and more restricted but increasingly uniform and stable populations.

Whatever the merits and demerits of this case, elsewhere there seems to be no doubt that speciation takes place most effectively in the disturbed niches so often invaded by colonizing species. At times, this may be seen particularly clearly on major *ecoclines*,[79] in which habitat conditions differ continuously along certain planes. Thus, in the perpetually instable ecoclinal chaparral shrub communities of Central California, Nobs[113] has differentiated several closely related and newly evolved species among the bush genus *Ceanothus*, which are infertile in experimental crossings. Similar ecotypic races or semispecies of yarrow (*Achillea millefolium*) and the cinquefoil *Potentilla glandulosa*[23, 24] have also been found in the same area. Briggs[14] has recorded seven mutually infertile new species of *Ranunculus lappaceus* in the 1520–1850 m (5000–6000 ft) zone of the Australian Kosciusko plateau, which are apparently maintained by very narrow ranges of ecological tolerance in an ecoclinal situation. Populations of the grass *Agrostis tenuis* in Central Wales show features which closely resemble this, though on a much smaller scale.[13] It is to be noted that all these species have evolved under conditions in which small populations are not, or were not, completely isolated from the gene pool of a larger population on the ecocline; in other words, these are situations close to the maximum requirements for species differentiation, as outlined by Sewall Wright.

Somewhat slower patterns of speciation may be seen on isolated islands, especially when they are invaded by small alien groups with limited gene pools. It has already been observed that on truly oceanic islands the number of endemics may be considerable. On Hawaii,[44] 90 per cent of the 1729 angiosperms, 3722 out of approximately 6000 insects, 77 birds, and 2 large families of land snails are not found naturally elsewhere. Of these, it is thought that the endemic plants and insects evolved from only about 275 ancestral immigrants, so that it is possible that only one successful introduction per 20 000 to 30 000 years was necessary for this to have taken place.[155] Similarly, only 7 ancient invasions could have given rise to the distinctive landbird fauna. On the Galapagos islands, 600 miles off the coast of South America, the 13 species of finches noted by Darwin have evolved from 1 ground-based species found in South America; yet only 6 now remain as ground-based birds, 6 more are tree finches, and 1 is more like a warbler than a finch, though its internal physiology places it among the finch group.[84] Similar examples to these may be found in many oceanic island groups, both on land and in off-shore waters, as detailed by Carlquist[20] and others. But the reasons as to why there should be such a large proportion of endemics on islands like these need careful consideration, for completely satisfactory answers to this problem have not yet been forthcoming. Part of the solution may be found in the very impoverishment of island ecosystems, which apparently offer less resistance to immigrants than may be found elsewhere. As a result, immigrant species often begin to occupy a wider range of niches and habitats than is possible in their source areas.[92] In itself, this can give rise to explosions of new evolutionary forms, as in Hawaii where non-parasitic bees have at times produced parasitic offspring, and the endemic drepaniid birds have developed major adaptive characteristics since the arrival of their ancestors, so that there are now sickle-billed nectar suckers, grosbeak-like seed crackers, heavy-billed fruit eaters, and sharp-billed insect catchers[155] among them. The number of endemics which form may also partly be a function of age and differential rates of evolution which, of course, can increase substantially if the island ecosystems are severely disturbed by man.

Other examples of speciation through geographical isolation are known from tropical mountains and on desert fringes. Thus, 80 per cent of the afro-alpine group of vascular plants are found only on the high mountains of Ethiopia and East Africa.[73] Twenty-one alpine species of *Senecio*, *Alchemilla*, and *Helichrysum* are endemic in

East Africa, seven generally, nine to Mount Kenya and the Aberdare Range, and five to Mount Kenya alone.[28] Speciation here may have resulted through isolation induced by the breakup of a former related forest as it moved upslope in the post-Pleistocene climatic amelioration, as Moreau[110] has postulated; but, in view of the fact that this would need a former depression of existing vegetation zones by ca. 1000 m, which is unlikely, it is perhaps more reasonable to envisage it as due to the chance long-distance dispersal of a few diaspores from populations with a very limited gene pool, as in the *Lobelia deckenii* group, the main pollinators here being a species of *Nectarina*. Less is known about the evolution of new species in areas isolated by deserts, though some examples are forthcoming, as in Western Australia, where Main[95] has recorded at least three migrations of *Crinus* frogs from the humid east since the early Tertiary period, each of which has given rise to new species.

As usual in natural ecosystems, the patterns of speciation are often nowhere near as simple as those outlined above. For one thing, enduring isolation need not necessarily result in the emergence of new species, for as Mayr has pointed out,[106] parapatric populations* may be formed, such as those present among birds in Australia, Africa, South America, and South-east Asia, and in *Miconia* plant groups of the Northern Range in Trinidad;[129] or hybrid zones may develop in which overlapping semispecies are fairly common. Moreover, recently evolved species may maintain themselves in ecosystems where they are apparently not isolated at all, as in the case of many closely allied groups in Europe, most of which emerged during the period of Pleistocene ice advance, when remnants of larger groups were driven into distinctly different regions (usually the Iberian and Balkan peninsulas) in their search for warmth, and then followed slightly different evolutionary pathways. This is true for the western and eastern nightingales of Central Europe (*Luscinia megarhynca* and *L. luscinia* respectively) which now overlap again in their distribution, though they show no sign of interbreeding or intense competition in that they prefer complementary niches—the one in deer woods and gardens, and the other in swampy woods.[17] However, the same is not true for the similarly evolved western and eastern newts (*Triturus cristatus* and *T. marmoratus*) which overlap in roughly the same area, but display signs of hybridization and inter-

* Parapatric populations are those in non-overlapping geographical contact, existing without interbreeding.

breeding, indicating that speciation is not yet complete. In addition, invasions of species into new habitats may not always create identical effects, and, here, the question of time often assumes importance. To illustrate this point, one may refer to the Canary islands, where the European chaffinch has arrived on two recorded occasions, producing, in the first instance, a new blue chaffinch (*Fringilla teydea*), but, in the second, forms which have not yet had time to be differentiated from the invader *F. coelebs*, except at a sub-species level.

One important additional and instantaneous genetic mechanism may also give rise to new species. This is termed *polyploidy*, and relates to the fact that, on occasions, the process of meiosis does not take place before reproduction, so that cells with twice the normal number of chromosomes may be produced. The result is that a normal haploid (n) cell may eventually fuse with an abnormal diploid ($2n$) one, so forming a triploid cell, with $3n$ chromosomes. Tetraploid ($4n$), pentaploid ($5n$), and other combinations may ultimately ensue. The condition of polyploidy appears to be unusual in animals, except in those species, such as rotifers and some crustaceans, which reproduce parthenogenetically; however, it is extremely common among plants, where about half of the known angiosperms are polyploid, with an even greater proportion among ferns.

The success of polyploidy depends, to a large extent, on the form which it takes. For instance, the triploid is normally sterile, since the chromosomes of one of its three sets will have nothing to pair with, so being randomly distributed among its daughter cells, and giving rise eventually to abnormal forms; however, it can still be reproduced vegetatively by mitosis. Similarly, tetraploids can only be crossed successfully with other tetraploids, or with themselves, for if they are mated with diploids the resultant hybrid will again have $3n$ chromosomes and all the disadvantages of triploids; in this case, since the tetraploid cannot produce fertile offspring when crossed with its parents, it may be classed immediately as a new species. As a general rule, polyploids whose pairs of chromosomes are both multiplied (producing $4n$, $6n$, $8n$, and so on, chromosomes) are fertile and may be reproduced sexually, while others with odd polyploid numbers are not. Occasionally, *aneuploidy* may occur as a result of the differential duplication of single or paired chromosomes, as in the cases of apples, pears, and quinces, all of which have 17 chromosomes derived from an original 7, of which 3 tripled and the remainder doubled. Wheat and rice are also aneuploid, with a basic number of 5 chromosomes, as are several other cultivated plants, including sugarcane, bananas, and Kentucky blue grass.[123] One must also distinguish between *autopolyploids* and *allopolyploids*, the former arising directly from similar parental species, or from a fertile hybrid of different races or species, and the latter when the duplication of a number of chromosomes in a sterile hybrid enables it to become sexually fertile again, which is essentially the case in a species hybrid which has become permanent in its form. Polyploidy of all kinds is generally associated with an increase in size of the species (*gigantism*) which may be best seen in cultivated plants; thus, potatoes, oats, alfalfa, grapes, tea, coffee, and many soft garden fruits and ornamental plants, are autopolyploids, whereas some *Prunus* species, rape, and particularly wheat, are allopolyploids. Herbs and weeds also show polyploidy, an important biogeographical consideration in that the polyploid forms may be better able to act as colonizers in a changing environment than their parents. This has been the case for many species in post-Weichselian Britain, in which diploid races of plants have tended to remain very close to their former southern refuges, whereas their related polyploid forms, which emerged during cold phases, have spread rapidly since the ice melted.[72] The development of polyploidy can induce a change from an annual habit to a perennial one, and also seems to stimulate hardiness in general, so enabling some temperate species to extend their range, more easily than might be expected, into mountain regions[68] or into areas whose climate is rigorous; the latter situation may be seen particularly clearly in the case of lady's smock (*Cardamine pratensis*) in England, the most common form of which in northern districts has a much more complex chromosome structure than that growing in the warmer south.[77]

It may be useful to close these sections on evolutionary processes by emphasizing again the essential differences between the genetic reaction of sexually-reproducing plants and animals to changing environments, since these contribute in no small measure to the functional and structural diversity of ecosystems which has been stressed throughout this book. *First*, many new species of plants are produced through rapid speciation or sudden chromosomal changes involving polyploidy, a condition rarely found in animals; in contrast, phyletic evolution may be much more important to many of the higher animals. *Second*, the tendency is for plants to produce ever more heterogeneous offspring through the mechanisms of genotypic variability, which leads to a great plasticity in phenotypes and increases the chances of survival of seeds which fall upon a wide range of environments. In contrast, animals, through their essential mobility, which enables them to seek out favourable environments, or by means of homeostatic* mechanisms which give them some freedom from environment, usually show much less variation and produce many fewer offspring. *Third*, although some plants are ephemeral, most have much longer lifespans than animals, especially in terms of their reproductive periods.

6.8 The problem of extinction

It is a cardinal rule in ecosystems that as new species evolve, others less fitted to survive become extinct, so as to keep an ecological balance. As in the case of speciation, extinction is more likely to occur when environments are changing very quickly, and populations become very small. Often, it is preceded by the development of a *relic* community, formed of old endemic species; however, this is not to say that all old endemics, or even all relic populations, may eventually become extinct, for some may increase their range later should environmental conditions change in their favour.

While the process of extinction is therefore a natural phenomenon, there is no doubt that recently, particularly following the innovation and dispersal of new intensive agricultural and industrial techniques in many parts of the world since the early nineteenth century, and the associated increased pressure on land as human populations have risen beyond former expectations, the rate of species extinction has been remarkably accelerated. Ziswiler[156] has noted that in the last 300 years, man has completely destroyed over 200 species of birds and animals, most of which were reasonably adapted to the environment in which they lived; and Bouillenne[12] suggests that 600 other species have been brought close to the point of extinction. At times, man may be directly responsible for these losses, as when billions of North American passenger pigeons (*Ectopistes migratorius*) were shot for sport and food from 1850 until the death of the last pigeon in 1914. The American buffalo (*Bison bison*) also came close to extermination through hunting: herds of several million were

---

* The term *homeostatic* refers, in this instance, to the ability of many animals to maintain a relatively stable internal physiological environment.

decimated after the building of railways onto the Great Plains of the USA in the eighteen-fifties, and by 1890 only a few dozen buffalo remained, the descendants of which are now carefully protected. Hunting has also caused a massive reduction in numbers of species such as the fin and sei whales (*Balaenoptera physalus* and *B. borealis*), and also the giant tortoises *Testudo elephantopus* and *T. gigantea*, found respectively on the Galapagos and Seychelles islands. The search for edible eggs, particularly around the shores of the North Atlantic, destroyed the great auk (*Alca impennis*) during the last century and, more recently, has led to a severe decline in populations of the related black guillemot (*Cepphus grylle*) and the puffin (*Fratercula arctica*). Many otters and seals are jeopardized, as also are the cheetah (*Acinonyx jubatus*) in India and the leopard (*Panthera panthera*) in Africa, due to the demand for furs and pelts; and the ostrich (*Struthio camelus*) narrowly escaped extinction in last century's fashionable vogue for its feathers. Many more similar instances may be cited, especially among birds and mammals.

Equally grave indirect results may follow man's interference with natural ecosystems, especially when environments are altered to serve his own ends. As forest was cleared in Central Europe in order to extend the area under agriculture, the auroch (*Bos primigenitus*) in the seventeenth century, and the bear (*Ursus arctos*), wolf (*Canis lupus*) and lynx (*Lynx lynx*) later, all lost their refuge areas, so as to become extinct in the region. Repeated fires in Philippine pine forests have resulted in the permanent removal of undergrowth shrubs and herbs, including *Nepenthes alata*, *N. ventricosa*, and *Lilium philippinensis*; and, in the same region, indigenous species of *Erythrina*, *Pandanus*, *Nauclea*, and others, have been sacrified in the draining of inland swampy and marshy areas.[118] Particularly on islands, evolutionary specializations are often so precise as to make some animal species totally unfit to survive if their immediate habitat is destroyed.[82] This was the case for huia bird (*Heteralocha acutirostris*) populations in New Zealand, whose male and female beak adaptations took very different forms, the male beak being short and tough, and the female long and slender. In their search for food, the male of any given huia pair would employ his beak to break through the bark of certain tree species, and the female would then utilize her beak to penetrate further into the softer, subbark layers so as to pick out palatable insect species which served as their main source of nourishment; accordingly, the significance of these beak dissimilarities lay in the fact that both were used as a single functional unit without which the huias could not survive. Once the tree-insect associations on which they depended were removed by European settlers, the huia beaks proved to be so overspecialized that the birds themselves could not adapt to life elsewhere, so that eventually they became extinct, last having been observed in 1907. Man's introduction of alien predators has also caused the elimination of species which were not capable of defending themselves against the new arrivals, and the effects of this may again most clearly be seen in faunally impoverished island habitats. Thus, pigs and rats brought by Europeans played havoc with the large ground-nesting pigeon of Mauritius, the dodo (*Rhaphus cucullatus*), which did not survive beyond 1681. Rats have helped to exterminate 7 of the 12 species of birds on Lord Howe island,[104] and have destroyed much of the native fauna of many other islands;[65] the introduced mongoose (*Herpestes burmanicus*) and certain aggressive ants have given rise to equally devastating consequences on several Pacific islands and in the Caribbean.

It is important, however, not to lose sight of the fact that man-induced extinction, alarming though it may be in terms of its effects on biological stability (see section 7.10), is superimposed upon naturally occurring patterns of species loss which may lie at the end of pathways of overspecialization, an overabundance of recessive genes, an overall loss in fertility, or a small gene pool from which new variants may be produced. Despite the fact that the relationships between the forces of natural selection and the fitness of an organism to survive are very difficult to evaluate, some generalizations concerning these may nevertheless be made. Usually, the *turnover* of forms is highest in groups of low taxonomic rank, so that extinction of species and genera is relatively common, whereas few orders and fewer classes have been removed; moreover, almost all phyla have survived from their time of origin to the present day. Generally, too, the turnover rates of birds and some mammals are often much greater than among plants, in view of the restricted variability of the former, and the great genetic plasticity of the latter. Moreover, rates of extinction can differ due to locational factors, and as might be expected in view of the small size of populations and the large numbers of minuscule niches, they seem to be relatively high in tropical rain forests. Crowded refuges may, in times of deteriorating climate, also experience a greater loss of species than many other areas, and Mayr and Phelps[107] and Martin and Wright[97] have suggested that this might be the case for bird populations in the mountain rain forest refuges of Guyana during the Pleistocene period. Extinction rates on oceanic islands appear to be particularly high, especially among birds, and because of this evolutionists, from Darwin to Simpson[127] and Darlington,[34] have speculated as to whether island species have genetic factors which presuppose an inability to survive for long periods, particularly when competing with invading species from outside their immediate environment. Other workers, such as Mayr,[103] have designated islands as *evolutionary traps*, in which small populations may become recessively homozygous and therefore much less suited to survival than those with a more randomly distributed gene pool. More recently, however, Mayr has revised this position on the grounds that there are too many exceptions to make this a hard and fast rule.[104] Whether these hypo-

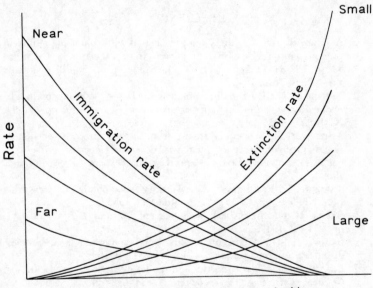

Fig. 6.14 The attainment of faunal equilibria (from an original diagram in R. H. MacArthur and E. O. Wilson, *Evolution*, Vol. 17, pp. 373–87, 1963). As the number of species increases, the immigration rate declines because of the decreasing number of available new immigrants, whereas the extinction rate increases because of the decreasing average population size. The number of species at equilibrium is obtained by dropping the intersection point of a given immigration and a given extinction curve to the abscissa. As the immigration rate increases, or the extinction rate decreases, the equilibrium number increases.

theses are valid or not, the fact remains that though the precise rates
may vary according to the number of species present (see Fig. 6.14),
a very rapid faunal turnover is characteristic of many islands. Thus,
despite New Caledonia's approximate age of 60 million years, there is
only one truly old bird species on the island, the kagu (*Rhynochetus
jubatus*), the others being either newly endemic, or only subspecific-
ally different from their Australian or Papuan neighbours, having
evolved or arrived within roughly the last 50 000 years.[105] Since there
is no reason why this island should not always have supported a
similar number of birds to those currently present, one must assume
that a continuous cycle of colonization, speciation, and extinction
has taken place there. Similar features have also been observed in
New Zealand[54] and on the isolated Pantepui highlands of Venezuela.

REFERENCES

1. ALLARD, R. W., Genetic systems associated with colonizing ability in
   predominantly self-pollinated species. In *The Genetics of Colonizing
   Species*, eds. H. G. BAKER and G. L. STEBBINS, pp. 50–78, Academic
   Press, New York, 1965.
2. ANDERSSON, G., The Swedish climate in the Late-Quaternary period. In
   Die Veränderungen des Klimas seit dem maximum der letzten Eiszeit, pp. 247–94,
   XI Inter. Geol. Congress, Stockholm, 1910.
3. ASHTON, P. S., Speciation among tropical forest trees: some deduc-
   tions in the light of recent evidence. In *Speciation in Tropical Environments*,
   ed. R. H. LOWE-MCCONNELL, pp. 155–96, Academic Press, London,
   1969.
4. AXELROD, D. I., Studies in late Tertiary paleo-botany, *Carnegie Inst.*,
   Pub. No. 590, pp. 1–323, Washington DC, 1950.
5. BAKER, H. G., *Melandrium* (Roehling em.), Fries., *J. Ecol.*, Vol. 35,
   pp. 271–92, 1947.
6. BAKER, H. G., Self-compatability and establishment after 'long
   distance' dispersal, *Evolution*, Vol. 9, pp. 347–9, 1955.
7. BAKER, H. G., Characteristics and modes of origin of weeds. In *The
   Genetics of Colonizing Species*, eds. H. G. BAKER and G. L. STEBBINS,
   pp. 147–72, Academic Press, New York, 1965.
8. BANNISTER, M. H., Variation in the breeding system of *Pinus radiata*.
   In *The Genetics of colonizing species*, eds. H. G. BAKER and G. L. STEBBINS,
   pp. 353–74, Academic Press, New York, 1965.
9. BARBER, H. N., Genetic polymorphism in the rabbit in Tasmania,
   Nature, Vol. 173, p. 1227, 1954.
10. BARBOUR, M. G., Patterns of genetic similarity between *Larrea
    divaricata* of North and South America, *Am. Mid. Nat.*, Vol. 81,
    pp. 54–67, 1969.

11. BOATMAN, D. J., *Mercurialis perennis* in Ireland, *J. Ecol.*, Vol. 44, pp. 587–96, 1956.
12. BOUILLENNE, R., Man, the destroying biotype, *Science*, Vol. 135, 1962.
13. BRADSHAW, A. D., Population differentation in *Agrostis tenuis*, III, Populations in varied environments, *New Phytol.*, Vol. 59, pp. 92–103, 1960.
14. BRIGGS, B. G., Hybridization in *Ranunculus*, *Evolution*, Vol. 16, pp. 372–90, 1962.
15. BROEKER, W. S., Isotope geochemistry and the Pleistocene climatic record. In *The Quaternary of the United States*, eds. H. E. WRIGHT and D. G. FREY, Princeton UP, 1965.
16. BUNTING, A. H., Some reflections on the ecology of weeds. In *The Biology of Weeds*, ed. J. L. HARPER, pp. 11–26, Blackwell, Oxford, 1960.
17. CAIN, A. J., *Animal Species and their Evolution*, Hutchinson Univ. Library, London, 1963.
18. CAIN, S. A., Concerning certain phytosociological concepts, *Ecol. Mon.*, Vol. 2, pp. 475–505, 1932.
19. CAIN, S. A., *Foundations of Plant Geography*, New York, 1944.
20. CARLQUIST, S., *Island Life*, Natural History Press, New York, 1965.
21. CASTETTER, E. F., The vegetation of New Mexico, *New Mexico Quart.*, Vol. 26, pp. 270–2, 1956.
22. CLAUSEN, J., *Stages in the Evolution of Plant Species*, Ithaca, New York, 1951.
23. CLAUSEN J., and HIESEY, W. M., Experimental studies on the nature of species, IV. Genetic structure of ecological races, *Carnegie Inst.*, Pub. No. 615, 312 pp., Washington DC, 1958.
24. CLAUSEN, J., KECK, D. D., and HIESEY, W. M., Experimental studies on the nature of species, III. Environmental responses of climatic races of *Achillea*, *Carnegie Inst.*, Pub. No. 581, 129 pp., Washington DC, 1958.
25. CLEMENTS, F. E., Plant succession: an analysis of the development of vegetation, *Carnegie Inst.*, Pub. No. 242, Washington DC, 1916.
26. CLEMENTS, F. E., Plant indicators: the relation of plant communities to processes and practice, *Carnegie Inst.*, Pub. No. 290, Washington DC, 1920.
27. CLEMENTS, F. E., The nature and structure of the climax, *J. Ecol.*, Vol. 24, pp. 252–84, 1936.
28. COE, M. J., *The Ecology of the Alpine Zone of Mt. Kenya*, The Hague, 1967.
29. COOPE, G. R. and SANDS, C. H. S., Insect faunas of the last glaciation from the Tame Valley, Warwickshire, *Proc. R. Soc. B*, Vol. 1965, pp. 389–412, 1966.
30. COOPER, W. S., The climax forest of Isle Royale, Lake Superior, and its development, *Bot. Gaz.*, Vol. 55, pp. 1–44, 115–40, 189–235, 1913.
31. COX, G. W., The role of competition in the evolution of migration, *Evolution*, Vol. 22, pp. 180–92, 1968.
32. DANSEREAU, P., *Biogeography: An Ecological Perspective*, John Wiley, New York, 1957.
33. DANSEREAU, P. and LEMS, K., The grading of dispersal types in plant communities, and their ecological significance, *Contrib. Institut. Botanique, Univ. Montreal*, No. 71, 1957.

34. DARLINGTON, C. D., *The evolution of genetic systems*, Cambridge, 1939.
35. DARLINGTON, P. J., The origin of the fauna of the Greater Antilles, *Quart. Rev. Biol.*, Vol. 13, pp. 274–300, 1938.
36. DARLINGTON, P. J., *Zoogeography*, New York, 1957.
37. DARWIN, C., *On the Origin of Species by Means of Natural Selection, or the Preservation of Favoured Races in the Struggle for Life*, London, 1859.
38. DAYHOFF, M. O., Computer analysis of protein evolution, *Scient. Am.*, Vol. 221, No. 1, pp. 86–95, 1969.
39. DEBACH, P., Some biological and ecological phenomena associated with colonizing entomophagous insects. In *The Genetics of Colonizing Species*, eds. H. G. BAKER and G. L. STEBBINS, pp. 287–306, Academic Press, New York, 1965.
40. DOBZHANSKY, T. H., *Genetics and the Origin of Species*, 3rd edn, New York, 1951.
41. DOBZHANSKY, T. H. and PAVLOVSKY, D., Indeterminate outcome of certain experiments on *Drosophila* populations, *Evolution*, Vol. 7, p. 198, 1953.
42. DOMIN, K., Is the evolution of the earth's vegetation tending toward a small number of climatic formations?, *Acta Bot. Bohemica*, Vol. 2, pp. 54–60, 1923.
43. DRUCE, G. C., *The Flora of Berkshire*, Oxford, 1897.
44. EGLER, F. E., Indigene versus alien in the development of arid Hawaiian vegetation, *Ecology*, Vol. 23, pp. 14–23, 1942.
45. EKMAN, S., *Zoogeography of the Sea*, London, 1953.
46. ELTON, C. S., *The Ecology of Invasions*, London, 1958.
47. EMILIANI, C., Pleistocene temperatures, *J. Geol.*, Vol. 63, pp. 538–78, 1955.
48. EMILIANI, C., Cenozoic climatic changes. In *Solar Variations, Climatic Change, and Related Geophysical Problems*, ed. R. W. FAIRBRIDGE, *Ann. N.Y. Acad. Sci.*, Vol. 95, pp. 1–740, 1961.
49. EMILIANI, C., Palaeotemperature analysis of the Caribbean cores A 254–BR–C and CP–28, *Bull. Geol. Soc. Am.*, Vol. 75, pp. 129–44, 1964.
50. EVERNDEN, J. F., SAVAGE, D. E., CURTIS, G. H., and JAMES, G. T., Potassium-argon dates and the Cenozoic mammalian chronology of North America, *Am. J. Sci.*, Vol. 262, pp. 145–98, 1964.
51. EYRE, S. R., *Vegetation and Soils*, London, 1963.
52. FEDEROV, A. A., The structure of the tropical rain forest, and speciation in the humid tropics, *J. Ecol.*, Vol. 54, pp. 1–11, 1966.
53. FISHER, R. A., *The Genetical Theory of Natural Selection*, Dover, New York, 1958.
54. FLEMING, C. A., History of the New Zealand land bird fauna, *Notornis*, Vol. 9, pp. 270–4, 1962.
55. FORBES, E., On the connexion between the distribution of the existing fauna and flora of the British Isles, and the geological changes which have affected their area, especially during the epoch of the Northern Drift, *Mem. Geol. Surv. G.B.*, 1., pp. 336–432, 1846.
56. FOSBERG, F. R., Plant dispersal in the Pacific. In *Pacific Basin Biogeography*, ed. J. L. GRESSITT, pp. 273–82, Honolulu, 1963.

57. GATES, D. M., Toward understanding ecosystems. In *Advances in Ecological Research*, Vol. 5, pp. 1–36, 1968.

58. GEER, G. de, A geochronology of the last 12 000 years, *C. R. XI Int. Geol. Congress (Stockholm)*, Vol. I, pp. 241–53, 1912.

59. GEORGE, W., *Animal Geography*, London, 1962.

60. GLEASON, H. A., The individualistic concept of the plant association, *Bull. Torrey. Bot. Cl.*, Vol. 53, pp. 7–26, 1926.

61. GLOCK, W. S., Principles and methods of tree-ring analysis, *Carnegie Inst.* Pub. No. 486, Washington DC, 1937.

62. GODWIN, H., *History of the British Flora*, Cambridge University Press, 1956.

63. GOOD, R., *The Geography of the Flowering Plants*, Longman, London, 1953.

64. GRAY, A., Comparative study of the floras of eastern North America and eastern Asia, *Am. Acad. Arts and Sci. Proc.*, Vol. 21, pp. 1–31, 1873.

65. GREENWAY, J. C., Jr., Extinct and vanishing birds of the world, *Am. Comm. Intern. Wildlife Protection, N.Y.*, Ser. Pub. No. 13, 1958.

66. GRESSITT, J. L. and YOSHIMOTO, C. M., Dispersal of animals in the Pacific. In *Pacific Basin Biogeography*, ed. J. L. GRESSITT, pp. 283–92, Honolulu, 1963.

67. GUPPY, H. B., *Plants, Seeds and Currents in the West Indies and the Azores*, London, 1917.

68. GUSTAFSSON, Å., Polyploidy, life-form and vegetative reproduction, *Hereditas*, Vol. 34, pp. 1–22, 1948.

69. HALDANE, J. B. S., The cost of natural selection, *J. Genet.*, Vol. 55, pp. 511–24, 1957.

70. HARPER, J. L., Establishment, aggression and cohabitation in weedy species, In *The Genetics of Colonizing Species*, eds. H. G. BAKER and G. L. STEBBINS, pp. 245–68, Academic Press, New York, 1965.

71. HARRIS, D. R., Recent plant invasions in the arid and semi-arid southwest of the United States, *Ann. Asso. Am. Geogr.*, Vol. 56, pp. 408–22, 1966.

72. HASKELL, G., Polyploidy, ecology and the British flora, *J. Ecol.*, Vol. 40, pp. 265–82, 1952.

73. HEDBERG, O., Evolution and speciation in a tropical high mountain flora. In *Speciation in Tropical Environments*, ed. R. H. LOWE-McCONNELL, pp. 135–48, Academic Press, London, 1969.

74. HESLOP-HARRISON, J., *New concepts in flowering-plant taxonomy*, Heinemann, London, 1953.

75. HESLOP-HARRISON, J., Forty years of genecology. In *Advances in Ecological Research*, Vol. 2, pp. 159–247, 1964.

76. HOWARD, W. E., Interaction of behaviour, ecology and genetics of introduced mammals. In *The Genetics of Colonizing Species*, eds. H. G. BAKER and G. L. STEBBINS, pp. 461–84, Academic Press, New York, 1965.

77. HUSSEIN, F., Chromosome races of *Cardamine pratensis* in the British Isles. In *British Flowering Plants and Modern Systematic Methods*, ed. A. J. WILMOTT, London, 1949.

78. HUTCHINSON, G. E., Copepodology for the ornithologist, *Ecology*, Vol. 32, pp. 571–7, 1951.

79. HUXLEY, J. S., Clines, an auxiliary taxonomic principle, *Nature*, London, No. 142, p. 219, 1938.

80. IVERSEN, J., *Viscum, Hedera* and *Ilex* as climate indicators, *Geol. Fören. Föhr*, Vol. 66, pp. 463–83, 1944.

81. JOHNSON, F., Half-life of radiocarbon, *Science*, Vol. 149, p. 1326, 1966.

82. KEAST, A., Competitive interactions and the evolution of ecological niches, as illustrated by the Australian honeyeater genus *Melithreptus* (Meliphagidae), *Evolution*, Vol. 22, pp. 762–84, 1968.

83. KENT, D. H., *Senecio squalidus* L. in the British Isles, *Proc. Botan. Soc. Brit. Isles*, Vol. 2, pp. 115–25, 1956; Vol. 3, pp. 375–9, 1960.

84. LACK, D., *Darwin's Finches*, Cambridge, 1957.

85. LADD, H. S., Origin of the Pacific island molluscan fauna, *Am. J. Sci.*, Vol. 258A, pp. 137–50, 1960.

86. LAMB, H., Britain's changing climate. In *The Biological Significance of Climatic Changes in Britain*, ed. C. G. JOHNSON, pp. 3–34, London, 1965.

87. LEMS, K., Population dispersal. In *McGraw-Hill Encyclopedia of Science and Technology*, New York, 1960.

88. LEWONTIN. R. C., Selection for colonizing ability. In *The Genetics of Colonizing Species*, eds. H. G. BAKER and G. L. STEBBINS, pp. 79–94, Academic Press, New York, 1965.

89. LIBBY, W. F., *Radiocarbon Dating*, 2nd edn, Chicago, 1965.

90. LOWE-MCCONNELL, R. H., Speciation in tropical freshwater fishes. In *Speciation in Tropical Environments*, ed. R. H. LOWE-MCCONNELL, pp. 51–75, Academic Press, London, 1969.

91. LYELL, C., *Principles of Geology*, London, 1830–33.

92. MACARTHUR, R. H., Patterns of species diversity, *Biol. Rev.*, Vol. 40, pp. 510–33, 1965.

93. MACARTHUR, R. H., Patterns of communities in the tropics. In *Speciation in Tropical Environments*, ed. R. H. LOWE-MCCONNELL, pp. 19–30, Academic Press, London, 1969.

94. MACARTHUR, R. H. and CONNELL, J. H., *The Biology of Populations*, New York, 1966.

95. MAIN, A. R., Ecology, systematics and evolution of Australian frogs, In *Advances in Ecological Research*, Vol. 5, pp. 37–86, 1968.

96. MANLEY, G., The range of variation of the British climate, *Geogr. J.*, Vol. 117, pp. 43–68, 1951.

97. MARTIN, P. S. and WRIGHT, H. E., Jr., *Pleistocene Extinctions: The Search for a Cause*, Yale University, 1967.

98. MATHER, K., Polygenic inheritance and natural selection, *Biol. Rev.*, Vol. 18, pp. 32–64, 1943.

99. MATTHEW, W. P., Climate and evolution, *Ann. N.Y. Acad. Sci.*, Vol. 24, pp. 171–318, 1915.

100. MATTHEWS, J. R., *Origin and Distribution of the British Flora*, Hutchinson's Univ. Library, London, 1955.

101. MAYR, E., Die Vogelwelt Polynesiens, *Mitt. Zool. Mus. Berlin*, Vol. 19, pp. 307–23, 1933.

102. MAYR, E., The origin and the history of the bird fauna of Polynesia, *Proc. 6th Pac. Sci. Congress, Calif., 1939*, Vol. 4, pp. 197–216, 1941.

103. MAYR, E., *Systematics and the Origin of Species*, New York, 1942.

104. MAYR, E., *Animal Species and Evolution*, Cambridge, 1963.

105. MAYR, E., The nature of colonization in birds. In *The Genetics of Colonizing Species*, eds. H. C. BAKER and G. L. STEBBINS, pp. 29–49, Academic Press, New York, 1965.

106. MAYR, E., Bird speciation in the tropics. In *Speciation in Tropical Environments*, ed. R. H. LOWE-MCCONNELL, pp. 1–18, Academic Press, London, 1969.

107. MAYR, E. and PHELPS, W. H., Jr., The origin of the bird fauna of the southern Venezuelan highlands, *Bull. Am. Mus. Nat. Hist.*, Vol. 136, pp. 269–328, 1967.

108. MENDEL, G., Versuche über Pflanzen-Hybriden, *Verhandlungen des naturforschenden Vereines in Brünn*, Vol. 4, pp. 3–47, 1865.

109. MITCHELL, G. F., Littleton Bog, Tipperary: an Irish Vegetational Record, *Geol. Soc., Am., Special Paper*, No. 84, pp. 1–16, 1965.

110. MOREAU, R. E., *The Bird Faunas of Africa and Its Islands*, Academic Press, London, 1966.

111. MORGAN, T. H., A biological and cytological study of sex determination in phylloxerans and aphids, *J. Exp. Zool.*, Vol. 7, pp. 239–353, 1909.

112. NICHOLS, G. E., A working basis for the ecological classification of plant communities, *Ecology*, Vol. 4, pp. 11–23 and 154–80, 1923.

113. NOBS, M. A., Experimental studies on species relationships in *Ceanothus*, *Carnegie Inst.* Pub. No. 623, Washington DC, 1963.

114. OOSTING, H. J., *The Study of Plant Communities*, 2nd edn, San Francisco, 1956.

115. PIGOTT, C. D., The vegetation of upper Teesdale in the north Pennines, *J. Ecol.*, Vol. 44, pp. 545–86, 1956.

116. PIGOTT, C. D. and WALTERS, S. M., On the interpretation of the discontinuous distributions shown by certain British species of open habitats, *J. Ecol.*, Vol. 42, pp. 95–116, 1954.

117. POLUNIN, N., *Introduction to Plant Geography*, London, 1960.

118. QUISUMBING, E., The vanishing species of plants in the Philippines. In *U.N.E.S.C.O. Symposium on the Impact of Man on Humid Tropics Vegetation*, pp. 344–9, Goroka, Papua, and New Guinea, 1960.

119. REID, C., *The Origin of the British Flora*, London, 1899.

120. RENSCH, B., *Die Geschichte des Sundabogens: eine tiergeographische Untersuchung*, Berlin, 1936.

121. RIDLEY, H. N., *The Dispersal of Plants Throughout the World*, Ashford, Kent, 1930.

122. SAURAMO, M., Studies on the Quaternary varve sediments in southern Finland, *Comm. géol. de Finlande Bull.*, No. 60, 1923.

123. SCHWANITZ, F., *The Origin of Cultivated Plants*, Harvard, 1967.

124. SCLATER, P. L., On the general geographical distribution of the members of the class Aves, *J. Proc. Linnean Soc. London, Zool.*, Vol. 2, pp. 130–45, 1858.

125. SETCHELL, W. A., Pacific insular floras and Pacific paleogeography, *Amer. Nat.*, Vol. 69, pp. 289–310, 1935.

126. SIMPSON, G. C., World temperatures during the Pleistocene, *Quart. Jl. R. Met. Soc.*, Vol. 85, pp. 332–49, 1959.
127. SIMPSON, G. C., *The Geography of Evolution*, Philadelphia, 1965.
128. SIMPSON, I. M. and WEST, R. G., On the stratigraphy and palaeobotany of a late-Pleistocene organic deposit at Chelford, Cheshire, *New Phytol.*, Vol. 57, pp. 239–50, 1958.
129. SNOW, D. W., Fruiting seasons and bird breeding seasons in the New World Tropics, *J. Ecol.*, Vol. 56, pp. 5–6, 1968.
130. STEBBINS, G. L., *Variation and Evolution in Plants*, Columbia UP, 1950.
131. STEBBINS, G. L., Longevity, habitat and release of genetic variability in the higher plants, *Cold Spring Harb. Symp. Quant. Biol.*, No. 23, pp. 365–78, 1958.
132. STEBBINS, G. L., Colonizing species of the native California flora, In *The Genetics of Colonizing Species*, eds. H. G. BAKER and G. L. STEBBINS, pp. 173–96, Academic Press, New York, 1965.
133. STEBBINS, G. L. and MAJOR, J., Endemism and speciation in the California flora, *Ecol. Mon.*, Vol. 35, pp. 1–35, 1965.
134. TANSLEY, A. G., The classification of vegetation, and the concept of development, *J. Ecol.*, Vol. 8, pp. 118–49, 1920.
135. TANSLEY, A. G., The use and abuse of vegetational concepts and terms, *Ecology*, Vol. 16, pp. 284–307, 1935.
136. TANSLEY, A. H., *The British Islands and their Vegetation*, Cambridge UP, 1939.
137. THORNE, R. F., Biotic distribution patterns in the tropical Pacific. In *Pacific Basin Biogeography*, ed. J. L. GRESSITT, pp. 311–54, Honolulu, 1963.
138. TURESSON, G., The scope and importance of genecology, *Hereditas*, Vol. 4, pp. 171–6, 1923.
139. USINGER, R. L., Animal distribution patterns in the tropical Pacific. In *Pacific Basin Biogeography*, ed. J. L. GRESSITT, pp. 255–61, Honolulu, 1963.
140. VAN STEENIS, C. G. G. J., Plant speciation in Malesia, with special reference to the theory of non-adaptive saltatory evolution. In *Speciation in Tropical Environments*, ed. R. H. LOWE-MCCONNELL, pp. 97–133, Academic Press, London, 1969.
141. WADDINGTON, C. H., Introduction to the symposium. In the *Genetics of Colonizing Species*, eds. H. G. BAKER and G. L. STEBBINS, pp. 1–7, Academic Press, New York, 1965.
142. WALLACE, A. R., *The Geographical Distribution of Animals*, London, 1876.
143. WALLACE, A. R., *Island Life: Or the Phenomena and Causes of Insular Faunas and Floras*, London, 1880.
144. WARBURG, M. R., The evolutionary significance of the ecological niche, *Oikos*, Vol. 16, pp. 205–13, 1965.
145. WATTS, D., Human occupance as a factor in the distribution of the California Digger pine (*Pinus sabiniana*), *Univ. of Calif., Unpublished dissertation*, 1959.
146. WEISZ, P. B., *Elements of zoology*, New York, 1968.
147. WEST, R. G., *Pleistocene Geology and Biology*, Longman, London, and John Wiley, New York, 1968.

148. WHITTAKER, R. H., A consideration of climax theory: the climax as a population pattern, *Ecol. Monog.*, Vol. 23, pp. 41–78, 1953.

149. WILMOTT, A. J., Evidence of survival of the British Flora in glacial times, *Proc. R. Soc.*, *B*, Vol. 118, pp. 197–241, 1935.

150. WILSON, E. O., The challenge from related species. In *The Genetics of Colonizing Species*, eds. H. G. BAKER and G. L. STEBBINS, pp. 8–28, Academic Press, New York, 1965.

151. WISEMAN, J. D. H., Deep sea caves as an aid to absolute dating in the Quaternary period, *Report VIth Int. Congr. on Quaternary*, Vol. I, pp. 743–66, 1961.

152. WRIGHT, S., Evolution of Mendelian populations, *Genetics*, Vol. 16, pp. 97–159, 1931.

153. WRIGHT, S., Breeding structure of populations in relation to speciation, *Am. Nat.*, Vol. 74, pp. 232–48, 1940.

154. ZAGWIJN, W. H., Zur stratigraphischen und pollenanalytischen Gleiderung der Pliozänen ablagerungen im Roertal-Graben und Venloer Graben der Niederlande, *Fortschr. Geol. Rheinld. u. Westf.*, Vol. 4, pp. 5–26, 1959.

155. ZIMMERMAN, E. C., Pacific basin biogeography: a summary discussion. In *Pacific Basin Biogeography*, ed. J. L. GRESSITT, pp. 477–84, Honolulu, 1963.

156. ZISWILER, V., *Extinct and Vanishing Animals*, New York, 1967.

7

Man in Ecosystems

7.1 Introduction

Although the origins of man are lost in the mists of prehistory, they may certainly be placed within the broad limits of the Mio-Plio-Pleistocene periods, when biological and environmental instability induced by mountain-building, vulcanism, climatic change, and relatively rapid species evolution were the general rule. His phylogenetic ancestors are little known, nor are the forms which were brought to extinction to make way for him. Probably his appearance coincided with the loss of a variety of other large but less adaptable primates, following which there might well have been a wide range of vacant niches which he could occupy. Once established and settled therein, he proved to be extremely well fitted to survive in the changing physical milieu, so much so that he has since become the dominant organism in most ecosystems. This fact, coupled with the biological and cultural complexity of his existence, is sufficient reason to accord him a separate chapter at the end of this book.

The patterns of development of the genus *Homo* from emergence to predominance are by no means easy to follow. Although the investigations by Leakey and his coworkers at Olduvai Gorge in Tanzania[98] have made it clear that early tool-using and tool-making man (*Homo habilis*) can be dated back to at least 1·75 million years ago, and it is probable, though by no means proved, that as a species he came from Africa, there are still considerable gaps in our knowledge of his subsequent history, particularly with respect to the modes of inception of his obvious polytypy, and the chronology of his dispersal (or repeated dispersal) onto other continents. It is also not yet

306

determined precisely how and when he was first able to transmit learning or culture, through which means he could gradually acquire more and more skills to safeguard his family, augment his numbers, and utilize his surroundings to make his life more comfortable. But, eventually, and at first very slowly, he gained a sufficient technological capability to assume a greater ecological preeminence, by steps which were always intricate and sometimes extremely obscure. As Sauer[126] has so explicitly put it, he began through the course of time to intervene in the environment 'with or without design, to increase and decrease, to expel or exterminate and to introduce, to modify and even to originate organic entities'. In general, he has profited from these disturbances, though, at times, catastrophic consequences have followed from his endeavours, as when severe soil erosion was induced or when floods devastated areas largely denuded of their natural vegetation. That these latter instances are more often regarded as 'acts of God' than 'acts of man' is proof of the ever-greater acceptance throughout history of an anthropocentric view of the biosphere, which grants recognition to the idea that there are few or no natural limits to man's economic and cultural activities. However, any such notion is inherently self-delusory in that it fails to take account of the all-important fact that man is not only a social animal, but also a biological one, whose ultimate bounds are those which control the lives of all other organisms within ecosystems; in other words, they are those of the genetic framework which he inherits and the physical and biological environment within which he lives. Since it is the contention of the author that one of the central themes of biogeography should be to encourage the integration of studies concerned with these two facets of man's heritage, the first part of this chapter will give some consideration to certain major environmental influences which are known to affect his well-being, and the second part to the cultural achievements which have contributed most substantially towards the modification of his habitat. The chapter, and the book, will then conclude with a brief résumé of the present ecological status of world ecosystems, and the current trends of change which may be detected within them.

SOME ENVIRONMENTAL RESTRAINTS

Man is a complex, multicellular polyploid organism. Within his body, he has roughly 135 million cells, each of which has the same basic structure as that outlined in chapter 3. In each cell nucleus, there are 46 chromosomes, though this number may be modified under certain conditions, all mammalian chromosomes being very difficult to fix in the process of meiosis. As is the case for most other animals, his rate of reproduction is strictly limited and his life relatively short, particularly when compared to that of many plants. His genetic structure ensures that he is one of the least specialized of organisms, and this has contributed towards his adaptability and his ability to maintain himself in a wide variety of environments. However, despite this, it is probable that during his early history, he survived only through utilizing the old animal trait of seeking favourable locales; then, with time, his increasing technology allowed the extension of his range through the more efficient use of tools and fire, and the creation of artificial living quarters focused on hearth and home. More recently, and particularly following the industrial and agricultural revolutions of the eighteenth and nineteenth centuries, the number and scale of these has been remorselessly augmented to accommodate an accompanying population explosion, notably in towns, cities, and conurbations. Man has thus become a cosmopolitan and self-protecting species on a scale never known before, and under these conditions it is even more imperative than formerly to understand the nature of the environmental constraints which might still affect him, or conceivably react against him.

7.2 Human origins and diversity

At present, one must turn to Africa for the best indications as to the patterns of man's emergence as a separate species. Here, fossil remnants of several distinctive forms of the family Hominidae (hominids) have been recovered from Lower and Middle Pleistocene strata. These are usually divided on a morphological basis into those located within the subfamily Homininae (members of the genus *Homo*) and those manlike primates which lie in the subfamily Australopithecinae (members of the genus *Australopithecus*). Several representatives of the family of anthropoid, or 'great' apes (the Pongidae) were also present in relatively close areal proximity to the hominids at this time.

The distinguishing features which set apart australopithecines from early man are most clearly seen in their skull structures. *Australopithecus* differed from *Homo* in having larger cheek teeth, a more massive cheekbone skeleton, and a much smaller brain capacity in a body which, at least for some forms, may well have

308

been very similar in size to that of early man. To date, the brain size of recovered adult australopithecine skulls has fallen within the range of 360 to 640 cm³, with an estimated mean value of about 500 cm³ as opposed to a mean of 800 cm³ in early man, and over 1400 cm³ in modern man. A further broad dissimilarity between the two subfamilies lay in the fact that, while both had an erect bipedal gait, that of *Homo* seems always to have been much freer, so enabling him to travel across much greater stretches of terrain than was ever the case for the australopithecines. In turn, the latter could be separated from the forest-oriented anthropoid apes in terms of their marked bipedalism, their somewhat larger brains and generally smaller bodies, and the reduced size of their front teeth.

In a recent study which tabulates, summarizes and classifies all the morphological data so far recovered from early hominid fossil fragments, Tobias[141] has argued that three definitive species of *Australopithecus* can be delimited within the subfamily Australopithe-cinae.* *A. africanus*, Dart, which is known from both eastern and southern Africa, may lie close to the ancestral line of hominids, in that it has relatively non-specialized features, including a cranium which is light in construction, a dental arrangement which comprises only slightly enlarged canines, and moderately enlarged cheek teeth. In contrast, *A. boisei*, Leakey, found only in Olduvai Gorge in East Africa,[98] where it was discovered in the same Lower Pleistocene stratum as *Homo habilis* (see p. 311), has much more massive cheek teeth and a more robust skull structure with a heavy cranial construction, inferring a rather more specialized form which has deviated from the main evolutionary line. The third species, *A. robustus*, Broom, which has only been recovered from the early Middle Pleistocene, also shows some intensive specialization, with massive cheek teeth, and a skull which, while still robust, is not quite

---

* It should be emphasized that the classification of Hominidae used within this chapter is not recognized by all authorities. Each of the three australopithecine species has, at times, been represented as a separate subgenus or genus: thus, *A. africanus*, Dart, has been referred to as *Australopithecus* sp.; *A. boisei*, Leakey as *Zinjanthropus* sp.; and *A. robustus*, Broom, as *Paranthropus* sp. Tobias's view that all should be placed inside one single genus was foreshadowed by a similar proposal, put forward at an earlier date by Washburn and Patterson,[148] which was based on a much smaller array of evidence than is currently available; this was later supported by Oakley,[112] Le Gros Clark,[34] and the Leakeys.[97] If Tobias's ideas come to be generally accepted, then, of course, there would be no further need to specify two sub-families (Australopithecinae and Homininae) within the Hominidae as presented herein.

so heavy as in the case of *A. boisei*, Leakey. Remains of *A. robustus*, Broom have been collected from widely scattered sites in both Africa and South-east Asia.

The cranial capacity of australopithecines suggests that they might have been able to develop a crude technology. Although there is no evidence that any of them could control fire, it is likely that some were able to utilize tools and weapons on a fairly large scale. Indeed, this trait in itself may not have been exceptional in that several other less intelligent animal groups are known to employ similar aids in a variety of ways. Among non-primates, one of the Galapagos finches (the cactus finch *Cactospiza pallida*) uses a stick or pebble grasped in its beak to smash shells more easily. In pongid communities, orangs may throw sticks to ward off intruders, and chimpanzees often carry water within leaves or break nuts with suitably-shaped rocks.[147] But while such activities are usually only marginal to the customary daily routine of these creatures, australopithecines seem, in contrast, to have depended on tools and weapons to a much greater extent and, perhaps, even needed them to ensure their survival in the somewhat dangerous open or semi-wooded environments in which they lived, where weaponry, in particular, was required both to kill animal predators and to facilitate the capture of animal food as a supplement to that obtained from plant sources.* From his examination of massive piles of broken debris on the living sites of *A. africanus*, Dart at Makapansgat in South Africa, in which he found very large numbers of discarded bones (and also horns, stalactites, sticks, and even well-shaped stones), the length and shape of which were so similar that their presence could only be explained by invoking their selection for some practical purpose, Dart[49] was the first to adduce a major tool-using capacity among these hominids. However, there are no indications that any australopithecine had the technological ability to *make* tools, so that one must assume that they lived off the land in a very primitive way, probably in small, relatively mobile groups, their total numbers being determined to some extent by the amount of edible food which they could collect or hunt (and then eat raw) within the confines of the territory which they occupied.

Even though the degree of australopithecine adaptation to a number of different ecological niches was quite high as compared to the anthropoid apes, it was not unlimited, and in any case was

* It is to be noted that, like man, australopithecines are now thought always to have had a largely omnivorous diet.

nowhere near as great as that of early *Homo*. The first identifiable fossil traces of the latter come from basal beds in Olduvai Gorge, Tanzania, where Leakey[98] in 1961 discovered a skull of a pygmy-sized hominized being which he termed *Homo habilis*, since it lay amid a debris of stone tools (small primitive choppers and flake tools) which showed positive signs of having been worked. Although this man was a contemporary of *A. boisei*, Leakey, having lived 1·75 million years ago, he appears always to have had a much greater capacity for innovation and learning, for, in addition to making tools, it is possible that, even at this early date,[96] he knew how to build small huts on fairly open sites. Indeed, once *Homo* had established himself, possibly on dry forest edges or in wooded savanna close to water, he proved himself able to take over a wide variety of empty niches, and, admirably fitted to survive in these, slowly assumed ascendency over the australopithecines, though the latter continued to maintain themselves on favoured sites for a considerable period.

It is now generally considered that the phylogenetic origins of hominid species can be traced back well into the Pliocene or even to the Miocene period. This idea has received some encouragement from the fact that a skull of 15 million years ago found in the Siwalik Hills of India (*Ramapithecus punjabicus*), and a Miocene skull from Africa (*Kenyapithecus wickeri*, Leakey), have both recently been shown by Simons[131] to have dental and facial features which are almost identical to *Australopithecus africanus*, Dart, a species which, as previously stated, may lie very close to the main evolutionary line of australopithecines. From this, one may deduce that the early protohominids of the Mio-Pliocene periods were very similar in appearance to the much later, relatively unspecialized, small-toothed australopithecines. Subsequent stages in the evolution of this family have recently been reexamined by Tobias.[141] He proposes that towards the end of the Upper Pliocene, or at the beginning of the Pleistocene, the group diverged into several distinctive lines or *clades*. This may not have been an unusual occurrence, in that the small, semi-isolated populations of the time could easily have fitted into a pattern which Sewall Wright has described as being an optimum situation for the evolution of new forms (see pp. 287 to 288); moreover, the changing environmental milieu which was characteristic of these geological periods might have stimulated the intensive and rapid selection of certain new clades at the expense of others. Accordingly, one can envisage the eventual emergence of a large-toothed species (*A. boisei*, Leakey) which, at a later date, might have given rise to the even more specialized *A. robustus*, Broom. In contrast, the main line of australopithecines would always have remained relatively stable and little changed. Divergence from this in other directions could easily have produced more manlike forms, from which *Homo habilis* was ultimately derived. One provisional representation of this possible sequence of events is depicted in Fig. 7.1.

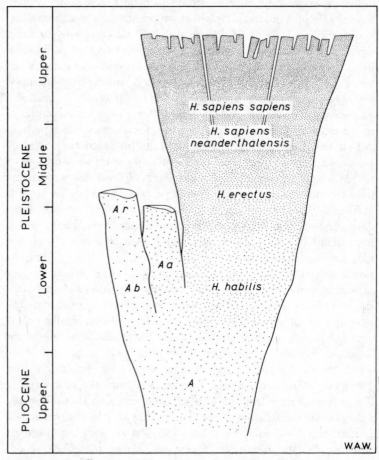

Fig. 7.1 A tentative scheme of hominid phylogeny from Upper Pliocene times to the Upper Pleistocene. Progressive degrees of hominization are represented by progressively darker shading. A, postulated ancestral australopithecine, comparable with *Australopithecus africanus* (Aa); Ab—*Australopithecus boisei*; Ar, *Australopithecus robustus* (from an original diagram in P. V. Tobias [141])

Once early man had become completely separated from the australopithecines,* the chance of further speciation then diminished in that the total gene pool of each subfamily was immediately reduced, although some minor facets of selection continued to operate, as, for example, in the choice of larger human phenotypes with ever greater cranial capacities. The latter could well have been encouraged particularly by the periodic emigration of groups outwards from Africa into other regions, once they had filled all the locally available ecological niches. It is still not certain how quickly this was achieved, though one may be sure that as a hunter, man must have been fairly mobile; indeed Birdsell[18] has estimated that successful lines of hominids and early man might have taken only 23 000 years to disperse from Africa to south-eastern Asia. As populations then became settled in radically different environments, they might easily have reacted subconsciously in the selection of better-suited forms, such as *Homo erectus* (Java man and Peking man), a tool-making, fire-using man of over 500 000 years ago. One or more of these eventually gave rise to the lineage of modern man (*Homo sapiens*), among whose European predecessors were Neanderthal man of 70 000 years ago, and Cro-Magnon man of 35 000 years ago.†

That minor patterns of selection still take place in human groups cannot be disputed, though often their precise mechanisms have not been fully demonstrated.‡ Some well-documented though scattered examples of the efficacy of negative selection in human populations have been traced, as in the case of the haemophilic gene. But the agencies of positive selection which have given rise to the major morphological phenotypes of man are often too complex, or their genotypic expressivity too variable, to allow for their full direct examination. Accordingly, one may often only impute a selective

* It is still not clear whether only one species of man evolved, as Darwin[50] postulated, or whether mankind arose from a multispecies structure, as Wallace[145] believed.

† Although past custom has assumed the existence of three distinct species of man in the evolutionary line from *Homo habilis* to *Homo erectus* to *Homo sapiens*, it is probable that these are distinguishable only in the light of their time-separation, and that in reality they are all merely minor phyletic variants of one broadly based species.

‡ It is unlikely that the present wide range of polymorphism in man can be explained through local mutation, for, although some populations such as those in Kerala, India (where soil radioactivity is particularly high) have been subjected for many years to above-normal rates of mutation,[20] it is only rarely that genes altered in this way can be passed on effectively to subsequent generations (see also p. 281).

313

basis for the apparently close relationships which exist between certain phenotypes and their environment. Thus, Bergman[15] and Allen[4] have postulated that genetically controlled 'ecological laws' can lower the ratio of a body's weight to its surface area in regions where, at least for certain times of the year, there is a need to eliminate excessive body heat, and Newman[109] has argued that this is particularly the case among Amerindian populations. As far as the major race variants of *Homo sapiens* are concerned, Cowles[43] has proposed that the highly pigmented skin of Negroes, Abysinnians, Hindus, and Polynesians may have been a result of natural selection operating in its favour under the intense radiation and heat income of tropical rain forest conditions, while, in contrast, the lighter pigmentation of northern populations is more suited to less intensive light rays, in that it allows these to penetrate the skin deeply enough to provide the body with adequate amounts of vitamin D which otherwise would not be forthcoming. Similarly, the long narrow noses of Caucasians are thought by some to be a genetic response to the need for extremely cold winter air to be warmed before passing into the lungs, whereas the short stubby noses of negroid peoples are characteristic of habitats where this is not necessary. However, other authorities see no evidence for any of these suppositions, and their frailty is emphasized by the fact that even the age of many phenotypes (and therefore the speed of selection) is contested, some researchers suggesting a very recent origin, and others, such as Coon,[40] a somewhat older one in most cases.

The mechanisms of selection in human populations, framed as they are within an immense variety of phenotypes, may, perhaps, best be clarified if one envisages the world as comprising a theoretical complex of differentially-sized swarms of people, most of whom are semi-isolated in the Sewall Wright sense (see pp. 287 to 288); these then would automatically tend to produce distinctively-adapted local phenotypes through genetic drift in relatively limited gene pools, while, at the same time, they would still be able to give rise to the maximum possible number of new variants. This view may form a close approximation of the facts of human existence, for in many parts of the world, even today, small town, village, or hamlet populations are, or have been until very recently, almost cut off from their neighbours through poor communications, so that they have become very tightly knit and ingrown social units. Other, more widely distributed sedentary groups, such as colonial settlers, have often behaved as semi-isolates in their tendency to marry within their own social class, and this is also true of many nomadic populations (e.g., gypsies). Usually, too, the three major racial lines (Caucasoid, Negroid, Mongoloid) and their subgroups (e.g., the Mediterranean, Alpine, and Nordic divisions of the Caucasoid line) act as disparate large-scale swarms which do not mix readily with each other. Within all of these, people may choose mating partners very selectively for complex psycho-social reasons, so creating further potential limitations to the gene pool. These facets of life might have been especially significant in prehistoric times, when all human groups were both numerically small and unlikely to have too close a degree of contact with each other; moreover, at this time, isolation was occasionally accentuated through the decimation of populations by disease and war. Some may even have come close to extinction and, upon recovery, could then have created populations with an especially restricted gene pool. From time to time, the repeated separation of smaller units from a larger group might also have been a very important subconscious procedural technique in the production of early variants, particularly when such units chose to move into radically new environments where they could multiply and evolve quickly. In this case, a scattered *mosaic distribution* of different new forms is always likely to follow. Hiernaux[77] has suggested that, in the past, this has been significant in many areas and, more recently, has accounted for much of the polymorphism found in Bantu tribes today. Of course, if individuals from any of these distinctive groups should intermingle on a reasonably large scale, many additional vigorous and successful hybrid lines will tend to emerge at a very rapid rate, as is the case currently in Hawaii and Trinidad, where Negroes, Mongoloids, and Caucasoids of different social backgrounds are interbreeding fairly freely.

A certain amount of evidence to support these contentions has already been collated. Thus, a very great degree of gene variability is known to have developed recently among several small groups which are now semi-isolated from each other, as in the case of the Dunkers of Pennsylvania.[63] Studies of antigene properties within blood streams, from which most of the proof of the mechanics of polymorphism in man has so far been obtained,[20] have tended to confirm these opinions. It is common knowledge that the relative frequency of each of the four major blood groups (A, B, AB, O) varies areally among populations of different phenotypes, both on a

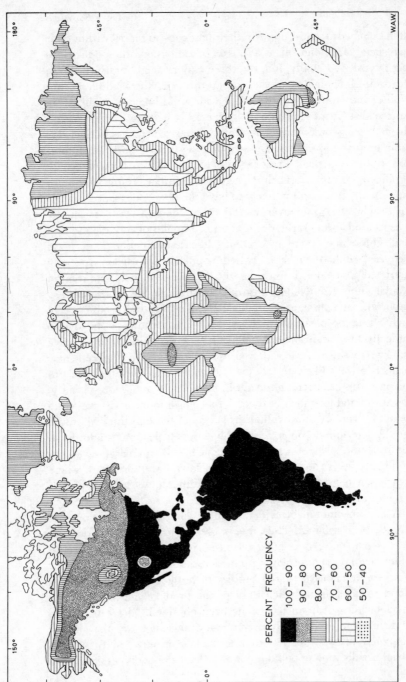

Fig. 7.2 The generalized distribution of the gene giving rise to blood group O (modified from an original diagram in A. E. Mourant, A. C. Kopec and K. D. Sobezak, The ABO Blood Groups, *Occasional Publication of the Royal Anthropological Institute*, No. 13, Blackwell

PERCENT FREQUENCY

100 − 90
90 − 80
80 − 70
70 − 60
60 − 50
50 − 40

continental or subcontinental scale and more locally. Over the world as a whole, most people have a mixture of O, A, and B alleles, though a majority of Amerindians from the MacKenzie valley to Tierra del Fuego belong only to blood group O. In contrast, Eskimos do not have particularly high O percentages, nor do any populations of the Old World, including tribes from north-eastern Asia, from which Amerindians are presumed to originate (see Fig. 7.2); in fact, some desert Bedouins and Copts have the highest Old World percentages of this group. Generally, the frequency of blood group B diminishes along a marked and continuous gradient from a peak in Central Asia and India, to zero in Australia, the Pacific, and parts of North and South America (Fig. 7.3). However, some curious anomalies exist in the Americas, and this is particularly the case among the Yaghan Indian tribes of Tierra del Fuego, South America, in which the highest world values of B by a large margin have been recorded. In Africa, the distribution of B alleles is confused and one finds both very high and very low frequencies here. Extreme values of blood group A (Fig. 7.4) are also a feature of the Americas, where some South American Indian populations (especially in the Argentine, Paraguay, and Peru) supply the only examples from which no A alleles have been recovered, while, in contrast, the highest known incidence of this group comes from the Blood, Blackfoot, and Piegan tribes of North America. A-allele distribution is further complicated by the fact that this blood group is usually divided into two (A_1 and A_2); Amerindians, Polynesians, and Eskimos have an A_1 allele only, while all other populations possess both, A_2 always being less abundant than A_1 (Table 7.1). Similar though

Table 7.1 Relative frequencies of A_1 and A_2 blood-group alleles in selected human groups (from A. S. Wiener, Blood groups and transfusion, 1943. Courtesy of Charles C. Thomas, Publisher, Springfield, Illinois)

Group	Ratio A_2/A_1	Group	Ratio A_2/A_1
American Indians	0·0	Russsians	0·31
Hawaiian Polynesians	0·0	Danes	0·34
Eskimos	ca. 0·0	Finns	0·37
English	0·23	USA (negroes)	0·37
Egyptians	0·28	Basques	0·39
USA (white)	0·30	Irish	0·46

less distinctive patterns of areal variation may also be seen if one studies the frequency of occurrence of M-N blood types, in which more than a 75 per cent frequency of the M allele is found among

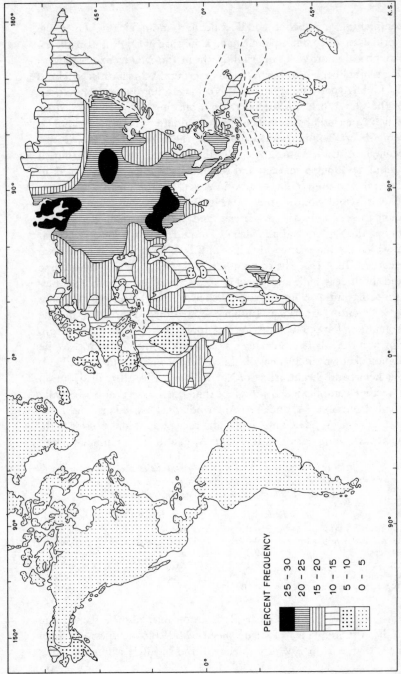

PERCENT FREQUENCY

⬛	25 – 30
	20 – 25
	15 – 20
	10 – 15
	5 – 10
	0 – 5

Fig. 7.3 The generalized distribution of the gene giving rise to blood group B (modified from an original diagram in A. E. Mourant,

K.S.

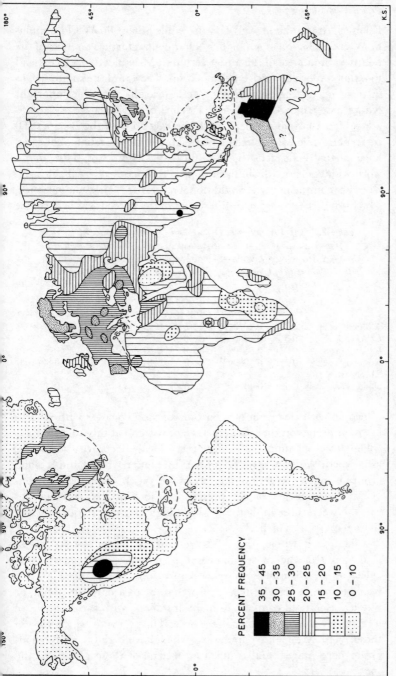

Fig. 7.4 The generalized distribution of the gene giving rise to blood group A (modified from an original diagram in B. Lundman[100])

PERCENT FREQUENCY

35 – 45
30 – 35
25 – 30
20 – 25
15 – 20
10 – 15
0 – 10

K.S.

Eskimos, Amerindians, and Hindus, while other Old World populations tend to have a much higher N frequency; further analyses of the
relative abundance of the anti-S serum in M-N blood types can lead
to a much more definitive delimination of certain human groups, as
in those occupying New Guinea and Australia. For the important
Rh-antibody blood types, the incidence of Rh-negative streams is
always less than 50 per cent, with the exception of the Basques (60
per cent), while Mongoloids, Polynesians, and Australian aborigines
have virtually no Rh-negative antibodies. Some of these major
worldwide genetic dissimilarities are summarized in Table 7.2.
They are complemented by additional patterns of genotypic variance,
which give rise to regional differences in the percentage occurrence

Table 7.2   Approximate gene frequencies in six serologically defined races
(from W. C. Boyd[20], The contributions of genetics to anthropology. In
Anthropology Today—Selections, ed. Sol Tax, copyright ©
The University of Chicago Press, Chicago, 1962, used with permission)

	Early European (Basque)	European (Caucasian)	African (Negroid)	Asiatic (Mongoloid)	Amerindian	Australian
A blood group	ca. 0·25	0·20–0·30	0·10–0·20	0·15–0·40	0–0·60	0·10–0·60
A_2/A_1 ratio	0·5	0·10–0·30	ca. 0·40	0	0	0
B blood group	0·01?	0·05–0·20	0·05–0·25	0·10–0·30	0	0
N blood type	0·5?	0·30–0·50	ca. 0·50	0·40–0·50	0·10–0·20	0·80–1·00
Rh-negative blood type	0·5?	0·40	ca. 0·25	0	0	0

of people with (for example) bass or soprano voices,[16] with whorls,
loops, or arches on their fingerprints, or with an ability to taste PTC
(phenylthiocarbamide).[136] More precise local effects are also known,
as in North Wales, where the proportion of men with O and B blood
groups is much greater among those with Welsh, as opposed to non-
Welsh surnames;[120] or, in Scotland, where positive correlations
exist between those individuals with a high O group frequency, and
those who speak Gaelic.[24]

Taken as a whole, these features suggest strongly two things.
First, some of the most important phenotypic differences which
characterize dissimilar human populations may, indeed, have
resulted from a certain degree of present or past cultural, social, or
physical isolation, coupled, in many instances, with the existence of
a relatively limited gene pool resource. This seems to be the case in
those Welsh and Celtic populations which have high O and B blood-
group percentages, and could also be true of those groups in the

Americas which show a remarkable incidence of extremely pure blood groups. The latter situation can possibly be explained by invoking the successive arrival from Asia over a long period of time of disparate populations whose size was initially very restricted, and whose isolation was subsequently preserved to a much greater extent than was ever the case in the Old World (with the possible exception of those native Australian and Pacific peoples who have no B group alleles). Taking this argument one stage further, one might then postulate that those Amerindian groups of Tierra del Fuego who possess very high B allele frequencies are descendants of some of the first migrants to North America, who had moved across the Bering Strait at a very early date, from an as-yet undetermined source in Asia, and that these were later pushed farther and farther towards the southern extremities of the Americas as newer, more aggressive migrants characterized by A, and then O alleles moved in to replace them.

Second, it is clear that blood group patterns in the Old World are much more confused than in the New because of a longer history of intensive intermixing. Some direct consequences of this may be seen in the case of the marked Eurasian gradient of blood group B, which represents the aftermath of fairly recent Mongol movements from Central Asia over a period of several centuries, and the mingling of their gene pool with that of conquered peoples in differential proportions mainly according to the distance from their homeland. The fact that more mongoloid features are not observed in these same populations today is due to the early breakup during interbreeding of those multifactorial gene combinations which determine the more obvious race variant characteristics, since these always tend to persist less completely than the more directly linked alleles which (for example) fix blood types.

At first sight, it is remarkable, and to some extent paradoxical, that such a wide range of polymorphism in man has been achieved without any apparent loss in his capacity to survive as a single species; but, as Dobzhansky[53] has pointed out, this might eventually have encouraged fitness once he had acquired even a limited technological capability, for then habitats were more often changed to suit human genes than vice versa, so that a new niche was rapidly created for each new phenotype. As population isolates are broken down by the inception of fast, modern communications, those selection processes which have given rise to the present variety of human forms might, in time, become less important, though at present there are no immediate signs that this is the case. Indeed, in some locales, the evolution of new forms may now be taking place faster than ever before, particularly in those areas where hybrid populations are beginning to emerge from an admixture of the old, as in many of our major cities.

7.3 Environmental limitations

Despite man's adaptability, he still has to survive within the confines of ecosystems which are not always favourable to him, and in an environmental complex of limiting factors which may affect him directly, or give rise to indirect nutritional or other physiological stresses. Any such reactions can modify his capacity for survival and reproduction, his efficiency, or his resistance to disease; often, they are felt most keenly at the extremes of life, during infancy or in old age, though he is not immune from them at other times, too. Emphasizing the overall genetic unity of mankind, there appears to be no easily differentiated variability in the general response to environmental stress of the major racial or subracial varieties, though, as one would expect in a species with a considerable range of polymorphism, very dissimilar degrees of individual tolerance to given situations may be observed; these are due not only to the patterns of physiological diversity, but are also a manifestation of disparate social mores, both of which make generalizations and detailed comparative analyses exceedingly difficult. Taken as a whole, our knowledge of those habitat factors which restrict man's activities is somewhat imbalanced, for as Bresler[23] has shown, most studies germane to this field of inquiry have been concerned with the consequences of severe external temperatures, nutritional inadequacies, and high-altitude settlement to human survival and work-effectiveness, so that we have relatively little information as to the significance of other environmental components, such as potential moisture deficiency, and know even less about dual or multifactor restraints. However, man clearly finds it difficult to live under certain extreme conditions, some of the most important of which are assessed below.

Temperature limitations may be broadly categorized into those of cold and heat. Many groups of men have fitted themselves successfully to active life in *cold* climates, for it is not only the Eskimo, the

Lapp, the northern Siberian tribes, and the Arctic Amerindian, who can exist in habitats where winter temperatures drop consistently to −40°C or below; Arctic and Antarctic explorers from temperate latitudes have also frequently adapted to these conditions for a few months or more. Elsewhere, Australian aborigines occasionally sleep nude in temperatures of −20°C, and the ability of Patagonians to stay unclothed in the cold dank climate of their region has long been recognized.[50] All groups who are constantly exposed to cold in this way undergo definable changes in their physiology, induced either by means of shivering, which may raise their general metabolism, or by allowing their skin and body temperatures to fall to slightly below normal values. Of course, life in cold climates can also be adjusted to by means of wearing suitably warm clothing, or by confining one's activities to an artificially-warmed habitat. However, as in the case of other organisms (section 4.4), sudden and severe environmental stress from cold will cause body damage to all, but particularly to those not normally expecting it, for then it will occur at considerably higher temperatures than, say, among polar *habitués* who have higher metabolic rates. At first, patterns of circulation change in response to the stress, but eventually shivering (the phasic contraction of muscles) sets in as a basic reaction; at this point, the general rates of metabolism may suddenly be raised by a factor of four to five times and muscle metabolism by up to twenty times. Despite this, body temperatures may still not be maintained, as the motion engendered by shivering gives rise to an increased loss of heat by convection and also to a decline in body insulation due to the stimulation of a greater flow of blood to peripheral areas, such as fingers, where heat loss is greatest.[105] Eventually, frostbite or hypothermia might set in, in which case death may follow.

The situation with respect to *heat* is much more complex, for it is very difficult to counteract the effects of excessively high temperatures in heat-generating organisms (*homeotherms*) such as man, where there must always be a net heat flux from the body to the environment. In view of man's supposed origins as a tropical animal, this might at first sight appear to be paradoxical, but it can be partly explained if one considers the fact that organisms evolved in such areas are much more likely to have been selected for cold rather than for heat tolerance, as they moved outwards from their primary habitat. Certainly, hyperthermia may often be a much more serious general threat to survival than hypothermia, and although populations which are indigenous to hot areas tend to develop some physiological and behavioural responses to counteract this possibility (see Table 7.3), even they are not always fully acclimatized to ex-

Table 7.3   *Adaptive changes to excessive heat, as detected in man (modified from R. H. Fox[61])*

Mechanism	Adaptation
Sweating	(a) Increased capacity
	(b) Quicker onset
	(c) Better distribution over body surface
	(d) Reduced salt content
Cardio-vascular	(a) Greater skin flow
	(b) Quicker response
Respiratory	Hyperventilation (panting)?
Heat storage	A lower resting body temperature
Anatomical	Change from short and stocky to long and thin?
Metabolic	Lower basic metabolism rate?

treme heat. For others, an acquired tolerance to heat is most likely to come with hard physical labour, which can give rise, in time, to a marked increase in heart-beat rates, along with an augmented production of sweat, which cools the skin through evaporation, and possibly to the eventual development of lower skin and deep body temperatures.[61] If these adaptations do not occur, heat stroke or heat prostration may result, with skin ailments, diarrhoea, or more extreme conditions, again leading to death.

In many parts of the world, *nutrient deficiencies* can also accentuate environmental limitations, even though, as Newman[110,111] has pointed out, man has a considerable degree of resistance to dietary inadequacies, and is able to produce small phenotypes with diminished energy needs or very slow maturation rates to counteract them. However, it is still possible for severe nutritional stress to lead to death from the prenatal stage onwards, or to less grave effects, such as the retardation of development, a reduction in physical or mental vitality, and an increase in susceptibility to certain diseases. Malnutrition, especially where it involves mineral and vitamin deficiencies, can cause congenital abnormalities, a general decline in fertility, or a more frequent and serious incidence of illnesses, such as rickets, scurvy, beri-beri, pellagra, and kwashi-orkor. The precise nutritional requirements for each individual depend to some extent on climate, for calorific intake must be increased to above mean values in cold regions, to allow for the greater use of energy expended in keeping warm. In cold climates, too, there is a greater specific need for vitamin D and ascorbic acid. In contrast, calorific demands decrease in the tropics, though mean protein consumption should be augmented here by a factor of 5 to 10 g/day,[106] and the diet modified to include more salt, iron, and liquids. In hot arid areas, additional quantities of fat may also be necessary for good health. Nutrient deficiencies in all parts of the world can not only occur from an inadequate total intake of calories, proteins, fats, carbohydrates, or vitamins, but also from more particular insufficiencies of micro-nutrient trace elements, such as copper, zinc, or manganese, all of which contribute towards the chemical well-being of the human body (see Table 3.2, p. 55). They might affect people throughout the year, or just within periods of seasonal hunger.

While man can exist for months or even years on a grossly-inade-quate diet, or live for some weeks completely without food, his need for *water* is much more acute, in that he can survive for only a few days without it. In this context, it should be remembered that al-though most men require a direct intake of only about two litres of water per day in order to keep themselves alive, the maintenance of the much greater supplies which support the food chain on which he depends may be more critical to him in the long run, and these (taking into account the transpiration rates of crops and the water consumption of animals which he eats) have been estimated to lie somewhere between 1100 and 9500 litres/day for a normal, naked, healthy, and well-nourished male.[21] Usually, the total quantity of freely circulating water present in any man varies a little, as when it is reduced by exercise, but it is usually close to two litres, with slightly more during the summer months;*[160] this is used to carry nutrients and oxygen between the brain and other organs, to pre-serve internal body temperatures, and to cool the skin through sweat-production. There is also a fairly large reserve of relatively static water inside cell structures, which can act as a supplementary supply. It is only when a body loss of between two to three litres of water has been effected that the mean minimal human requirements for survival have been approached, though at this stage the individual usually remains fully active in drawing upon his cell reserves; however, deterioration is rapid once his weight has been reduced by between 8 to 10 per cent, and death may follow soon after this point has been reached.[1, 92]

* In addition, there are usually considerable amounts of water in the human gut, which do not, however, contribute towards the osmotic balance of the body.

Physiological difficulties can also be encountered when man attempts to settle, temporarily or permanently, at *high altitudes*. These usually arise from one or more of the following restraints. Some result from the greater incidence of cosmic and ultraviolet radiation which, as already indicated (section 4.3), might have deleterious or even fatal mutational effects on certain cellular structures. Others develop from anoxic anoxia, or oxygen deficiency, caused by the 'thin' atmosphere which, in turn, creates an above-normal sensitivity to carbon dioxide. Pugh[117] has estimated that close to 10 million people living at elevations of 3600 m (12 600 ft) or above might be affected to some extent by this impediment, even though many of them can eventually become partially acclimatized to the conditions through an augmentation of the haemoglobin content of the blood, a greater cardiac output, the development of larger lung volumes, and an elevation of pulmonary arterial pressure due to the low O_2 tension. However, despite these body adjustments, an unexpected loss of acclimatization (termed *Monge's disease*) can still take place fairly quickly from time to time among such populations, following which severe pulmonary hypertension occurs, and occasionally heart failure, unless patients are removed immediately to lower altitudes. This sequence of events has been noted especially among mining populations who work at over 4100 m (14 000 ft) in the South American Andes. More general signs of physical deterioration, such as dehydration and starvation, can intermittently affect almost every group which lives at high altitudes where cold is an additional limitation to existence; moreover, some women suffer from an unexplained temporary or permanent loss of fertility.[107]

These few instances of physio-environmental limitations to human existence could, if space permitted, be followed by many more. They serve to show that, while man can adapt to most environmental situations, there may always be extreme conditions with which even his present genetic fluidity cannot cope. The creation of an increasingly large number of artificial niches which he can fill does not necessarily mean that the ultimate physical restraints of his habitat have been lessened, and this is perhaps emphasized by the continuance of nutritional stress in many parts of the world, together with the explosive emergence of new 'illnesses of civilization' in our towns and cities. Clearly, he is still to a large extent an organism marshalled by the confines of his biospheric milieu.

7.4 The web of disease

The closeness of man's ties with his physical and biological environment can also be exemplified by a consideration of the patterns of human disease. From time to time throughout history, human populations have been decimated by the pandemic illnesses of cholera, plague, and typhus, and other severe maladies, such as tubercolosis, syphilis, and malaria. On a worldwide scale, Banks and Hislop[9] have suggested that the effects of these reached their peak in the eighteenth century, at which time, furthermore, parasitic complaints within human bodies were the rule rather than the exception, and an extremely lethal form of infant diarrhoea killed many children at a young age. More recently, while the treatment of old afflictions has been considerably improved, the incidence of new illnesses, such as cancer, coronary thrombosis, and leukaemia, has greatly increased. Often, both old and new are maintained by means of specific symbiotic relationships between man and other organisms, arranged into a web pattern, as the title of this section implies.

These characteristics may perhaps be made more explicit through the medium of a few case studies. To begin with some fairly simple situations, it is commonplace that *infectious* diseases are induced by the growth and dispersal inside the human body of microorganisms which are harmful to it; a third organism (the *vector*) could also be involved, the precise role of which is to transmit the microorganisms from one human host to another. Thus, the plague virus is spread by the intermediary agents of rats, lice, and fleas,[163] whose ability to move easily from place to place has aided the dissemination of the disease itself. Yellow fever has as its common vector the mosquito *Aedes aegyptii*, and this, through the ease with which it travelled across oceans in the water barrels of European ships of exploration and trade, has helped in no small measure to extend the illness from its primary habitat in tropical Africa to South America, where it has found a second subsidiary vector in the form of certain forest monkeys. Malaria, the single most important killer disease of mankind, is normally carried by several species of *Anopheles* mosquitoes, all of which are found in regions where summer temperatures do not fall below 15°C (ca. 60°F), and where there is a reasonable amount of moisture. When feeding on the blood of infected people, mosquitoes ingest malaria parasites (one of four species of *Plasmodium*— *P. malariae*, *P. falciparum*, *P. vivax*, and *P. ovale*), which then undergo

327

a sexual cycle of development before being passed on, by further feeding, to infect another human host (see Fig. 7.5). Following this, they enter the red cells of the blood stream, which they often

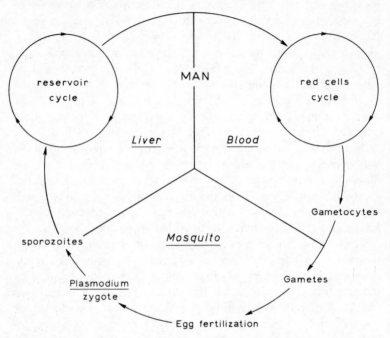

Fig. 7.5 A diagrammatic representation of the life cycle of the malaria parasite (modified from an original diagram in J. D. Smyth[133])

partially destroy, and subsequently produce toxins which give rise to an intermittent fever and great debility. Usually, a considerable reservoir of *Plasmodium* species is found in all affected persons.

Examples such as this, which occur fairly widely, involve the intra-specific transmission of a disease from one individual of a host species to another, by means of one vectorial agent. But more complex situations can also arise, in which dispersal is accomplished by vectors showing little specificity in their choice of host. This is seen in the case of the common and dangerous nematode (roundworm) *Trichinella spiralis*, which is frequently found in men, pigs, rats, cats, dogs, walrus, whales, and polar bears (Fig. 7.6). In these instances, infection results from the consumption of contaminated meat, such as pork, after which the development of larval worms rapidly

328

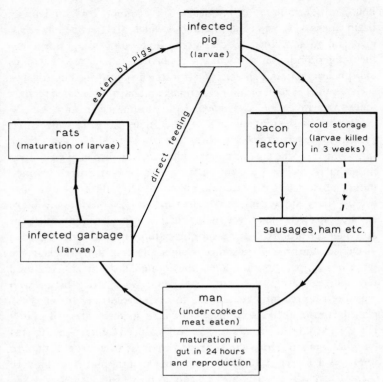

Fig. 7.6 A diagrammatic representation of the life cycle of *Trichinella spiralis* (modified from an original diagram in J. D. Smyth[133])

follows. Upon reaching maturity, sexual activity takes place and then the male worms die. However, each female subsequently produces up to 1500 larvae, which then, in turn, penetrate the intestinal wall of the host to make their way into the blood stream. Eventually, they come to rest in muscle fibres, where they form cysts, particularly around the jaw, tongue, ribs, diaphragm, larynx, and eyes.[58] The resulting illness of trichinosis is serious in man, and can gravely affect the efficiency of respiration and other bodily functions.

Often, the full relationships between host organism, vector, and *pathogen* (the agent which causes the disease) are even more intricate and involved than this. However, although the pathogens which affect human hosts are extremely diverse, most can be readily placed within one of the groups of microorganisms already noted in chapter 3 (pp. 83 to 84); thus *viruses* cause smallpox, influenza, measles,

329

polio, rabies, mumps, yellow fever, the common cold, and many brain diseases; *bacteria*, the smallest of which are classified as *germs*, give rise to sore throats, diarrhoea, and pus-forming infections, including diphtheria, tuberculosis, gonorrhea, some forms of syphilis and pneumonia, and cholera; *fungi* frequently lead to skin diseases, such as athlete's foot, or lung infections; and harmful *protozoa* include many parasites, such as *Plasmodium*, certain amoeba, and the trypanosomes of sleeping sickness. In addition, parasitic worms and various other arthropods, such as the tropical chigger flea (*Tunga* spp.) which lays its eggs underneath human skin or toenails and lives immediately below the skin surface, are the active agents which induce other relatively common afflictions. Not all of these require intermediary organisms for their transmission, and not all are equally effective; moreover, while relationships between pathogens and man are usually of some antiquity, differences in their degree of virulence, and in the resistance to their effects of individuals in the same host population are sometimes quite substantial.[13] At times, too, the association of particular elements within the disease web may be so apparently tenuous as to be inferred only through the close examination of seemingly unrelated factors. Thus, Burkitt[25] has shown that certain carcinogenically-induced tumours (malignant lymphomata) in Africa are restricted to areas where the minimum temperatures exceed 15°C in the coldest month, and has therefore suggested that the illness may be caused by a virus which some unknown vector, possibly a mosquito, can transfer only within the confines of this climatic milieu.

The complexity of these webs is further compounded by the knowledge that there is always a reaction against pathogens within the bodies of host organisms, usually of a biochemical nature. Particularly in the case of viruses, this takes the form of antibody production. Usually, a balance between antibody and pathogen is eventually reached, when the disease is brought under control, but should the pathogens be sufficiently virulent or numerous (as in some smallpox outbreaks), or the release of antibodies be abnormally slow, then the host may be killed before antibodies are present in adequate numbers to preserve it. For certain illnesses (e.g., in mild outbreaks of yellow fever, measles, and smallpox), the host will continue to produce antibodies once it has survived, so creating permanent *immunities* to them, whereas for other complaints (e.g., the common cold) antibody formation is only temporary, so allowing reinfection to occur. It is, of course, an understanding of these circumstances which has led to the successful development of techniques of vaccination and immunization against diseases such as smallpox and poliomyelitis, and to more recent attempts to find substances which can react permanently against several of the more recurrent ailments, such as influenza and the common cold itself.

It can also be assumed that the efficacy of certain human diseases has varied considerably with the passage of time. Dubos[55] has suggested that in the early stages of man's prehistory, when he maintained himself through hunting and collecting in small roving bands, he coexisted with pathogenic organisms (particularly parasites) in a balanced way which encouraged local ecological stability. But once he had learnt the use of fire for cooking, and then begun to establish semi-permanent living quarters centred around the twin foci of hearth and home, these initial patterns of enforced cooperative symbiosis might have commenced to break down for two reasons: *first*, a rapid consequential growth in human populations, which followed the widespread adoption of a better diet (see pp. 334 to 335); and, *second*, a more spasmodic though still massive increase in the number of those disease vectors (e.g., rats, fleas, and mosquitoes) which were well suited for life in the artificial microenvironments which man created.[10] A predictable result was the emergence of disequilibria between men and pathogens, in which the former were occasionally decimated when the latter assumed dominance. However, the best opportunities for the development of such imbalances may not have arisen until the Mesolithic and Neolithic periods, when contacts between even larger though still quite isolated human groups of vastly different antibody production and capability first became relatively frequent. During these periods and subsequently, disease vectors came to be dispersed along well-established trading routes, which in some regions retained their pre-eminence for several thousands of years; indeed, throughout history, the ties between trade and the spread of disease have been remarkably close. Thus, in historic time, Venice has had the unenviable record of 63 outbreaks of plague in the past 600 years due to its long-standing commercial associations with the Far East where this illness is endemic;[123] and when plague struck the city of London in 1666 its only other occurrence in Britain was in the quiet Derbyshire village of Eyam over 200 miles away (320 km), to which a bundle of infected cloth had been sent.

After the expansion of intercontinental trading links to the New World in and following the sixteenth century, many Old World diseases extended their range considerably to become cosmopolitan or pantropical in nature, though the rate at which this was achieved varied substantially. Those which were transmitted fairly directly from host to host moved quickly. This was so in the case of smallpox which arrived in Mexico with Cortez in 1519, and which, in ravaging a population with little antibody resistance to it, contributed towards the overthrow of the ruling Aztec régime. Measles reached the New World shortly afterwards and had a similar effect. On the other hand, illnesses such as yellow fever and malaria usually did not travel from the Old World until their vectors had preceded them into new territory, a process which was also often unwittingly encouraged by man. In particular, the transfer of malaria from one continent to another has been hastened both by an increase in the number of irrigated moist habitats,[99] and by the planting of crops (such as cacao in Trinidad[60]) which have provided shade for mosquito vectors. Man's activities can further the diffusion of intestinal diseases as well. One instance of this is the rapid arrival of Bancroftian filariasis (a disease induced by parasitic worms of the genus *filaria*) in the Tokelau islands of Polynesia. The filaria pathogen (*Wucheraria bancroftii*) has as its chief vector a mosquito, *Aedes polynesiensis*, the larvae of which live in a variety of sites, including crab holes, natural and manmade tree holes, but most commonly by far in the fermented fluid left in rat-chewed coconuts.[94] Since neither the coconut nor the rats (*Rattus exulans*) are native to these islands, it is clear that the extension of this disease in this locality could not have taken place so quickly were it not for the preliminary introduction by man of these two common species.*

This section has presented a general analysis of some of the biological interrelationships which exist between man, his environment, and several of the more common major long-established diseases. It has not tried to deal with the many minor ones, the nutritional illnesses, or the recently incurred stress and ageing complaints of western civilization. Nor is it concerned with the control of disease. Its main purpose is to demonstrate that the essential characteristics

---

* In contrast to these examples, it should be remembered that certain illnesses are unable to affect organisms outside of their endemic areas, in that their vectors are incapable of moving elsewhere; thus, sleeping sickness and its vectors (*Glossina* spp.) are confined solely to Africa.

of most, if not all, diseases are to some extent a reflection of the close bonds which man has had, and still has, with his physical and biological environment, and the adaptations—in this case phenotypic, in antibody production—which have followed his attempts to come to terms with it. That the latter are not always successful is shown by the occasional decimation of world or regional populations by particularly virulent diseases. Today, there is a growing awareness that states of health in man are ultimately dependent upon the maintenance of an equilibrium with all other organisms within the biosphere and with the environment which sustains them. It follows from this that there must be a limit to the influence which he can exert upon them, beyond which the balance breaks down, with potentially disastrous results for all.

MAN AS AN AGENT OF ECOSYSTEM CHANGE

Notwithstanding the conclusions reached in the previous sections of this chapter, it is clear to even a casual observer that man has been by far the most successful organism in the biosphere, in terms of his ability to mould ecosystems for his own use. Thanks to his exceptional mental qualities, he began to free himself from dependence upon his immediate natural habitat in lower Pleistocene times. His lack of physiological specialization, particularly with respect to his omnivorous diet and his acquisition of culture and technology, enabled him to explore on foot most parts of the earth at an early date. At first, perhaps because of the very paucity of ideas circulating at the time, cultural attainments were crude and slowly envisaged. Moreover, they were apparently similar over large areas, one manifestation of this being the great number of almost identical artifacts of Palaeolithic age found in many parts of the world. But even so it is here that, as Wilkinson[157] has termed it, the human transmogrification of the earth's surface really commenced. Initially, man's alteration of his habitat was minimal, but as he learnt how to make fire, and subsequently, during Mesolithic and Neolithic times, to domesticate plants and animals, increasingly definitive modifications to world ecosystem structure took place. These were later accentuated as more sophisticated means of shaping environments favourable to him spread through diffusion from major cultural and civilization hearths such as the Lower Indus Valley, parts of China, Egypt, and the Near East. Since the establishment of these early civilizations, the pace of man-induced biospheric change has been relatively rapid, though regionally varied, and hindered from time to time through the existence of dark ages in which the will to experiment and initiate has been temporarily stilled. Currently, the rate of technical innovation is exceedingly fast, having risen alarmingly in recen years to a point where the inherent stability of ecosystems everywhere is being challenged.

7.5 The use of fire

All the available evidence suggests that the utilization of fire by man comes at quite a late stage in his cultural evolution. Certainly, it follows the development of a tool-making capacity and the acquisition of speech patterns and a primitive linguistic ability, the origins of which may be associated with the beginnings of group living and territorial behaviour. Leakey and Leakey[97] have denied the existence of any signs of fire-charring among the older living sites at Olduvai Gorge, and in summarizing present information, Oakley[113] states that no traces of fire have been found in any of the australopithecine encampments in Africa, despite the fact that naturally induced fires are very frequent there,[38] so that hominids must have been aware of them. Indeed, the first known indications of the use of fire, which have been detected from charred remains in caves used by Peking man (*Homo erectus*), date back only approximately 500 000 years.[108] In Africa, the four earliest fire-hearths so far recorded are even more recent in age, being correlated with the very end of Acheulian, Middle Pleistocene times, at just over 100 000 years ago.[81] The infrequency of these occurrences suggests that, at this stage, man was still not very adept in his control of fire and, perhaps, was a *fire-user* rather than a *fire-maker;* in other words, he would attempt to maintain naturally-occurring fires for his own use, but would not have the wherewithal to set them himself. Even today, there is still some evidence of this among primitive tribes; those individuals who have difficulty in igniting plant materials will often direct fire to plant roots, in order to keep them burning slowly underground for several days.[90] Probably it took a considerable amount of time to develop the skills needed to make fire and, although the chronological sequence of events leading up to this has not yet been defined in detail, the much greater number of hearths which are distributed widely among Upper Paleolithic groups, such as Neanderthal and Cro-Magnon man, at roughly 70 000 to 35 000 years ago, is strongly suggestive of the fact that the skills had been gained by this time, that is within the confines of the last glacial period.

Once man knew both how to initiate and regulate fire, his way of life must have changed appreciably. No doubt it encouraged home routine, particularly with regard to the taking of decisions by family groups centred around the hearth; and, as Clark[32] has suggested, it might also have aided the evolution of more sophisticated languages from their early primitive forms. It certainly helped to improve man's hunting techniques, and increase the speed with which he could mould tools and weapons; in time, it led also to more complex skills, such as boat-building. It enabled him to move much more freely than ever before in cold climates which formerly had been largely unacceptable, and gave him some degree of protection against predatory animals or other humans. A further, extremely important consequence was that, as he learnt expertise in cooking, the range of foods which he could eat was substantially augmented, for many plant proteins and starchy products can only be digested easily after they have been prepared by heating.

As his familiarity with the use of fire grew, so did his capacity to alter the biological environment. Often, he took little trouble to extinguish fires; indeed, Stewart[138] has compiled records which indicate that, even today, a wide variety of travellers from Australian aborigines to modern campers leave unattended fires when they move on, and these can easily escape and grow to burn off vegetation, sometimes over a wide area. Even before he could make fire, man may have consciously directed it to clear dense woodlands which hampered his search for game,[7] but once it could be set, it was certainly very often employed by late-Pleistocene man to force large lumbering animals over cliffs in *fire drives* so as to kill them and add to his sources of meat. Sauer[125] and Eiseley[56] have suggested that drives such as these were utilized effectively 20 000 years ago on the North American Great Plains to decimate local populations of (now extinct) bison, mammoths, ground sloths, giant beavers, mastodon, camel, horse, tapir, and cave bear; and a similar removal of big game by Solutrean hunters took place contemporaneously in Europe and Asia. Fire was also an early means of selectively improving the production of desired species (e.g., grasses) and of clearing the ground in order to collect protein-rich seeds, nuts, and acorns more easily.[137] Evidence for these last two types of activity, which have been continued to modern times in some parts of the world, has accumulated steadily, and is summarized in studies by Kuhnholtz-Lordat,[91] Sauer,[126] Stewart,[138] and Daubenmire,[51] all of whom support the view that fire was, indeed, the first great force employed by man in the modification of his habitat.

But the full effectiveness of fire as a tool in the shaping of ecosystems varies according to its intensity and frequency of occurrence and with the nature of the environment in which it is set. There might also be considerable differences in the efficacy of natural and man-created fires. The former can be caused either by lightning, volcanic activity, spontaneous combustion or other factors, all of which have been widely observed throughout the world.[88] Although it has been postulated from time to time that fires induced by these phenomena play an important part in the moulding of ecosystem structure, Braun-Blanquet[22] expressed some scepticism of this view as early as 1932, in proposing that major naturally-established fires were the exception rather than the rule; and this latter point has recently been reemphasized by Stewart,[137] who suggests that, in particular, lightning-derived fires are usually accompanied by rainfall, so that they are extinguished fairly quickly. However, there may still be areas, as in the dry western mountains of the USA and Canada, in which lightning fires can be effective, though it is still not known definitively whether they occur in the same place sufficiently often to substantially change the vegetation. If this is not the case, one is left with the probability that many of the great fires which affect ecosystems are man-induced, and, of these, several types have been delimited. Occasional and usually unplanned severe crown fires, reaching temperatures (in Africa) of up to 1150°C,[86] tend to devastate tree and all other vegetation, and leave behind a wasteland, at times destroying the organic soil itself. Even hotter fires than this might arise if such conflagrations are accompanied by a back-blow against the direction of the prevailing wind. But more usually man-set fires are not so extreme as this, being designed only to clear forest undergrowth, ground cover, or dead organic material. In these instances, the temperatures rarely rise above 720°C,[103] so that more mature trees, and particularly those which are fire-resistant (e.g., species with a thick, corky, insulating bark, such as redwoods, cork oak, and some pines) can survive. If such fires recur often enough, they can prevent undergrowth from regenerating, so giving rise to an open woodland in which herbaceous plants with protected roots and rhizomes, or freely seeding annuals, gain a competitive advantage in the ground layer, with significant accompanying modifications to the fauna. In time, the forests themselves could disappear through old age, leaving a treeless landscape in which herbs and grasses are dominant, and soil profiles, patterns of energy and chemical element exchange, and dependent populations are very different to those formerly present. If fire is then utilized to maintain the herbs and grasses, it will be much less intense in view of the limited amount of fuel available, rarely exceeding 100°C[51] at the ground surface. Fires such as these are still set in many parts of the world at the present time, so they can be regarded as a continuing ecological force.

The local conditions of climate, and the lie of the land can both modify the ultimate consequences of these processes. In year-round humid ecosystems, such as the tropical rain forest, or where the ground is saturated for most of the year, as in parts of the high Arctic, burning is often ineffectual; moreover, the potential intensity of fire can be restricted in arid climates, in view of the sparseness of vegetation there. In contrast, sub-boreal forests are easily devastated by fire, since the slow rate of decay of dead organic material in these areas means that there is enough timber on the ground surface to provide an abundant fuel supply, but once an area has been burnt off, the equally slow growth rates ensure that the forest takes many years to grow back. Repeated fires can alter ecosystems appreciably in any seasonally arid climate, where the new growth of each wet season adds to the potentially large tinder supply of the dry; but in this situation they are probably most significant in areas of flat land, where they are frequently wind-driven for many miles. Thus, on the western slopes of the Californian Sierra Nevada, fires set by Indians in pre-European times have swept most of the flattish foothills clear of trees to leave grass-dominated communities. However, beyond the first major breaks in slope, species such as the digger pine (*Pinus sabiniana*) and blue oak (*Quercus douglassii*) have maintained themselves in sheltered hollows and valleys along with many chaparral shrubs, in an open, probably still fire-affected community, which is nevertheless protected from the severest effects of burning in that most of the generated heat at ridge-top altitudes never penetrates downwards beneath this general level (see Fig. 7.7).[149] But perhaps the best examples of *fire-climaxes* on flat, seasonally-arid land are located in the extensive grassland regions of the world (notably within mid-latitudinal *steppes* and tropical *savannas*), from which there has been collated an increasing body of evidence suggestive of the fact that these are mainly fire-induced, and not as first supposed—and still affirmed by some botanists (e.g., Weaver[153])—a climatic climax.* Often, the very sharpness of forest/savanna or forest/steppe boundaries, which is the exact antithesis to the ecocline expected from

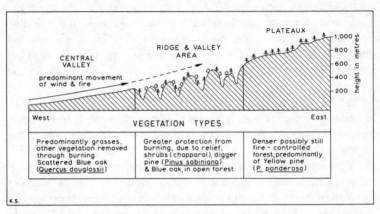

Fig. 7.7 The effects of burning on the vegetation of the slopes of the California Sierra Nevada

climatic control,[79] is indicative of artificiality and human interference; indeed, in many instances, these limits also coincide closely with divisions between hilly and flat land, giving point to a contention of Sauer that 'grasslands occupy plains, woody growth dominates rough terrain',[126] the implication being that forest species have been removed from such plains by frequent fires which have not affected the adjacent hill country so rigorously. Further support for these views has been forthcoming from an exceedingly broad spectrum of past and present opinion, among which may be mentioned the contributions of Marsh,[102] Gleason,[64] and Stewart[138] *vis-a-vis* North America; Denevan[52] and Parsons,[114] with respect to Nicaraguan savannas; Wymstra and van der Hammen,[159] concerning selected savannas in Columbia and Guyana; Schmeider,[127] for the Argentinian pampas; Cooper,[42] for Australia; Cumberland,[44] for post thirteenth century New Zealand; West,[155] for African savannas; and

---

* It is also the case that some grasslands are currently edaphic climaxes which are maintained by conditions of soil and drainage, as in the case of certain Venezuelan and Guyanan savannas;[159] while others might be a legacy of past pediplanation, as suggested by Cole [36] initially for Brazil, and later for Australia.

Bartlett[11] and Batchelder and Hirt,[12] within the tropics as a whole. In North America at least, such arguments are also enhanced by the realization that on plains now protected from fire, tree and shrub species have often begun to recolonize, so confirming that there is no natural environmental reason for their absence in these locales. Thus, mesquite and other leguminous shrubs have spread widely in recent years among the Pennyroyal grasslands of Kentucky,[124] and an explosive invasion of mesquite, sagebrush, and dwarf white oak has taken place on flat land in much of the American south-west.[30, 70]

7.6 The domestication of plants and animals, and the ecological status of agricultural systems

After man's emergence as a fire-maker, the next essential steps prerequisential to the growth of advanced cultures and civilizations took place within the Mesolithic and Neolithic periods, when he acquired the skills of plant and animal domestication. These paved the way for a considerable expansion in the sedentary way of life, and also made possible a major increase in world population, for while a hunter or gatherer needed the resources of approximately 20 km^2 of land to maintain himself on a year-round basis, the same area under reasonably intensive cultivation could, if fertile, easily sustain as many as 6000 people.[128] The immense expertise of the early domesticators may be judged by the fact that modern agriculturalists have been able to improve only slightly on the plant and animal materials which they chose, and have succeeded only rarely in adding to the general stock of cultivated and domesticated species.

It is thought that cultivated plants originated through mutation, or by means of selective advantage, on disturbed land around the hearths of food-gatherers and collectors, where the clearing of woodland or forest for home sites and the accumulation of nutrient-rich human wastes created a constantly changing and extremely artificial type of micro-ecosystem. One other preliminary was usually required: the presence of a sufficient variety of wild plant forms to encourage hybridization on a large scale, and so increase the intensity of selection of potential cultigens. This latter condition can best be fulfilled in hilly areas, where the existence of a number of very different habitats, or ecoclines, can result in the production of many divergent phenotypes. Those which were chosen for domestication frequently displayed polyploidy and gigantism, especially in their leaf size.

Through gigantism, total rates of photosynthesis per plant and the associated production of nutritional organic compounds were augmented, so leading, in turn, to the formation of more or larger fruits, seeds, and roots as compared to those of their wild progenitors. In the process of domestication, other phenotypic modifications tended to give rise to a loss of non-essential or potentially harmful characteristics, as when bitter-tasting compounds, such as saponin and betaine, were removed from wild sea beet (*Beta maritima*) during its conversion to cultivated beet (*Beta vulgaris*), or when thorns were replaced by smooth stems (compare wild and cultivated spinach), or when the general patterns of non-simultaneous ripening in wild plants were changed to a much more predictable response. Once the basic selection of cultigens had been completed, there was usually an extremely rapid emergence of additional phenotypes as their ranges came to be extended by man into unaccustomed territory, although the degree of variety in each was always largely dependent upon the breeding system, and this, as Hutchinson[83] has indicated, was far from uniform. Thus, some species which had a relatively free gene interchange through wind-pollination (e.g., maize) became highly variable, while those having a much more restricted means of cross-fertilization showed less differentiation (cotton, wheat, barley, oats), and others, such as bananas, were phenotypically and genotypically very similar, being reproduced almost wholly by vegetative means.

It is still a matter of some controversy as to where and when the first agriculture originated. Sauer's well-known treatise of 1952,[126] Darlington's study of chromosome botany,[47] Hutchinson's collection of papers on crop-plant evolution,[83] a comprehensive review by Harris,[71] a recent summary of a symposium on the domestication and exploitation of plants and animals by Ucko and Dimbleby,[142] and additional works by Zukhovskij,[165] Schwanitz,[128] and Baker,[8] together present a vast array of material evidence of a botanical, archaeological, and geographical nature, all of which has been interpreted in several ways. One of the current major dichotomies of opinion lies between those (e.g., Hawkes[74]) who argue that the domestication of plants and animals must have begun in several places independently of each other, possibly at different times, and those who propose that agricultural practices could only have arisen in one favoured region, from which they spread later, by diffusion, into other parts of the world. Supporters of the latter view need to define the locality in which one might expect the skills of agriculture to have developed most easily, and this has also proved to be a difficult task. Some, such as Sauer,[126] have postulated that South-east Asia forms an ideal area for this, in that it includes a wide variety of habitats (and, accordingly, plant and animal forms), ranging from well-watered fertile valleys to relatively high mountain ranges, in which asexual agriculture, oriented to the cultivation of vegetatively-reproduced plants (*vegeculture*), could easily have emerged from a former, far-from-primitive sedentary way of life in which fish-collecting, the utilization of plant-derived fish poisons, and fibre preparation were everyday skills, used within the general context of a hunting-gathering economy. So far, very little archaeo-logical evidence has come to light in verification of this premise, and perhaps this is not surprising in view of the prevalence of wooden artifacts and construction materials in the earliest cultures of the region, all of which would have decayed quickly under the charac-teristically hot and moist climatic conditions. Nevertheless, it has received some confirmation from the discovery that the most ancient forms of many asexual cultivated plants (as determined by their chromosome pattern), a considerable number of which are well suited to irrigated or alluvial land agriculture, have been traced back here or to adjacent parts of India. This is certainly the case for bananas, ginger, taro, the greater yam, sago palm, pandan, bamboo, sugarcane, breadfruit, and many citrus species. In contrast to this approach, other authorities (e.g., Darlington[47]) contend that South-east Asia could not possibly have been a good area for agricultural innovation, on the somewhat philosophical grounds that it did not offer a sufficient challenge to stimulate early domesticators; instead, they seek the origins of domestication in a nuclear *seed-planting* zone of the Old World, ranging from Anatolia and Iran to Syria, in which grain agriculture is known to have been practised before 9000 BP. One may conclude from the existence of these two irreconcilable views that it is as yet too early to consider in detail the full implica-tions of many of the spatial and temporal interrelationships of early agriculture; however, despite this, certain broad inferences can still be made about them, on the basis of existing information.

One of the first scientists to search for world 'centres of origin' of cultivated plants was the Russian, N. I. Vavilov, who assumed that one might correlate these with districts in which a particularly wide variety of cultivated plant species and varieties were located. Follow-ing a worldwide search for such *areas of maximum diversity*, he proposed that the first seed-planting centres had originated independently in hilly regions, mainly within the northern hemisphere. These included (a) North China; (b) Indo-Malaysia; (c) Central Asia; (d) Southwest Asia; (e) the Mediterranean region; (f) Ethiopia; (g) Central Mexico; and (h) Peru.[143] But these suppositions have had to be revised in the light of subsequent research, which has established that areas of maximum diversity are more a secondary consequence of seed-plant domestication than a primary indicator of it: for example, emmer wheat is now known to have been first sown not in Ethiopia, where there are more of its wild relatives than anywhere else, but in Syria;[68] and cultivated cottons evolved in Africa[82] rather than in India, as Vavilov thought. Indeed, the most recent archaeological and palynological evidence suggests that the origins of seed agriculture can now be firmly placed in one area alone—the Near East. Hole and Flannery[80] have put forward the notion that here, in western Iran, a period of semi-nomadic hunting and gathering which lasted until approximately 12 000 to 10 000 BP gave way to dry farming practices dominated by the cultivation of emmer wheat (*Triticum diccocum*) and two-row hulled barley (*Hordeum distichum*), and the rearing of sheep and goats, in a climate which was essentially similar to that of the present day. After 7500 BP these were replaced by a more advanced form of agriculture in which bread wheats (*Triticum aestivum*), six-row hulled or naked barley (*Hordeum vulgare*) lentils, linseed, and grass-peas were also grown, and domestic cattle and pigs were raised, in addition to sheep and goats; dogs were common as house animals, and early concepts of irrigation had been put into effect. The beginnings of an increasingly sophisticated society were eventually to emerge from this, as well as the evolution of walled towns, at about 5000 BP.

Such radical alterations to the patterns of life of early man need some further explanation. It is possible that they were stimulated either by climatic change or as a result of a rapid and substantial growth in population density. Binford[17] has argued persuasively for the idea that even if climates in the prehistoric Near East were very different to those of the present day (which they were not), they still could not possibly have been modified sufficiently frequently to account for all the cultural innovations in the area; and if one accepts this, one is left with the possibility that such innovations are in some measure associated with a major rise in the population densities of certain hunting-gathering groups. Since it has been already established (pp. 339 to 340) that the pre-agricultural human carrying capacities of even favourable localities will be low, one may postulate that the upper asymptote levels throughout this region might have been approached relatively quickly under these circumstances; and that following this, cultural change (and agricultural innovation) might well be the inevitable consequence should populations then maintain their tendency to grow. This theory is further encouraged by a consideration of the wide range in quality of sites available for human occupance in the Near East, some of which are excellent and others less propitious. Bearing this in mind, Flannery[59] considers that the origins of agriculture in this region can only be satisfactorily explained if they are examined in the light of even earlier modifications to the patterns of life, which took place in the Upper Palaeolithic period of approximately 22 000 years BP. Hunting and gathering groups which were antecedent to this existed off a relatively narrow range of meats dominated by ungulates, but, thereafter, this diet was supplemented by an increasing quantity of additional seasonally-available foodstuffs, such as fowl, fish, mussels, snails, and plants. Among the plants, freely-growing protein-rich wild cereal grasses were by far the most important. Flannery suggests that these dietary changes took place as population spilled out from those parts of the Near East, where human carrying capacities at a hunting-gathering level were particularly good, into adjacent districts of a much lower potential, and that, probably, the innovations commenced on the poorer, marginal sites, being transferred back later to more favourable ones. They may have encouraged a well-defined division of labour within families, for while women and children could collect seeds, the men were still able to hunt; and they might also have contributed substantially in themselves towards the adoption of a semi-sedentary way of life.

Taking this argument one stage further, one can envisage that, once populations dependent upon wild cereal crops approached the limits of their land's carrying capacity, a second major migration into less favourable areas for wild grains was initiated, and this, in turn, led to a further innovation: the beginnings of plant and animal domestication. Zohary[164] has ably demonstrated that one can live quite well off those patches of wild wheat which still remain; thus, there seems to have been no good reason why settlers in optimum areas of wild wheat growth should have instigated cultivation practices which involved very hard work, when adequate amounts of

343

naturally-produced food were already to hand. Furthermore, the archaeological record strongly suggests that the first cultivation did indeed take place outside the optimum area for wild wheats, as, for example, in the Khuzistan steppe, where good yields of grain could only be obtained by the use of domesticated varieties. Once adopted in these less propitious regions, the dispersal of agricultural techniques to a wide range of other sites might then have taken place fairly quickly.*

Current evidence places the first seed agriculture in Iran and adjacent areas of the Near East, at shortly after 10 000 BP. Certainly, by 9000 BP fairly large agricultural villages had become established, such as those at Haçilar and Çatal Hüyük in Turkey.[104] By 8000 BP, similar settlements had been formed in Salonika,[122] from whence others spread at a later date throughout Europe. Towards the east, the picture is more confused, although it is thought that seed agriculture in India was derived by the diffusion of ideas from a South-west Asian source, following which it was modified to suit local conditions by the inclusion of rice, sorghum, and soybeans as staple crops. Harris[71] has pointed out that the only certain signs of ancient seed-planting in Africa (dating from about 7000 BP) are located in the

---

* The idea that agricultural innovation is stimulated by a rapid increase in population density to a level close to the carrying capacity for any given economy, has also been invoked in attempts to explain more recent, historical changes in agricultural practices. Thus, Boserup[19] has proposed that population growth is the one independent and autonomous variable which brings about social and economic change. She divides present and past world land-use patterns into five types, each of which become increasingly labour-intensive, and therefore capable of supporting a larger population:

1. Forest-fallow cultivation (swidden, or slash-burn). Plots are cleared in the forest, cropped for a year or two, and then left fallow for a longer period (6–20 years). Secondary forest grows back. Ashes add nutrients to the soil.
2. Bush-fallow cultivation. Fallow period between crops is shorter (6–7 years). Secondary 'bush' grows back. Cultivation periods are from 1–2 years to 6–8 years.
3. Short-fallow cultivation. Fallow is only 1 year, or at the maximum a few years. Wild grasses grow back; a grass-fallow cultivation.
4. Annual cropping. Fallow is only for a few months. Often, systems of annual rotation are included.
5. Multicropping. Very intensive, with two or more crops per year. Short or negligible fallow. Relatively rare.

In case 5, patterns of labour-use can, at times, become excessively complicated or 'involuted' if there is a need to accommodate a very high population density within them. This could reverse the general sequence of increasing labour-intensiveness and efficiency which is implied in Boserup's scheme. It is an exceptional case, and perhaps, best exemplified among the wet-rice cultivators of Indonesia.[62]

extreme north-east, where they are also closely linked to South-west Asia. However, the recovery of polished stone hoes and grindstones, and the known long-term existence of sedentary peoples elsewhere in this continent, as in Ethiopia, Kenya, northern Tanzania, and (possibly) West Africa, might presuppose the former presence of a similar old type of seed-agriculture in these areas, too.[32] Early seed-planting in the Americas has been traced back to beyond 9000 BP, for remnants of pepper (*Capsicum annuum*), pumpkin (*Cucurbita pepo*), and bottle gourd (*Lagenaria siceraria*)* have all been recovered from caves of this age in the Mexican states of Tamaulipas and Puebla,[101] while in contrast the more advanced cultivation of maize (*Zea mays*), amaranth (*Amaranthus leucocarpus*), squash (*Cucurbita moschata*), and common lima beans began later, at intervals between 7000 to 4000 BP. Information from other parts of the New World is more scanty, though it may be that subsidiary centres of domestication developed subsequently between 6000 and 4000 BP in Peru and Venezuela.

Some interesting differences in diet arose relatively quickly between the seed-planters of the Old and New Worlds. The diet of the New World, dominated as it was by maize, beans, and squash, was extremely protein rich and more than adequate in carbohydrate, fats, and vitamins, but that of Old World small-grain agriculture was much less so, necessitating a greater reliance on meat as a protein source. This meant that Old World farmers always placed a considerable emphasis on the herding of multipurpose animals, such as goats, sheep, pigs, and cattle. Thus, the earliest indications of domesticated sheep recovered from the Iraquian village of Zawi Chemi[118] date back to 10 800 BP, and these animals were certainly also present in North Africa (Cyrenaica) at least by 6800 BP, and possibly as early as 8400 BP.[78] However, sheep of this period would have been somewhat different in aspect from those of the present day, in that they had little wool in their fleece. In certain districts of the near East, as in Iranian Khuzistan, the rearing of goats could well have preceded that of sheep, and traces of 'house' pigs and dogs, which existed at 9000 BP, have been found at Çayönü, in south-western Anatolia.† Probably the last of these animals to be brought

---

* A plant of possible Old World origins.
† The dog is a particularly interesting animal in that it was present in North America (Idaho) at 10 000 BP, or over 1000 years earlier than its first known occurrence in the Near East.[95]

under control were cattle and, then horses, for, while the former were utilized in Greek Thessaly by 8500 BP and in Macedonia by 8100 BP, there is at present no evidence that the horse was tamed before 4000 BP, when steppe tribesmen (e.g., those of the Tripolje culture) who valued this creature were first becoming prominent. The domestication of other useful beasts, such as camels, followed later. At a much more recent stage of cultural evolution, the demand for meat and milk became such that in parts of Eurasia and Africa where a general or seasonal shortage of browse or grazing material was common, as in the drier grasslands or on desert margins, more specialized groups of pastoral nomads, such as the Arab and Bedouin, eventually broke away to concentrate solely on animal-herding at the expense of seed-planting.

Until very recently, no similar evidence suggestive of an early start to *vegecultural* practices had been forthcoming from the tropics of either the Old or the New World.* However, within the last few years, Solheim[134] has recovered primitive cord-marked pottery, flaked tools (possibly hoes), and piles of freshwater or seawater shells and animal bones among the debris of former Hoabinhian tribes on mainland South-east Asia, from which he infers that a relatively advanced hunting-gathering economy existed in this region several millenia ago. He considers, furthermore, that primitive forms of agriculture could easily have emerged from such a cultural milieu. Analyses of pollen cores taken from the bottom of Sun-Moon Lake in Taiwan also indicate the local presence of a well-developed slash-burn type of horticulture (with fruit and root crops) as far back as 11 000 BP. If one assumes that there must have been a long period of experimentation before agricultural practices could become as thoroughly organized as the Taiwan data implies, the date of initial plant domestication in South-east Asia can be put back at least to 12 000 BP. On the basis of this, and his own extensive knowledge of the prehistory of the region, Solheim has come to agree with Sauer's original contention that this part of the world was the first to provide the ideas from which all subsequent forms of agriculture were ultimately derived. He proposes, moreover, that South-east Asia remained a progressive area of agriculture until it was replaced in significance by the grain-producing regions of China in the fourth millenium BP, and that there is some evidence that substantial trad-

---

* Although it has long been known that certain animals were first domesticated in these regions (e.g., the chicken in South-east Asia).

ing contacts had encouraged a massive interchange of ideas between South-east Asia and many other parts of the world before this latter date. Thus, the well-known Dongson bronzeware designs, found in this area at 4500 BP, have also been retrieved from early iron age sites among the Caucasus Mountains, and more generally among bronze age encampments throughout Europe. At this time, many South-east Asians were excellent sailors and travelled widely along trading routes which ranged from Easter Island to Madagascar, and it is probably the case that they were especially important as agents

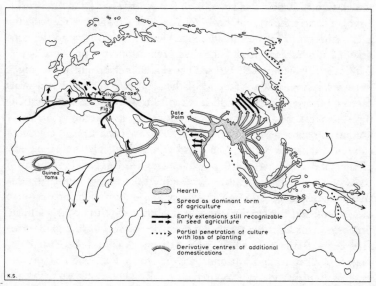

Fig. 7.8 The origins and spread of early asexual agriculture: a scheme suggested by C. O. Sauer (from an original diagram in C. O. Sauer, *Agricultural origins and dispersals*, American Geographical Society, 1952[126])

of commodity exchange within and around the shores of the Indian Ocean. Certainly, there seems to be no good reason why early techniques of agriculture could not have been dispersed fairly quickly from South-east Asia into other parts of the Old World, so leading to an extension of domestication and, eventually, to the adoption of other plant materials for local cultivation in favoured areas (see Fig. 7.8 as an illustration of the way in which this might have been achieved). Carter[29] considers that such skills were also selectively transferred to the New World, where an additional centre of vege-culture (based on species such as manioc and sweet potato) was

347

formed, possibly on the northern flanks of the Andes in Columbia and Ecuador.[126] But so far very little has been ascertained concerning these latter possibilities, except that certain crops (the sweet potato[8,126] and some cottons[84]) are known to have been extensively raised in both Old and New Worlds in pre-Columbian times, and it is thought by Heyerdahl[75,76] that this might have been the case for several other species as well.

The consequences to world ecosystem structure of man's cultural progression from a predator to a domesticator and farmer, and the association diffusion of agricultural concepts throughout the world, have, of course, been profound. While a hunter, and even a collector and gatherer, he usually lived conservatively, according to the confines of his habitat, and with few pretensions towards ecological dominance, except where he used fire relatively frequently. He might not even have been at the end of the food chain, for often he was eaten by more powerful predators than himself. But as a domesticator, he manifestly altered this situation by modifying plant and animal associations, simplifying the patterns of energy and chemical element exchange within them, and entrenching himself at the end of the food chain, if at all possible. Despite this, the agricultural ecosystems which he produced remained for many centuries relatively inconspicuous, though gradually his removal of forest, drainage of swamps and restriction of natural plant growth from Mesolithic times onwards assumed greater importance in the biosphere. Eventually, in the fertile river valleys of South-east Asia and the Near East, major 'hydraulic civilizations' emerged which were based on the intensive cultivation of irrigated land. Elsewhere, forest-clearance slowly grew in scale to extend the area of non-irrigable agriculture and to cater for an increased demand for timber to build shelters, houses, and ships, and for use as charcoal in early industry. In Europe, the history of forest removal from the Mesolithic period onwards has been detailed by a host of authors, including Iversen,[85] Godwin,[65] Clark,[33] and Darby.[45,46] Here, the net result has been that while today less than a quarter of the land lies under forest cover in western and central Europe, close to 95 per cent would have done so before the arrival of man.[146] Moreover, human interference has encouraged wholesale modifications in the structure and composition of those forests which remain. It has been suggested that the present dominance of oakwood in several parts of Europe is artificially derived, in that it dates from mediaeval planting practices which

348

favoured this species over any other; and a more recent corollary can be seen in the spread of those conifers favoured by modern foresters at the expense of deciduous hardwoods. Similar instances to these are found in many other parts of the world.

Following forest clearance, the extent to which ecosystem inter-relationships are further changed depends largely on the type of agricultural system adopted and on the human population density; the effects of both of these influences can vary widely. In regions which practice traditional, tropical vegeculture, an apparently-confused tangle of vegetation greets the untrained eye; this consists of a generous mixture of tree crops, and shrub and herb elements, so that frequently very little bare soil is detected. Usually, there is no particular harvesting time, the farmer or peasant taking in crops when he wishes, and the land is only used for a few years before forest is allowed to regenerate, the precise period being determined largely by the intensity of population pressure (see the footnote on p. 344). New land is then cleared through felling or burning to be cultivated rotationally in its turn. This *slash-burn* type of agriculture is extremely *conservative* in its approach, in that, consciously or unconsciously, it seeks to preserve the structure of a natural ecosystem, and rarely over-strains the chemical and energy resources of the environment, although some reduction in productivity may be noticed if population density becomes very great and the rotational patterns have to be compressed so as to meet the increased human food requirements. Closely similar conservative patterns of cultivation are often seen elsewhere; for example, they exist, albeit in a much more modified form, within temperate-latitude systems which allow for well-balanced crop rotations, since these leave time for the land to occasionally lie in fallow, when natural fertilizer from grazing animals can be added to replenish nutrients taken out in the production of demanding crops such as cereals. Both approaches ensure that no more is removed from the ground than is put back in, so maintaining an environmental equilibrium, particularly in terms of chemical element transfer, despite the artificiality of the ecosystems involved.

In contrast, exploitative and seriously *destructive* agricultural systems have been recently extended in scale by the widespread adoption of ever more intensive methods of food-crop production in many parts of the world. Their inception and retention frequently give rise to a major decline in the physico-chemical resources of the soil, a general loss of fertility, and eventually soil loss and soil erosion

349

(see also pp. 86 to 90 for a further discussion of some of the biological implications of these possibilities). Sometimes, their ultimate consequences are extremely severe and disastrous, as when soil was almost completely removed by windblow from certain semi-arid Dust Bowl areas of the American Great Plains in the 'thirties and from some districts of Soviet Kazahkstan in the 'fifties, following a breakdown in its structure due to the overproduction of small grains. However, an equally grave general reduction in soil quality without conspicuous erosion is a much more common result, such as that formerly observed on numerous cotton and tobacco plantations of the southern USA[102] and on many sugar-producing estates in Caribbean islands such as Barbados.[151]* Usually, this sequence of events most frequently attends the non-rotational or monocultural planting of demanding crops, and the effects are seen most clearly in those areas which have very intensive precipitation, or a marked seasonal imbalance in water availability (e.g., in areas of semidesert or a Mediterranean climate), or a combination of both.

By way of illustration, one may cite the case of the eastern Mediterranean. Here, mature forests of evergreen and deciduous oaks, beech, pines, firs, and cedars were distributed widely even as late as Homeric times (approximately the 9th century BC). Despite a climatic régime which allowed for only limited and often variable quantities of incoming precipitation, contemporary amounts of soil water were always sufficient to support small-grain agriculture, which within the pre-Greek period had expanded in a conservative way for several millenia. Outside the major river basins, there was little irrigation, most crops being raised by scratch-ploughing techniques, or on terraces similar to those found in the area today; moreover, a common use of sheltering tree species, such as grapes or olives, was entertained wherever possible. This was a balanced agriculture with little or no loss of soil through erosion. The initial steps towards environmental deterioration commenced when population densities increased, following Greek and Roman colonization (especially between the 5th century BC and the 4th century AD), for then large quantities of timber began to be removed for the first time, so that some forests came to be replaced by shrub com-

---

* It is true that visible erosion is also present in all these areas, though on a much smaller number of landholdings. The consequences of soil erosion and soil deterioration can both, for a while, be concealed by a heavy application of artificial fertilizer, but this temporary measure can never counteract the harm already done.

munities. As a result, the rate of rainfall interception was reduced, runoff rates were augmented, and a general diminution in the reserves of soil water took place. At the same time, the colonizing powers demanded ever greater quantities of locally produced cereals for their own use, and in so doing overtaxed the chemical and water resources of the soil so severely as to give rise to greatly inferior crop yields, and, eventually, to the subsequent curtailment or abandonment of grain agriculture in many areas, notably from 300 AD onwards. This was especially true of the coasts of North Africa from Tripoli to Cairo, where sedentary populations began to leave distressed areas, to be replaced by nomadic Berber tribes with their flocks of close-grazing Merino-type sheep; these effectively chewed off much of the remaining, or regenerating palatable woody and herb vegetation, and contributed towards a further general reduction in the quantity and quality of graze and browse. The somewhat later introduction into the herding milieu of hardy goats, some of which had a climbing ability, led in certain places to the removal of the last vestiges of tree and shrub species and the further general impoverishment of ground communities; in short, they gave rise to the final stages of degradation from the former forest to the present very restricted grass, herb and weed communities which are characteristic of so much of this region today (see Fig. 7.9). It is to be noted that, although cycles of dry years do occur in the Mediterranean, these events appear to have taken place independently of any patterns of climatic change and, thus, are solely an anthropic effect; indeed, there is no climatic reason why relatively complex woodland assemblages cannot still be supported on many sites, as witnessed by those which have persisted (e.g., on islands such as Cyprus), wherever agricultural and grazing pressures have not been so intense.

7.7 Human migration, and its biogeographical consequences

It has already been made clear that man's planned or inadvertent transfer of organisms from one region or continent to another can considerably alter the structure of certain ecosystems. Presently-cosmopolitan animal pests, such as rats and bats, and widely-distributed diseases and their vectors have frequently migrated along his trading routes. His commercial activities have also encouraged indirectly the range extension of many birds, weeds, and

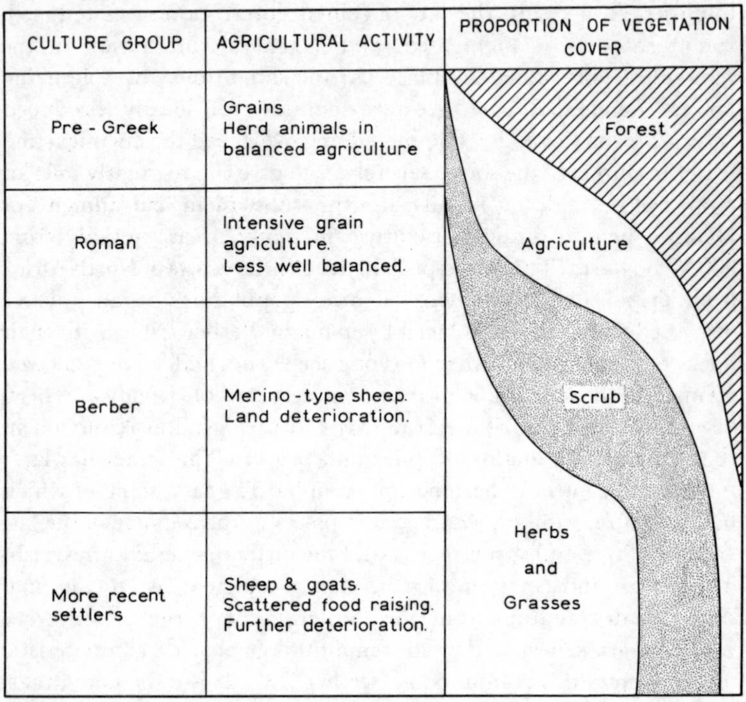

CULTURE GROUP	AGRICULTURAL ACTIVITY	PROPORTION OF VEGETATION COVER
Pre - Greek	Grains Herd animals in balanced agriculture	Forest
Roman	Intensive grain agriculture. Less well balanced.	Agriculture
Berber	Merino-type sheep. Land deterioration.	Scrub
More recent settlers	Sheep & goats. Scattered food raising. Further deterioration.	Herbs and Grasses

Fig. 7.9 Patterns of change in land use and vegetation of the eastern Mediterranean

other wild plant and animal species. He is more directly responsible for the diffusion of cultigens and a large number of domesticated animals. Indeed, his individual and group tendency to travel long distances can be accounted as a major potential cause of change within the biosphere as a whole.

Particularly since his organization into family, tribal, or broader cultural groups, man has, whenever necessary, removed himself and his dependents into new settlement areas for a variety of reasons. He might do this in order to escape intergroup warfare or the possibility of slavery; to seek more fertile, or leave less favourable or excessively overcrowded habitats; to flee from especially virulent diseases or pests which have adversely affected him, his domesticated animals, or his cultivated plants; to counteract seasonal or longer resource deficiencies, as in the case of transhumance, pastoral nomadism, semi-nomadism, or a bush-fallow rotational system (see footnote,

p. 344); or temporarily or permanently to undertake religious pilgrimages to cities such as Jerusalem, Benares, and Mecca. In fact, major religious shrines often attract their faithful from all parts of the world (see Fig. 7.10, for the case of Mecca). These long-standing causes of migration have been considerably augmented since European trade expansion began in the sixteenth century, by an increasingly rapid conveyance of plants, animals, and men from one continent to another; and these trends, in turn, have been supplemented more recently, following the Industrial Revolution of the eighteenth and nineteenth centuries, by complex, large-scale social migrations on a scale never known before. Usually, the latter have been initiated by a desire to improve living standards, and/or obtain better employment, as instanced in the attraction of North America for many European settlers in the last two centuries, the equivalent drift of West Indians and Asians to Britain since the Second World War, and the widespread shift of populations from rural areas to city regions. As families have increased their standard of living, new leisure migrations have also developed, both on a small scale ('weekending'), and also on a much broader basis, as exemplified in the emergence of resort areas in the Caribbean and the Mediterranean; and these have been encouraged by a parallel improvement in travel facilities.

Recognition of this enhanced scale of human mobility can be seen in the general extension of stringent customs regulations along the borders of many countries or regions, the aim of which is to restrict or prevent the dispersal of those potentially harmful live organisms which follow in the wake of man. One of the severest of these controls is found on the frontiers of the USA, where fresh fruit is often confiscated against the possibility that it may harbour hidden pests. But despite these precautions, which are never entirely effective, many creatures still succeed in becoming established in new territory as a direct or indirect result of man's movement from place to place; and they then make their presence felt in a number of intricate ways. Thus, within the present century, the Asiatic chestnut blight (*Endothia parasitica*), a fungus which does not affect trees adversely in its area of origin, first decimated sweet chestnut (*Castanea dentata*) woodlands following its accidental introduction to North America; and then, subsequent to its carriage by man into Italy and Southern Switzerland in 1938, began to attack European chestnuts (*Castanea sativa*), too.[162] Travelling in the reverse direction, the

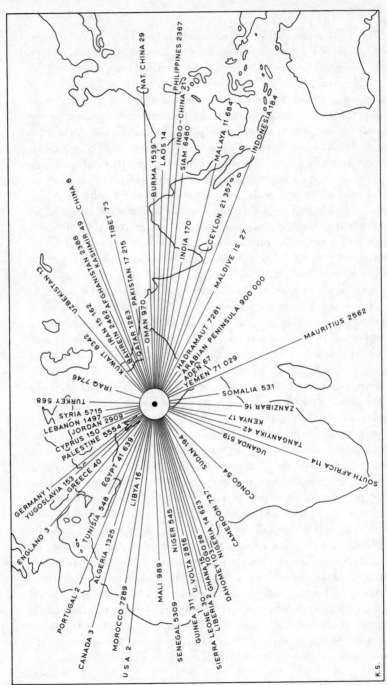

Fig. 7.10 The convergence of pilgrims on Mecca and Medina (from an original diagram in R. M. Prothero, *Migrants and Malaria* Longman Group, 1965. Based on a map prepared by the World Health Organization)

K.S.

European fungus disease *Cerastomella ulmi* destroyed many majestic elms (*Ulmus americana*) once it had been brought unwittingly to North America, although, in this case, the effect was less immediate in that the disease could not be implanted into the elms before it had found suitable vectors. Three species of bark beetle were used for this purpose, two of which were of American origin; but the involved nature of this particular ecological reaction can be judged by the fact that the third, and ultimately the most common vector (*Scolytus multistratus*), proved also to be a European immigrant, having crossed the Atlantic at an even earlier date than the fungus itself.[37]

Occasionally, the introduction of a major crop by man can lead to unforeseen biological repercussions in a wide range of ecosystems. When new immigrants first planted the potato in Colorado, USA, in 1859, the now infamous Colorado beetle (*Leptinotarsa decemlineata*), which feeds off this crop, was stimulated to move outwards from its native habitat in the Rocky Mountains until it had reached the eastern seaboard of America by 1874. It then crossed the Atlantic in a ship's cargo to become established in France in 1920, since when it has disseminated itself to most other parts of mainland western Europe.[57]

Unanticipated ecological results can also attend the conscious transfer by man of other plant and animal species which are deemed to serve a useful or pleasurable purpose. In 1891, several pairs of European starlings (*Sturnus vulgaris*) were brought as ornamental birds into central New York City, from whence they have spread into other towns throughout the United States,[39] often competing with the native bluebird (*Sialis sialis*) and the flicker (*Colaptes auratus*) to their detriment. The West African green monkey (*Cercopithecus sabeus*), introduced as a house pet into some West Indian islands by seventeenth-century French settlers, then escaped into the wild, where, after scarcely surviving for over two centuries, its descendents have now grown explosively in numbers to become a major pest. Also in the West Indies, the mongoose, transported from Calcutta in 1872 as a means of controlling rat and snake populations, has itself developed into a thorough nuisance.[3,129] The devastations of the European rabbit in Australia are notorious, yet the fact that it was originally imported by colonists as a game animal is not advertised. An additional example is that of the coypu, a fur-bearing South American rodent, which, within the last few years, has extricated itself from fur farms in eastern England to cause immense havoc to adjacent riverbank ecosystems through its burrowing activities. Among plants, the setting of the tall American estuarine grass *Spartina alterniflora* in Britain during the nineteenth century, and its later hybridization with the English *S. maritima* to produce *S. townsendii*, gave rise to a completely unexpected reaction in that, while both of its progenitors displayed ecological stability, the new cross, first noted in Southampton Water in 1870, proved to be an aggressive colonizer, and in a short space of time had become conspicuous in a large number of tidal esturies elsewhere in the country.

Many more similar instances have been recorded throughout the world, and the reader is referred, in particular, to Sir Charles Elton's *The ecology of invasions*[57] for a more complete examination of these.

One must also reconsider briefly, in this context, the planned movement of cultigens by man from one region to another. It has previously been inferred that this has occurred on a particularly intensive scale from the sixteenth century onwards as European trade contacts expanded widely into other continents. The result has been a massive dislocation of many cultivated plants and their associated weeds, a few details of which are given in Table 7.4, which emphasizes the fact that transplanted species have often

Table 7.4   Tropical crop production

Crop	Area of origin	Area of current maximum production
Sugarcane (*Saccharum* spp.)	S.E. Asia	W. Indies and S. America
Rubber (*Hevea brasiliensis*)	Tropical S. America	Malaysia
Pineapple (*Ananas comosus*)	Tropical America	Hawaii
Coffee (*Coffea arabica*)	Ethiopia	Central and S. America
Cocoa (*Theobroma cacao*)	Tropical S. America	Ghana
Citrus (*Citrus* spp.)	S.E. Asia	USA
Banana (*Musa* spp.)	S.E. Asia	W. Indies and American tropics
Oil palm (*Elais guineenais*)	W. Africa	Malaysia
Nutmeg (*Myristica fragrans*)	Moluccas	Grenada, W. Indies
Sisal (*Agave sisalana*)	Mexico	E. Africa

given better yields in their new environments than in their old. Many of the crop transfers of the last four centuries were originally sponsored by state botanical gardens in Europe, and Harris,[71] quoting Kraus,[89] has noted that the precise records of these are such that one can often pick out distinctive biogeographic patterns which signify the relative importance of particular regions to plant exploration at different points in time. A slightly modified presentation of the views of these two authorities suggests that between 1560 and 1620 most species received in Europe had been collected in the Near East. Between 1620 and 1772, they were largely derived from the Americas and the Caribbean, and (subsequent to 1687) from the Cape of Good Hope region. Between 1772 and 1820, they came mainly from Australia and the Pacific, and between 1820 and 1930, from Japan and west China. Once catalogued, a selection of these plants was then sent overseas from Europe for relocation in suitable colonies elsewhere.

Although the biogeographical consequences to ecosystem composition and structure of man-induced plant and animal dispersal have obviously been considerable in the artificial biological habitats (cleared land, gardens, fields, etc.) which he has created, it has still not yet been fully ascertained how extensive or permanent they may be in those more natural communities (forests, swamps, etc.) which still remain. Studies in this field have been hampered to date by the surprisingly small amount of detailed information which has been accumulated, both in respect of the chronology and the pathways of transfer of plant and animal species between particular regions, and the degree of success of alien species colonization. So far, the most complete evidence has been gleaned from areally confined milieus, such as small oceanic islands, where the patterns of ecological change can often be seen with a greater clarity than elsewhere. Thus, Allan[2] and Ridley[119] have been able to delimit several specific stages of alien plant invasion in New Zealand. Harris[69] has shown both that it is possible to distinguish the major source areas from which plants and animals were brought by British settlers to Antigua, Anguilla, and Barbuda in the West Indies (see Fig. 7.11), and that some correlation exists between the relative success of alien plant colonization and the successive types of land use adopted in these islands. Furthermore, in an analysis of the results of alient plant introduction in Barbados, the West Indies, Watts[150,152] has remarked on the fact that many native plants, and two native plant associations have proved to be very capable of resisting the ecological challenge of alien species competition, despite the maintenance of a consistently high population density, a severe and intensive pattern of agriculture and animal husbandry, and, accordingly, an extremely disturbed environmental situation for over 300 years. Indeed, it seems more and more

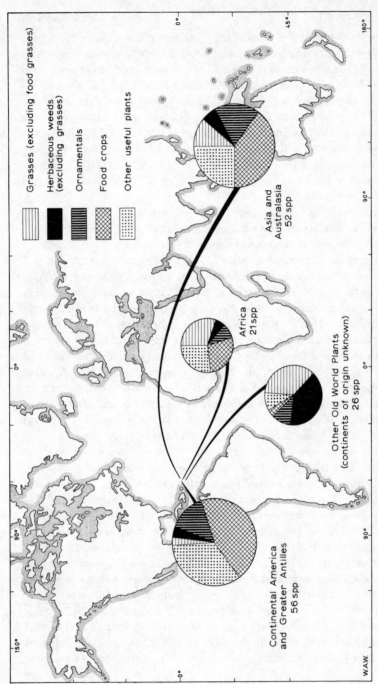

Fig. 7.11 Areas of origin of alien plants introduced into Antigua (from an original diagram in D. R. Harris[71]. Reprinted by permission of the Regents of the University of California)

likely that many plant aliens are able to find suitable niches in new habitats only for as long as the latter continue to be interfered with by man, and, even then, certain native species can adapt to a radically altered niche structure extremely quickly, and (in some instances) perhaps more completely than the aliens themselves.

7.8 Environmental pollution

The point has previously been made (specifically in sections 3.9 to 3.12) that man's alteration of his habitat has not always been beneficial to it. Instances of general environmental deterioration can be traced back to the beginnings of agriculture in Mesolithic times, and more recently specific patterns of pollution have been initiated, and subsequently seriously augmented by the release of large quantities of noxious agents into the atmosphere, soil, and water-bodies of the earth's surface. Both pollution, and the continuing decline in environmental quality which is characteristic of some parts of the world, are the indirect results of an explosive growth in population, whose origins can be dated back to the late seventeenth century (Table 7.5),

Table 7.5   Estimates of world population, 1650–1970, after A. M. Carr-Saunders[28] and United Nations reports

Date	World population (in millions)
1650	545
1750	728
1800	906
1850	1171
1900	1608
1950	2550
1970	3700

coupled with an associated rapid expansion of urban and industrial areas in the last two centuries. All large towns, and particularly the new urban-industrial complexes, tend to feed off neighbouring or even distant regions, to which they give little if anything of a biological nature in return; they therefore emphasize any biospheric imbalance which might exist in the agricultural ecosystems which they dominate and, in the long term, can contribute massively towards the rundown of chemical and energy resources held therein. The inherent artificiality of such areas also means that organisms which lead to the breakdown of human and other wastes for recycling are often absent, or present in very reduced numbers, so that new and

359

contrived ways of refuse disposal must be adopted. That these are not always successful is witnessed by the spread of pollution, first, in the immediate vicinity of cities, and, later, on a much wider scale. Some of the ways in which these comprehensive patterns of environmental impairment can be instigated are indicated below, in a brief discussion which seeks not to duplicate the information given in chapter 3, but rather to extend it, and place it within a broader perspective.

Until 1800, the population of most cities had remained below 100 000 due to the inbuilt limitations of their economy.* Even so, many were far from being in biological equilibrium with their surroundings, as Marsh[102] has recorded in the case of imperial Rome, whose demands for wheat, olives, dried fish, and pottery led eventually to deforestation and soil erosion locally, and extensive land and river spoilation farther afield (see also p. 350). But the potentialities for disequilibrium have been compounded many times over within the last two centuries, as agricultural regions adjacent to developing urban areas have been subjected to an ever greater pressure and intensity of land use. This has often been particularly the case around the larger conurbations, especially where they have partially merged to form megalopoles, such as that present on the eastern seaboard of the USA between Washington, DC and Boston, which has been so well described by Gottman.[66] Moreover, within the confines of such conurbations, one can expect substantial biospheric imbalances to form not only as a result of the considerable output of industrial pollutants (see section 3.12), but also because of the vast extent of brick and concrete microhabitats in which plant growth is severely restricted or even prevented entirely. One result of this is that the overall production of photosynthetic materials per unit area is reduced to much below normal values, so leading to the emergence of urban climates whose atmospheres have greatly modified totals of O_2 and CO_2. In addition, the frequency of bright sunshine, cloud, good visibility, fog, and precipitation can be very different to that in surrounding rural regions;[93] and the close proximity of many heated homes and office buildings usually induces a slow general rise in temperature towards the heart of a built-up area, a phenomenon which has been termed a heat-island effect by Chandler.[31]

* In the Europe of 1800, only London (959 000), Paris, and Constantinople exceeded 500 000 in population, there being 18 other cities with populations in excess of 100 000.

Equally grave consequences in the form of extensive water pollution can also arise from the inevitably huge output of human and industrial wastes which is characteristic of conurbations; and this is augmented by that emitted from smaller industrial sites as well. Water may be contaminated through the release of too many industrial effluents; by toxic agents such as phenols, oils, and detergents; by radioactive wastes and pesticides; through an excess of organic materials in sewage; as a result of mine discharges; or by too great an accumulation of salt, as in some irrigation water.[87] These various substances can be placed within one or more of four major categories of material contributing to water pollution: organic wastes; dissolved inorganic wastes; solid or semisolid refuse; and heat effluents.[139]

Of these, *organic wastes*, such as human sewage or other animal residues, are all readily oxidized by dissolved oxygen existing in the water, but problems can arise when they are present in excessive quantities. In this case, the demand for oxygen by microorganisms of decay might exceed the available supply, so that the breakdown of wastes is only partial, leaving some coagulated debris and the characteristically unpleasant smell of accumulated hydrogen sulphide. The question of *dissolved inorganic wastes*, which are frequently emitted from mining or quarrying industries, is more intractable, for even if these are reduced by dilution, they can still create severe pollution effects. Thus, the sulphides and sulphates in many industrial effluents act as inorganic reducing agents which can diminish oxygen supplies in river water; and ferrous salts carried in solution from mine adits or metal shops might give rise to noxious riverbed deposits, such as the red ochre of the Don Valley which lies downstream from the steel mills of Sheffield in England. *Solid or semisolid refuse*, which is either rolled along the floor of a river, or transported in suspension, also results largely from mining or quarrying. Large-scale hydraulic placer mining, such as that practiced in the eighteen-fifties in the mother-lode gold zone of the California Sierra Nevada foothills, can destroy fluvial ecosystems for many miles downstream through the sheer amount of rock, gravel, and sand debris amassed therein. Even the finer refuse can kill plant and fish life, or reduce water visibility so greatly that fish can no longer see their sources of food. Thus, the outwash from china clay (kaolin) quarries has greatly disturbed river and inshore marine fish populations along the southern coast of Cornwall in the UK. In fact,

Patrick[115] has postulated that any sedimentation at above normal rates is potentially able to annihilate the habitats of many river species. Within the last few years, *heat effluents* from power stations have also given rise to some pollution, especially where they have resulted in an augmentation of river temperatures by a factor in excess of $5°C$, for this has proved sufficient to cause the death of trout and other organisms in mid-latitude rivers. The rates of growth of sewage bacteria could also be raised to unusually high values by heat effluents, so causing an increase in the rate of oxidation and the relative abundance of CO_2 in the water, which in itself might have a deleterious effect on other forms of aquatic animal life; and, moreover, it is probable that the toxicity of any poisons which are present in the water might be increased by a rise in temperature.

Altogether, in recent years, pollution by these various means has been responsible for lowering the quality of 600 to 700 km^3 of water per annum, one-third of which will need treatment before it can be used again by humans; and this figure is growing monthly. Major rivers, such as the Rhine, Mississippi and St Lawrence, and many minor ones, are now little more than open sewers, and, through these, impurities are beginning increasingly to affect inshore waters and enclosed seas, such as the Baltic, in which fish are generally tainted. Similar effects have been observed in the Great Lakes of North America, where many aquatic organisms have been unable to survive in the changed conditions. In the deeper oceans, pollution substances are further augmented by oil discharges and the dumping of toxic and radioactive chemicals at sea. No one can as yet foresee what the end result of the introduction of so many potential poisons into aquatic and other ecosystems may be, though it is certain that in the long run it cannot be advantageous to most forms of life.

THE PRESENT STATUS OF WORLD ECOSYSTEMS

A systematic examination of the functional mechanisms of ecosystems has been presented in the foregoing chapters. It has been shown that the evolution of all ecosystems is primarily determined by the amounts of energy, received mainly from the sun, which flow through them in chains or webs. They are further maintained by intricate patterns of chemical cycling, their growth is encouraged or hampered by other parameters of the natural environment, and the populations which live within them can be considerably affected by the demographic restraints of competition and niche structure. Every ecosystem is modified with the passage of time, as the challenge of habitat leads to the emergence of new life forms through natural selection or to modified biogeographical interrelationships between organisms, as their ranges are extended, contracted, or qualified in subtle ways. It has also been emphasized that, although man is still very much dependent upon his environment, he has been able to dominate it to a degree far in excess of any other organism, and so considerably change it to suit his own requirements.

Throughout, the main thesis has been that under natural conditions, ecosystems have, of necessity, been in a state of ecological equilibrium. But within the last few millenia, and particularly in recent years, as their essential characteristics have been altered by man more comprehensively than ever before, signs of a severe imbalance, or, at the very least, a declining functional efficiency, are beginning to be observed in many of them. If these trends towards increasing instability are sustained in future years, there is no doubt that their ultimate consequences will prove to be extremely deleterious, or even fatal, for many organisms. It is therefore imperative that man should closely examine the ecological *status* of all ecosystems, with a view to profiting from and correcting the past errors of judgement which have led to disequilibrium, and preserving for all time that stability which remains.

7.9 The continued impoverishment of world ecosystems

Despite the possibility that minor fluctuations in emitted energy have taken place from time to time, it appears to be the case that the solar constant has remained relatively invariable, at least since the Cambrian period. Within the same timespan, the earth's atmosphere, ocean, and crust have been comprised of gases and chemical elements which are identical, and have been present largely in the same proportions to those found today. One can accordingly assume that the worldwide availability of energy and chemical elements for ecosystem growth has not been appreciably modified during the last several hundred million years. While specific environments might have been altered through mountain-building, vulcanism, glaciation, and similar phenomena, the overall balance of producers, consumers, and decomposers in the biosphere has also stayed approximately the same; and, although the total quantities of living biomass could theoretically have shown some considerable variation, even these, and the numbers of component species, do not seem to have been qualified to any great extent under natural conditions, certainly since the commencement of the Tertiary, and, perhaps, for very much longer than this.[132]

Although the initial disturbance of this basically stable milieu coincided with man's emergence as a thinking organism, it is likely that he, too, lived largely according to the confines of his natural habitat when he was a hunter, gatherer, and collector, for then his material resources—wood for fuel and shelter, wild fruits, roots, fish and game—were consumed so slowly in view of his limited numbers that they could be regarded as being essentially renewable. Probably the first major inroads into non-renewable resources occurred when he began to set fire and, then, to produce foodstuffs as opposed to merely collecting them. A further result of these changes in his daily and annual routine was that certain natural plant and animal assemblages came to be replaced by artificial ones designed to meet his own ends; and, since then, the shift from *wild* to *tame* ecosystems has been vastly accelerated. The consequences of these activities have been noted in full elsewhere in this book, though it is perhaps worthy of repetition to say that through them, in the course of time, man has simplified energy exchange patterns, reduced the qualitative and quantitative passage of chemical elements through the biosphere, imposed himself at the end of the food chain, and seriously interfered with the intricate interrelationships of many organisms, so as to induce those patterns of ecosystem impoverishment and disequilibrium which are found today in so many areas. These are manifested, for example, in the reduction of species numbers in secondary, as opposed to primary forest; in the widespread replacement of woodland by grassland; in the devastation of formerly good, fertile agricultural or grazing land through over-intensive use; in a general loss of biological productivity; in the massively augmented extinction rate of species as their habitats are disturbed; and in the instigation of pollution.

7.10 The need to maintain biological diversity

One of the most disturbing features of ecosystem impoverishment is a reduction in the total numbers of component species. This can be induced by means of niche destruction in particular habitats, as a result of direct extermination through hunting, or by excessive competition in those refuge areas to which species resort if their normal ranges are endangered (see also section 6.8). In all cases, the over-riding effect is to severely alter the balance of nature, for, while species loss under natural conditions takes place only when organisms have become unfit to survive in given environments and are ready to be superseded by newly evolved and better-adapted forms, man is quite capable of initiating the removal of plants and animals which are still well suited to play important roles in the biospheric tranfers of energy and nutrients, in which case the functional efficiency of those assemblages in which they formerly lived will be greatly lowered.

Moreover, once the species complexity of ecosystems has been greatly reduced, those populations which remain usually begin to show some considerable signs of disequilibrium. It was Gause who first provided mathematical and, later, experimental proof that single-species populations could be expected to fluctuate greatly in numbers, no matter what the state of their habitat might be. He also determined that the stability and, therefore, the potentiality for survival among all organisms is much enhanced in multispecies populations. Indeed, one can presume that the greater the species diversity in any plant-animal assemblage, the better the chances of a balanced interrelationship between each organism will be.

There are striking practical demonstrations of the efficacy of this premise in many parts of the world. It is instanced in the great instability of faunally- and floristically-impoverished oceanic islands, which has already been described; it is also exemplified in the massive fluctuations of population size occurring among a wide range of organisms whose habitats have been very much simplified by man, this being particularly the case for insect groups where monoculture is employed. It is further supported by the demographic stability which is so characteristic both of species-rich tropical rain forests, and of areas of traditional tropical slash-burn agriculture, whose complicated and interracting patterns of planting and harvesting are such that insect outbreaks seldom occur, and pests are rarely a problem. A naturally maintained ecological balance is also prevalent in many temperate-latitude communities, as in Wytham oakwoods near Oxford,[57] where the numbers of each of the 225 different insect species which live therein vary only slightly from year to year. It can be inferred from these examples that man might be wise to set in motion attempts to reverse the present trends of

365

ecosystem impoverishment by building up the species complexity of plant and animal communities everywhere. That this is a difficult task is not to be denied, for the biological and environmental inter-relationships which it involves are still poorly understood. However, even if entirely favourable results are not immediately forthcoming, the possibilities for ecosystem equilibrium will undoubtedly be enhanced by this course of action in the long run, to the eventual benefit of all life forms.

7.11 The need for conservation

The philosophical ideas which gave birth to practices of conservation are of ancient origin. Nevertheless, they were rarely expressed until the second half of the nineteenth century, when, for the first time, it was generally realized that man's old and basically stable relation-ships with his environment were rapidly deteriorating on a world-wide scale. They then reemerged almost as a by-product of the destructive extravagances of the industrial revolution, grew partly as a child of repulsion against the wholesale decimation of animals such as the American passenger pigeon and the Great Plains buffalo, and have since matured increasingly to evolve into a broadly based reac-tion against the wastemakers of modern society. Today, concepts of conservation vary widely, though it is agreed that the conservation movement undoubtedly seeks to preserve in balance the maximum rates of biological productivity, energy transfer, and chemical element exchange in all ecosystems, while at the same time upholding the quality of the environment and all suited organisms which live therein. To some, it is also explicitly involved in the protection of natural, semi-natural, or aesthetically-pleasing landscapes; to others, it is invoked in the innovation of planning schemes which set precise limits to zones of urban and industrial growth. It should also gener-ate an attitude of mind which encourages man to adopt a constructive and harmonic relationship with his environment, though, as we have seen, such attitudes are, as yet, not those which are followed by human populations as a whole.

Initially a North American concern, modern views of conserva-tion became a practical reality with the creation of the first National Park at Yellowstone in 1872, partly in order to provide a refuge for the scattered remnants of former Great Plains buffalo herds. At first, attempts were made merely to maintain existing habitats and their constituent organisms in almost static situations; but much

more sophisticated and dynamic techniques of landscape management are now customarily followed, so as to improve the quality of
ecosystems wherever it is possible to do so. However, the extent to
which this aim is eventually fulfilled depends upon a large number
of economic, social, and political considerations. At best, *nature
reserves* are set aside, in which hunting, and the collection of plants
and animals is forbidden or placed under strict surveillance, species
are kept in balance by trained conservationists, and the entire
landscape is sustained in a positive way. These reserves can be
extremely large, as in the case of the huge game parks of Africa and
North America; but they may also be envisaged as much smaller
units, such as those of the British Nature Conservancy, some of
which are only a few acres in size. Where the pressure of land use is
too great to permit the founding of reserves, *conservation areas* are
sometimes delimited instead. These are different in concept in that
they usually include within their boundaries a wide variety of
assorted human groups (e.g., wild fowlers in marshes, and the sheep-
farmers of heaths or moorlands) who still make their living off the
land and retain a considerable degree of overall control of land management; indeed, the activities of these people might be an absolute
prerequisite to the maintenance of the intrinsic character of such
areas. Nevertheless, gradually the administration and direction of all
protected sites is coming to fall more and more under the aegis of
national governmental agencies with a strong conservation bias,
supported by international organizations like the Conservation
Foundation and the World Wild Life Fund.

It should, however, be remembered that the adoption of a con-
servationally inspired philosophy of land management is necessary
to the retention of stability in all environments, and not just in those
specified as reserves, parks, or conservation areas. In practice, this
might mean, for example, the introduction of new, non-harmful
organisms to agricultural ecosystems so as to increase their prospects
of ecological equilibrium; the strict control of pollution substances;
or even simply the protection of milieus in which wild species have a
chance to survive. In respect of the last possibility, many natural-
ists in Britain and northern France have stressed the significance of
hedgerows, roadside verges, old railway embankments, and disused
canal sides as microhabitats which can support a diverse flora and
fauna, and have often succeeded in persuading local authorities to
keep these small refuge sites in a semi-natural state. Where their

removal is essential, as, for instance, in the augmentation of field sizes to pave the way for mechanized agriculture, farmers have at times been prevailed upon to compensate for this by setting copses on rough unused land, which effectively take their place. Similar habitats are catered for in some of the land diversification schemes prepared for parts of the USA. The extension of these ideas into other regions of the world, and the preservation of those semi-wild tracts of land which remain, might, in the long term, be absolutely indispensable to the future welfare of mankind and all other organisms, bearing in mind the desirability of encouraging, keeping, and imitating the inbuilt variety and functional efficiency of natural ecosystems in the increasingly artificial world in which we live.

REFERENCES

1. ADOLPH, E. F., *Physiology of Man in the Desert*, Interscience, New York, 1947.
2. ALLAN, H. H., Indigene versus alien in the New Zealand plant world, *Ecology*, Vol. 17, pp. 187–93, 1936.
3. ALLEN, G. M., Mammals of the West Indies, *Bull. Mus. Comp. Zoology, Harv.*, Vol. 54, pp. 175–263, 1911.
4. ALLEN, J. A., The influence of physical conditions in the genesis of species, *Radical review*, Vol. 1, pp. 108–40, 1877.
5. AMES, O., *Economic Annuals and Human Cultures*, Cambridge, Mass., 1939.
6. ANDERSON, E., *Plants, Man, and Life*, Boston, 1952.
7. AUBREVILLE, A., *Climats, forêts et désertification de l'Afrique tropicale*, Paris, 1949.
8. BAKER, H. G., *Plants and Civilization*, Belmont, California, 1964.
9. BANKS, A. L. and HISLOP, J. A., Sanitation practice and disease control in extending and improving areas for human habitation. In *Man's Role in Changing the Face of the Earth*, ed. W. L. THOMAS Jr., pp. 817–30, Chicago, 1956.
10. BARNICOT, N. A., Human nutrition: evolutionary perspectives. In *The Domestication and Exploitation of Plants and Animals*, eds. P. J. UCKO and G. W. DIMBLEBY, pp. 525–30, 1969.
11. BARTLETT, H. H., Fire, primitive agriculture, and grazing in the tropics. In *Man's Role in Changing the Face of the Earth*, ed. W. L. THOMAS Jr., pp. 692–720, Chicago, 1956.
12. BATCHELDER, R. B. and HIRT, H. F., Fire in tropical forests and grasslands, *US Army, Natick Lab. Techn. Report*, 67–41–ES, 1966.
13. BATES, M., Man as an agent in the spread of organisms. In *Man's Role*

in Changing the Face of the Earth, ed. W. L. THOMAS Jr., pp. 788–806, Chicago, 1956.

14. BATES, M., *Man in Nature*, Engelwood Cliffs, New Jersey, 1961.

15. BERGMANN, C., Uber die Verhältnisse der Wärmeökonomie der Thiere zu ihrer Grösse, *Göttinger Studien*, No. 8, 1848.

16. BERNSTEIN, F., *Variations und Erblichkeit-statistik*, Berlin, 1929.

17. BINFORD, L. R., Post-Pleistocene adaptations. In *New Perspectives in Archeology*, eds. L. R. BINFORD and S. R. BINFORD, Chicago, 1968.

18. BIRDSELL, J. B., Some population problems involving Pleistocene man, *Cold Spring Harb. Symp. Quant. Biol.*, Vol. 22, pp. 47–69, 1957.

19. BOSERUP, E., *The conditions of agricultural growth*, London, 1965.

20. BOYD, W. C., The contributions of genetics to anthropology, In *Anthropology today*, ed. A. L. KROEBER, pp. 488–506, Chicago and London, 1953.

21. BRADLEY, C. C., Human water needs, and water use in America, *Science*, Vol. 138, pp. 489–91, 1962.

22. BRAUN-BLANQUET, J., *Plant Sociology*, New York, 1932.

23. BRESLER, J. B., *Environments of Man*, Reading, Mass., 1968.

24. BROWN, E. S., Distribution of the ABO and Rh(D) blood groups in the north of Scotland, *Heredity*, Vol. 20, ph. 289–303, 1965.

25. BURKITT, D., A children's cancer dependent on climatic factors, *Nature*, Vol. 194, pp. 232–4, 1962.

26. BUTZER, K. W., *Environment and archeology*, Chicago, 1964.

27. CARLSON, L. D. and HSIEH, A. C. L., Cold. In *The Physiology of Human Survival*, eds. O. G. EDHOLM and A. L. BACHARACH, pp. 15–52, London, 1965.

28. CARR-SAUNDERS, A. M., *World Population: Past Growth and Present Trends*, Oxford, 1936.

29. CARTER, G. F., Movement of people and ideas across the Pacific. In *Plants and the Migration of Pacific Peoples*, ed. J. BARRAU, pp. 7–22, Honolulu, 1961.

30. CASTETTER, E. F., The vegetation of New Mexico, *New Mexico Quart.*, Vol. 26, pp. 270–2, 1956.

31. CHANDLER, T. J., The changing form of London's heat-island, *Geography*, Vol. 46, pp. 295–307, 1961.

32. CLARK, J. D., The spread of food production in sub-Saharan Africa, *Journ. of African Hist.*, Vol. 3, pp. 211–28, 1962.

33. CLARK, J. G. D., Forest clearance and prehistoric clearing, *Econ. Hist. Rev.*, Vol. 17, pp. 45–51, 1947.

34. CLARK, W. E. LE GROS, *The fossil evidence for human evolution*, 2nd edn, Chicago, 1964.

35. CLAYTON, W. D., Secondary vegetation and the transition to savanna near Ibadan, Nigeria, *J. Ecol.*, Vol. 46, pp. 217–38, 1958.

36. COLE, M. M., Cerrado, caatinga and pantanal: the distribution and origin of the Savanna vegetation of Brazil. *Geog. Journ.*, Vol. 126, pp. 168–79, 1960.

37. COLLINS, C. W., Two elm scolytids in relation to areas infected with the Dutch elm disease fungus, *J. Econ. Ent.*, Vol. 31, pp. 192–5, 1938.

38. COOKE, H. B. S., MOREAU, R. E. and LEAKEY, L. S. B., Transcript of discussions. In *African Ecology and Human Evolution*, eds. F. C. HOWELL and F. BOURLIERE, p. 621, London, 1964.

39. COOKE, M. T., The spread of the European starling in North America, *Circ. US Dept. Agric.* No. 40, pp. 1–9, 1928.

40. COON, C. S., *The Origin of Races*, New York, 1962.

41. COON, C. S., GARN, S. M., and BIRDSELL, J. B., *Races: A study of the Problems of Race Formation in Man*, Springfield, Illinois, 1950.

42. COOPER, C. F., An annotated bibliography of the effects of fire on Australian vegetation, *Mimeo. Soil Cons. Auth. of Victoria, Australia*, 1963.

43. COWLES, R. B., Some ecological factors bearing on the origin and evolution of pigment in the human skin, *Am. Nat.*, Vol. 93, pp. 283–93, 1959.

44. CUMBERLAND, K. B., Man's role in modifying island ecosystems in the south-west Pacific. In *Man's Place in the Island Ecosystem*, ed. F. R. FOSBERG, pp. 187–205, Honolulu, 1963.

45. DARBY, H. C., The clearing of the English woodlands, *Geography*, Vol. 36, pp. 71–83, 1951.

46. DARBY, H. C., The clearing of the woodland in Europe, In *Man's Role in Changing the Face of the Earth*, ed. W. L. THOMAS Jr., pp. 183–216, Chicago, 1956.

47. DARLINGTON, C. D., *Chromosome Botany and the Origins of Cultivated Plants*, 2nd edn. New York, 1963.

48. DARLINGTON, C. D., The silent millenia in the origin of agriculture. In *The Domestication and Exploitation of Plants and Animals*, eds. P. J. UCKO and G. W. DIMBLEBY, pp. 67–72, 1969.

49. DART, R. A., The osteodontokeratic culture of *Australopithecus prometheus*, *Transv. Mus. Mem.*, No. 10, 1957.

50. DARWIN, C., *The Descent of Man*, London, 1871.

51. DAUBENMIRE, R., Ecology of fire in grasslands. In *Advances in Ecological Research*, ed. J. B. CRAGG, Vol. 5, pp. 209–66, 1968.

52. DENEVAN, W. M., The upland pine forests of Nicaragua, *Univ. of Calif. Publs. in Geog.*, Vol. 12, pp. 251–320, 1961.

53. DOBZHANSKY, T., Genetic entities in hominid evolution. In *Classification and Human Evolution*, ed. S. L. WASHBURN, pp. 347–62, New York, 1963.

54. DROWER, M. S., The domestication of the horse. In *The Domestication and Exploitation of Plants and Animals*, eds. P. J. UCKO and G. W. DIMBLEBY, pp. 471–8, 1969.

55. DUBOS, R. *Mirage of Health*, New York, 1959.

56. EISELEY, L. D., Man the fire-maker, *Sci. Am.*, Vol. 191, pp. 52–7, 1954.

57. ELTON, C. S., *The Ecology of Invasions*, London, 1958.

58. FIENNES, R., *Man, Nature and Disease*, New American Library, New York, 1965.

59. FLANNERY, K. V., Origins and ecological effects of early domestication in Iran and the Near East. In *The domestication and Exploitation of Plants and Animals*, eds. P. J. UCKO and G. W. DIMBLEBY, pp. 73–100, 1969.

60. FONAROFF, L. S., Man and malaria in Trinidad: ecological perspectives

of a changing health hazard, *Annals of the Association of American Geographers*, Vol. 58, pp. 526–56, 1968.

61. Fox, R. H., Heat. In *The Physiology of Human Survival*, eds. O. G. Edholm and A. L. Bacharach, pp. 53–80, Academic, London, 1965.

62. Geertz, C., *Agricultural Involution: The Processes of Ecological Change in Indonesia*, Berkeley and Los Angeles, 1966.

63. Glass, B., Genetic changes in human populations, especially those due to gene flow and genetic drift, *Adv. Genet.*, Vol. 6, pp. 95–139, 1954.

64. Gleason, H. A., The vegetational history of the Middle West, *Ann. Ass. Am. Geogr.*, Vol. 12, pp. 39–85, 1932.

65. Godwin, H., Neolithic forest clearance, *Nature*, Vol. 153, p. 411, 1944.

66. Gottman, J., Megalopolis, or the urbanization of the Northeastern Seaboard, *Econ. Geog.*, Vol. 33, pp. 189–200, 1957.

67. Harlan, J. R., A wild harvest in Turkey, *Archaeology*, Vol. 20, pp. 197–201, 1967.

68. Harlan, J. R. and Zohary, D., Distribution of wild wheats and barley, *Science*, Vol. 153, pp. 1074–80, 1966.

69. Harris, D. R., Plants, animals and man in the Outer Leeward islands, West Indies, *Univ. of Calif. Publ. in Geogr.*, Vol. 18, 1965.

70. Harris, D. R., Recent plant invasions in the arid and semiarid southwest of the United States, *Ann. Ass. Am. Geogr.*, Vol. 56, pp. 408–422, 1966.

71. Harris, D. R., New light on plant domestication and the origins of agriculture: a review, *Geogr. Rev.*, Vol. 57, pp. 90–107, 1967.

72. Harris, D. R., Agricultural systems, ecosystems and the origins of agriculture, In *The Domestication and Exploitation of Plants and Animals*, eds. P. J. Ucko and G. W. Dimbleby, pp. 3–16, 1969.

73. Hatcher, J. D., Acute anoxic anoxia. In *The Physiology of Human Survival*, eds. O. G. Edholm and A. L. Bacharach, pp. 81–120, London, 1965.

74. Hawkes, J. G., The ecological background of plant domestication. In *The Domestication and Exploitation of Plants and Animals*, eds. P. J. Ucko and G. W. Dimbleby, pp. 17–30, 1969.

75. Heyerdahl, T., *American Indians in the Pacific; the theory behind the Kon-tiki expedition*, London, 1952.

76. Heyerdahl, T., Prehistoric voyages as agencies for Melanesian and South American plant and animal dispersal to Polynesia. In *Plants and the migrations of Pacific peoples*, ed. J. Barrau, pp. 23–36, Honolulu. 1961.

77. Hiernaux, J., Some ecological factors affecting human populations in sub-Saharan Africa. In *African Ecology and Human Evolution*, eds. F. C. Howell and F. Bourliere, pp. 534–46, London, 1964.

78. Higgs, E. S., Early domesticated animals in Libya. In *The Background to Evolution in Africa*, eds. W. W. Bishop and J. D. Clark, pp. 165–73, 1967.

79. Hills, T. L. and Randall, R. E., eds., The ecology of the forest-savanna boundary, *I.G.U. Humid Tropics Commission Symposium, 1964*, Montreal, 1968.

80. HOLE, F. and FLANNERY, K. V., The pre-history of southeastern Iran: a preliminary report, *Proc. Prehist. Soc.*, Vol. 33, pp. 147–206, 1967.

81. HOWELL, F. C. and CLARK, J. D., Acheulian hunter-gatherers of sub-Saharan Africa. In *African Ecology and Human Evolution*, eds. F. C. HOWELL and F. BOURLIERE, pp. 458–533, London, 1964.

82. HUTCHINSON, J. B., History and relationship of the World's cottons, *Endeavour*, Vol. 21, pp. 5–15, 1962.

83. HUTCHINSON, J. B., ed., *Essays on Crop Plant Evolution*, London, 1965.

84. HUTCHINSON, J. B., SILOW, R. A., and STEPHENS, S. G., *The Evolution of Gossypium*, Oxford, 1947.

85. IVERSEN, J., *The Influence of Prehistoric Man on Vegetation*, Copenhagen, 1949.

86. IWANAMI, Y. and IIZUMI, S., On the relations between the burning temperature and the amounts of fuels in natural grassland, 1. Measurements of burning temperatures with varying amounts of *Miscanthus sinensis, Tohoku Univ. Sci. Dept. Res. Inst (D)*, No. 17, pp. 27–33, 1966.

87. KLEIN, L., *Aspects of River Pollution*, London, 1957.

88. KOMAREK, E. V., The natural history of lightning, *Tall Timbers Fire Ecol. Conf.*, Vol. 3, pp. 139–83, 1964.

89. KRAUS, G., *Geschichte der Pflanzeneinführungung in die europäischen botanischen Gärten*, Leipzig, 1894.

90. KROEBER, A. L., Culture element distributions, XI: Tribes surveyed, *Univ. of Calif. Anthropological Records*, Vol. 1, pp. 435–40, 1939.

91. KUHNHOLTZ-LORDAT, G., La terre incendée, Nimes, 1939.

92. LADELL, W. S. S., Water and salt (Sodium chloride) intakes. In *The Physiology of Human Survival*, eds. O. G. EDHOLM and A. L. BACHARACH, pp. 235–91, London, 1965.

93. LANSBERG, H. E., The climate of towns. In *Man's Role in Changing the Face of the Earth*, ed. W. L. THOMAS Jr., pp. 584–606, Chicago, 1956.

94. LAIRD, M., Rats, coconuts, mosquitoes and filariasis. In *Pacific Basin Biogeography*, ed. J. L. GRESSITT, pp. 535–42, Honolulu, 1961.

95. LAWRENCE, B., Early domestic dogs, *Z. Saugetieck*, Vol. 32, pp. 44–59, 1967.

96. LEAKEY, L. S. B., Facts instead of dogmas on man's origin. In *The Origin of Man*, ed. P. L. DEVORE, pp. 3–16, Chicago, 1965.

97. LEAKEY, L. S. B. and LEAKEY, M. D., Recent discoveries of fossil hominids in Tanganyika: at Olduvai and near Lake Natron, *Nature*, Vol. 202, pp. 5–7, 1964.

98. LEAKEY, L. S. B., EVERNDEN, J. F., and CURTIS, G. H., Age of Bed I, Olduvai Gorge, Tanganyika, *Nature*, Vol. 191, pp. 478–9, 1961.

99. LEARMONTH, A. T. A., Some contrasts in the regional geography of malaria in India and Pakistan, *Institute of British Geography, Transactions and Papers*, Vol. 23, pp. 37–60, 1957.

100. LUNDMAN, B., Geography of the human blood groups, *Evolution*, Vol. 2, pp. 231–7, 1948.

101. MACNEISH, R. S., The origins of New World civilization, *Sci. Am.*, Vol. 211, pp. 29–37, 1964.

102. MARSH, G. P., *Man and Nature*, New York, 1864.

103. MASSON, H., Temperature du sol au cours d'un feu de brousse au Senegal, *2nd Inter-African Soils Conf.*, Leopoldsville, 1954.

104. MELLART, J., A neolithic city in Turkey, *Sci. Am.*, Vol. 210, pp. 94–104, 1964.

105. MERYMAN, H. T., Tissue freezing and local cold injury, *Physiol. Rev.*, Vol. 37, pp. 233–51, 1957.

106. MITCHELL, H. H. and EDMAN, E., *Nutritional and Climatic Stress*, Springfield, Mass., 1951.

107. MONGE, C., *Acclimatization in the Andes: The Historical Confrontation of Climatic Aggression in the Development of the Andean Man* (trans. by D. F. BROWN), Baltimore, 1948.

108. MOVIUS, H. L., The lower Palaeolithic cultures of southern and eastern Asia, *Trans. Am. Phil. Soc. N.S.*, Vol. 55, pp. 379–420, 1949.

109. NEWMAN, M. T., The application of the ecological rules to the racial anthropology of the aboriginal New World, *Am. Anthrop.*, Vol. 55, pp. 311–27, 1953.

110. NEWMAN, M. T., Biological adaptation of man to the environment: heat, cold, altitude and nutrition, *Ann. New York Acad. Sci.*, Vol. 91, pp. 617–33, 1961.

111. NEWMAN, M. T., Ecology and nutritional stress in man, *Am. Anthrop.*, Vol. 64, pp. 23–33, 1962.

112. OAKLEY, K. P., The earliest tool-makers and the earliest fire-makers, *Antiquity*, Vol. 30, pp. 4–8 and 102–07, 1956.

113. OAKLEY, K. P., On man's use of fire, with comments on tool-making and hunting. In *Social Life of Early Man*, ed. S. L. WASHBURN, pp. 176–93, New York, 1961.

114. PARSONS, J. J., The Miskito pine savanna of Nicaragua and Honduras, *Ann. Ass. Am. Geogr.*, Vol. 45, pp. 36–63, 1955.

115. PATRICK, R., A proposed biological measure of stream conditions, *Proc. Acad. Nat. Sci., Philadelphia*, Vol. 101, pp. 277–341, 1949.

116. PROTHERO, M., *Migrants and Malaria*, London, 1965.

117. PUGH, L. G. C. E., High altitudes. In *The Physiology of Human Survival*, eds. O. G. EDHOLM and A. L. BACHARACH, pp. 121–52, London, 1965.

118. REED, C. A., The pattern of animal domestication in the prehistoric Near East. In *The Domestication and Exploitation of Plants and Animals*, eds. P. J. UCKO and G. W. DIMBLEBY, pp. 361–80, 1969.

119. RIDLEY, H. N., *The Dispersal of Plants Throughout the World*, Ashford, Kent, 1930.

120. ROBERTS, J. A. F., Blood group frequencies in North Wales, *Ann. Eugenics*, Vol. 11, pp. 260–71, 1942.

121. ROBINSON, J. T., Adaptive radiation in the australopithecines, and the origin of man. In *African Ecology and Human Evolution*, ed. F. C. HOWELL and F. BOURLIERE, pp. 385–416, London, 1964.

122. RODDEN, R. J., An early Neolithic village in Greece, *Sci. Am.*, Vol. 212, pp. 82–92, 1965.

123. SAND, R., *The Advance to Social Medicine*, London, 1952.

124. SAUER, C. O., *Geography of the Pennyroyal*, Frankfort, 1927.

125. SAUER, C. O., Environment and culture during the last deglaciation, *Proc. Am. Philosoph. Soc.*, Vol. 92, pp. 65–77, 1948.

126. SAUER, C. O., *Agricultural Origins and Dispersals*, American Geographical Society, 1952.

127. SCHMEIDER, O., The pampa, *Univ. of Calif. Publ. in Geogr.*, Vol. 2, pp. 255–70, 1927.

128. SCHWANITZ, F., *The Origin of Cultivated Plants*, Harvard, 1967.

129. SEAMAN, G. A., The mongoose and Caribbean wild life, *Trans. 17th North Am. Wildlife Conf.*, Vol. 17, pp. 188–97, 1952.

130. SEARS, P. B., *Life and Environment*, New York, 1939.

131. SIMONS, E. L., The early relatives of man, *Sci. Am.*, Vol. 211, pp. 51–62, 1964.

132. SIMPSON, G. G., *The Geography of Evolution*, Philadelphia, 1965.

133. SMYTH, J. D., *Introduction to Animal Parasitology*, London, 1962.

134. SOLHEIM, W. G., Southeast Asia and the West, *Science*, Vol. 157, pp. 896–902, 1967.

135. STEARN, W. T., The origin and later development of cultivated plants, *J. Roy. Hort. Soc.*, Vol. 90, pp. 279–91 and 322–40, 1965.

136. STERN, C., *Principles of Human Genetics*, 2nd edn, San Francisco, 1960.

137. STEWART, O. C., Burning and natural vegetation in the United States, *Geogr. Rev.*, Vol. 41, pp. 317–20, 1951.

138. STEWART, O. C., Fire as the first great force employed by man. In *Man's Role in Changing the Face of the Earth*, ed. W. L. THOMAS Jr., pp. 115–33, Chicago, 1956.

139. THOMAS, H. E., Changes in quantities and qualities of ground and surface water. In *Man's Role in Changing the Face of the Earth*, ed. W. L. THOMAS Jr., pp. 542–66, Chicago, 1956.

140. TOBIAS, P. V., *Australopithecus, Homo habilis*, tool-using and tool-making man, *S. Afr. Acheol. Bull.*, Vol. 20, pp. 167–92, 1965.

141. TOBIAS, P. V., The cranium and maxillary dentition of *Australopithecus (Zinjanthropus) boisei*, Vol. 2 of *Olduvai Gorge*, ed. L. S. B. LEAKEY, Cambridge, 1967.

142. UCKO, P. J. and DIMBLEBY, G. W., eds., *The Domestication and Exploitation of Plants and Animals*, London, 1969.

143. VAVILOV, N. I., The origin, variation, immunity and breeding of cultivated plants (trans. by K. S. CHESTER), *Chronica bot.*, Vol. 13, 1949–50.

144. VIOSCA, P., Spontaneous combustion in the marshes of southern Louisiana, *Ecology*, Vol. 12, pp. 439–42, 1931.

145. WALLACE, A. R., Geological climates and the origin of species, *The Quart. Rev.*, Vol. 126, pp. 359–94, 1869.

146. WALTER, W., Die Waldveranderungen in Mitteleuropa in historische zeit, *Grundlager der Pflanzenverbreitung, II, Arealkunde*, Stuttgart, 1954.

147. WASHBURN, S. L., An ape's-eye view of human evolution. In *The Origin of Man*, ed., P. L. DEVORE, pp. 89–96, Chicago, 1965.

148. WASHBURN, S. L. and PATTERSON, B., The evolutionary importance of the South African 'man-apes', *Nature*, Vol. 167, pp. 650–1, 1951.

149. WATTS, D., Human occupance as a factor in the distribution of the

California Digger Pine (*Pinus sabiniana*), *Unpublished M.A. dissertation, University of California at Berkeley*, 1959.

150. WATTS, D., Man's influence on the vegetation of Barbados, 1627 to 1800, *Occasional Papers in Geography*, University of Hull, No. 4, Hull, UK, 1966.

151. WATTS, D., Origins of Barbadian cane-hole agriculture, *Barbados Mus. Hist. Soc. J.*, Vol. 32, pp. 143–151, 1968.

152. WATTS, D., Persistence and change in the vegetation of oceanic islands, *Can. Geogr.*, Vol. 14, pp. 91–109, 1970.

153. WEAVER, J. E., *North American Prairie*, Lincoln, Nebraska, 1954.

154. WEIDENREICH, G. *Apes, Giants and Man*, Chicago, 1946.

155. WEST, O., Fire and vegetation, and its use in pasture management, with special reference to tropical and subtropical Africa, *Commonw. Bur. Pastures and Crops*, England, 1965.

156. WIENER, A. S., *Blood Groups and Transfusion*, Springfield, Illinois, 1943.

157. WILKINSON, H. R., Man and the natural environment, *Occasional Papers in Geography*, University of Hull, No. 1, Hull, UK, 1963.

158. WITTFOGEL, H., *Oriental Despotism and Hydraulic Society*, New Haven, Conn., 1957.

159. WYMSTRA, T. A. and HAMMEN, T. VAN DER, Some palynological data on the history of tropical savannas in northern South America, *Leid. Geol. Meded.*, Vol 38, pp. 71–90, 1966.

160. YOSHIMURA, H., Seasonal changes in body fluids, *Jap. J. Physiol.*, Vol. 8, pp. 165–9, 1958.

161. YUDKIN, J., Archaeology and the nutritionist. In *The Domestication and Exploitation of Plants and Animals*, eds. P. J. UCKO and G. W. DIMBLEBY, pp. 547–54, London, 1969.

162. ZIMMERMAN, R. C., Chestnut blight in the forests of southern Switzerland, *Geogr. Rev.*, Vol. 55, pp. 99–104, 1965.

163. ZINSSER, H., *Rats, Lice and History*, Boston, 1935.

164. ZOHARY, D., The progenitors of wheat and barley in relation to domestication and agricultural dispersal in the Old World. In *The Domestication and Exploitation of Plants and Animals*, eds. P. J. UCKO and G. W. DIMBLEBY, pp. 47–66, London, 1969.

165. ZUKHOVSKIJ, P. M., *Cultivated Plants and Their Wild Relatives* (abridged trans. by P. S. HUDSON), Commonw. Agric. Bur., Farnham Royal, Bucks., 1962).

Glossary

Most of the technical terms utilized herein have been explained at appropriate points in the text; and these can be located by referring to those page numbers which are *italicized* in the index. A comprehensive glossary is therefore considered to be unnecessary. However, for reasons of space or convenience, it has not always proved possible to give precise definitions of every term; consequently, a few whose usage might be unfamiliar to certain readers have been selected for inclusion in the limited glossary which follows.

ADSORPTION. The concentration of a substance at the interface between two dissimilar materials, e.g., the adherence of a liquid, or a dissolved substance at a solid surface.

ALGAE. Marine and freshwater chlorophyll-producing organisms.

AMINO-ACID. An organic acid with both basic and alkaline affinities. There are 39 known amino-acids, many of which play important roles in protein synthesis.

ANGIOSPERMAE. The 'flowering plants': those vascular plant species which bear their seeds in a hollow ovary (the fruit).

APHID (APHIDOIDEA). Plant-lice, which do a great deal of damage by feeding off the sap of leaves, and exuding a sticky substance from their abdominal tubes.

ATMOMETER. An instrument designed to measure the evaporative potential of the atmosphere at the earth's surface.

CATENA. A group of closely-associated soil types, which, nevertheless, differ in their essential characteristics because of local variations in relief and drainage. Catenas are often well developed on hillslopes.

COLLOID. A non-crystalline semi-soluble substance whose molecules have not completely disbanded as in a true solution, but have stayed grouped together to form solute 'particles'. Colloidal substances can usually take in large quantities of liquid.

CULTIGEN. A cultivated plant species.

376

DIATOMS. Microscopic, unicellular or colonial algae which comprise the class Bacillariophyceae of the phylum Chrysophyta.

ECESIS. The process of plant and animal colonization.

ENDOGENOUS RHYTHM. A type of biological rhythm which coincides in period roughly with the length of the solar day or year, or the lunar month.

ENTROPY. A term referring to the extent to which the energy of any system has been degraded. The higher the entropy, the more energy *unavailable* to work in the system.

EUPHAUSIACEA. A group of oceanic shrimplike organisms which serve as a major source of food for whales.

EUSTATIC CHANGE. A change in sea level induced by the formation, or by the melting of polar ice caps.

EVAPOTRANSPIROMETER. An instrument which measures the amount of water lost in the process of evaporation and transpiration within unit area and unit time.

GUTTATION. The natural exudation of water from an uninjured plant surface.

GYMNOSPERMAE. Those vascular 'naked-seed' plant species whose seeds are produced without an ovary and lie in an exposed position on the plant. The conifers are the most conspicuous and most numerous of existing gymnosperms.

HAEMOGLOBIN. A pigment in the blood of vertebrates (and some other organisms), which allows respiration. It is bright red when oxygenated, and bluish-red on other occasions.

HOMEOSTACY. The maintenance of a relatively stable internal thermal environment in higher animals; or of stable conditions in in animal groups.

HOMEOTHERM. A warm-blooded animal which maintains a relatively uniform body temperature regardless of what external environmental temperatures might be.

HYDROCARBON. One of a large class of organic compounds which contain only carbon and hydrogen. They occur frequently in combustible fuels, such as coal, natural gas, and petroleum.

HYGROSCOPIC FORCE. The force which holds moisture firmly onto the surface of soil particles.

HYPERTHERMIA. An above-normal increase in body temperatures among certain animals (e.g., man).

HYPOTHERMIA. A below-normal cooling of body temperatures among animal species.

ION. An atom, or group of atoms, which carry a strong electric charge when combined in a molecule, as a result of having gained or lost one or more electrons. ANIONS are negatively-charged ions; CATIONS are positively-charged.

LYSIMETER. An instrument which measures the amount of water percolating through a given depth of soil.

MEGALOPOLIS. An amalgamation (or potential amalgamation) of major conurbations, e.g., those on the eastern seaboard of the USA between Boston and Washington DC.

MOTILE. Capable of movement.

MYCOPHAGE. An organism which eats off fungi.

MYCORHIZZA. The symbiotic association of fungi with roots of certain seed plants.

OLEFIN. An open-chain hydrocarbon.

OSMOTIC PRESSURE. The pressure of a solvent passing through a permeable membrane.

PEPTIDE. A chain of linked amino-acids. Some are extremely complex.

PHOTOOXIDATION. The oxidation of chemical compounds which is encouraged by the presence of light.

PLANKTON. A number of plant and animal species which live close to the surface of water bodies. They are usually microscopic in size, consisting mainly of unicellular algae, protozoa, small crustacea, and the larvae of other invertebrates. Usually the algae form the base of the food chain upon which the life of all other marine organisms depends.

POIKILOTHERM. An organism with a variable body temperature which is usually slightly warmer than its external environment (e.g., a frog).

PROTEIN. Complex nitrogenous compounds comprised of numerous amino-acid molecules which have been linked together. A basic constituent of all life forms. The permutations of possible amino-acid combinations are almost endless, so that it is hardly surprising that relatively few proteins have been thoroughly analysed.

PYRHELIOMETER. An instrument which measures the total quantity of incoming solar energy, as received at the earth's surface at normal incidence.

RADIOMETER. An instrument which is capable of measuring electromagnetic radiation received from the sun.

ROTIFER (ROTIFERA). Minute unsegmented acquatic animals which are usually limited to fresh, or slightly-brackish water. Many have very precise ranges of tolerance.

RUDERAL. Pertaining to rubbish heaps.

TRANSHUMANCE. The seasonal movement of livestock (especially sheep) between valley and mountain pastures.

TRICLAD (TRICLADIDA). Free-living flatworms, in which the gut is divided into three main sections.

Author Index

Alexander, P., 96, 97, 98
Alexander, W. B., 225
Allan, H. H., 357
Allard, H. A., 133
Allard, R. W., 285
Allee, W. C., 206, 207, 218, 220, 229
Allen, J. A., 314
Allen, S. E., 64, 66, 67, 88
Alway, F. J., 153
Andersson, G., 267, 268
Andrewartha, H. G., 221, 222, 223
Ashton, P. S., 289
Asprey, G. F., 130
Axelrod, D. I., 287

Baker, H. G., 340
Ball, R. C., 21
Banks, A. L., 327
Bartlett, H. H., 339
Basilevic, N. I., 44, 53, 105, 106, 108, 109
Batchelder, R. B., 339
Bates, M., 209
Bergmann, C., 314
Bertrand, D., 57
Billings, W. D., 130, 132, 137, 172
Binford, L. R., 342
Birch, H. F., 90
Birch, L. C., 84, 85, 216, 221, 222, 223, 227
Birdsell, J. B., 313
Birge, E. A., 28
Black, R. F., 136
Blackman, G. E., 134
Boserup, E., 344

Bouillenne, R., 294
Bourliere, F., 44, 45
Boyd, W. C., 320
Braun-Blanquet, J., 336
Bray, J. R., 37, 38, 42, 230
Breidenbach, A. W., 94
Bresler, J. B., 322
Briggs, B. G., 289
Brougham, R. W., 38
Bryant, F. J., 99
Budyko, M. I., 15, 162
Burges, A., 74
Burkitt, D., 330

Cain, A. J., 283
Cain, S. A., 3, 261
Candolle, A. L., de., 157
Carlisle, A., 66
Carlquist, S., 290
Carr-Saunders, A. M., 359
Carson, R., 21
Carter, G. F., 347
Chadwick, M. J., 147
Chandler, T. J., 360
Chapman, R. N., 206
Chapman, V. J., 172, 181
Chitty, D., 200
Chitty, H., 200
Choudhuri, G. N., 181
Clark, J. D., 335
Clark, J. G. D., 348
Clark, W. E. Le Gros, 309
Clarke, D. P., 84, 85
Clarke, G. L., 177, 178
Clements, F. E., 227, 257, 259

379

Cloudsley-Thompson, J. L., 147
Cole, L. C., 217
Cole, M. M., 338
Coleman, D. C., 100
Coombe, D. E., 183
Coon, C. S., 314
Cooper, C. F., 338
Cooper, J. P., 37
Cooper, W. S., 127
Cowles, R. B., 314
Crick, F. H. C., 63
Crisp, D. T., 88, 89
Crombie, A. C., 227
Cumberland, K. B., 338
Cunningham, W. J., 207

Dansereau, P., 3, 249, 251, 253, 255, 257.
Darby, H. C., 348
Darlington, C. D., 283, 297, 340, 341
Darlington, P. J., 256
Dart, R. A., 310
Darwin, Charles, 2, 82, 214, 234, 243, 275, 290, 297, 313
Daubenmire, R. F., 131, 153, 336
Davies, P. W., 224
Dawson, F. R., 69
Dayhoff, M. O., 277
Deevey, E. S., 202, 203
Dempster, J. P., 20
Denevan, W. M., 338
Dimbleby, G. W., 340
Dobzhansky, T. H., 283, 286, 288, 321
Doering, C. R., 202
Domin, K., 260
Douglas, I., 87
Dowdeswell, W. H., 200, 216
Dubos, R., 331

Edwards, C. A., 93
Eidman, F. E., 149
Eiseley, L. D., 335
Ekman, S., 256
Elton, C. S., 20, 23, 24, 198, 225, 230, 231, 233, 252, 356
Emerson, A. E., 229
Emiliani, C., 262, 265
Eriksson, E., 102
Evans, C. G., 66
Evans, F. C., 39
Evans, L. T., 182
Eyre, S. R., 260

Farrow, E. P., 187
Federov, A. A., 289
Firbas, F., 102
Fisher, R. A., 200
Flannery, K. V., 342, 343
Fogg, G. E., 14, 80
Forbes, E., 263, 269
Ford, E. B., 200, 214, 215, 216
Ford, H. D., 200, 214
Fosberg, F. R., 254
Fowells, H. A., 88
Fox, R. H., 324
Franklin, B., 2
Frenkiel, F. N., 105
Frost, L. C., 183
Funk, H. B., 90

Gaertner, E. E., 151
Galston, A. W., 56
Garfine, V., 37
Garner, W. W., 133
Gates, D. M., 10, 12, 13, 19, 282
Gause, G. F., 206, 228, 365
Geer, G. de, 266
Geiger, R., 138, 148, 149
George, W., 246, 252
Gerloff, G. C., 42, 43
Gessner, F., 37
Gibb, J., 226
Gilbert, O., 228
Gleason, H. A., 260, 338
Glinka, K. D., 179
Glock, W. S., 267
Godwin, H., 271, 348
Golley, F. B., 16
Good, R., 249, 250, 251
Gorham, E., 42
Gottman, J., 360
Green, F. H. W., 163
Gressitt, J. L., 253
Grime, J. P., 128
Grinnell, J., 230, 231
Grisebach, A. H. R., 157
Gross, A. O., 210
Guppy, H. B., 253, 254

Haagen-Smit, A. J., 105
Haeckel, E., 121
Hardin, G., 228
Hare, F. K., 160, 162, 163, 164, 179
Harris, D. R., 340, 344, 356, 357, 358
Harvey, H. W., 25
Hawkes, H. E., 58, 59

Hawkes, J. G., 340
Heyerdahl, T., 348
Hiernaux, J., 315
Hirt, H. F., 339
Hislop, J. A., 327
Hodges, J. D., 38
Hofmann, G., 146
Hole, F., 342
Hooper, F. F., 21
Hoppe, E., 149
Hoshino, M., 38
Howard, L. E., 225
Hughes, A. P., 127
Humboldt, A. von, 2
Hunt, L. A., 37
Hutchinson, G. E., 71, 231
Hutchinson, J. B., 340
Huxley, J. S., 226

Iversen, J., 140, 348
Ivlev, U. S., 40

Jackson, R. M., 68, 72, 73, 77, 81
Jahnke, L. S., 37
Jarvis, C. S., 150
Jeffries, R. C., 154
Jensen, J., 102
Jensen, P. B., 198
Jones, J. R. E., 23
Jones, R. L., 146
Jorgensen, C. A., 102
Juday, C., 28
Junge, C. E., 102

Kendrew, W. G., 145
Kettlewell, H. B. D., 200
Kinne, O., 124
Kittredge, J., 147, 148
Kleiber, M., 46
Klopfer, P. M., 230
Kopec, A. C., 316, 318
Köppen, W., 157, 158, 159, 160
Koutler-Andersson, E., 102
Kramer, P. J., 131
Kraus, G., 356
Krulwich, T. A., 90
Kuenzler, E. J., 21
Kuhnholtz-Lordat, G., 336
Kulp, J. L., 98
Kuriowa, S., 37

Lack, D., 221, 222, 223, 226, 227
Lamb, N., 264

Langenheim, J. H., 121
Larsen, E. B., 122
Lavoisier, A. L., 27
Lawrence, D. B., 37
Leakey, L. S. B., 306, 309, 311, 334
Leakey, M. D., 311, 334
Lee, C. H., 151
Lems, K., 253, 254, 255
Leslie, P. H., 200, 216
Levitt, J., 123
Lichtenberg, J. J., 94
Leibig, J., 53, 121
Lincoln, F. C., 200
Lindemann, R. L., 27, 28
Linnaeus, C., 2, 276
Litav, M., 69
Livingston, B. E., 121
Loach, K., 131
Lotka, A. J., 198, 211, 221, 228
Lounamaa, J., 105
Lundman, B., 319
Lyell, C., 263

Macarthur, J. W., 230
Macarthur, R. H., 230, 297
McCabe, L. C., 103
McDougall, B. M., 84
MacFadyen, A., 100
McWhirter, K. G., 216
Main, A. R., 291
Malmer, N., 67
Malthus, T., 2, 204
Manley, G., 264
Marples, T. G., 21
Marsh, G. P., 338, 360
Martin, P. S., 296
Mason, B., 53, 57
Mason, H. L., 121
Mather, J. R., 162, 164, 165, 166,
    167, 168
Mather, K., 283
Matthews, J. R., 271
Mattson, S., 102
Mayr, E., 256, 284, 286, 291, 296,
    297
Meiklejohn, J., 81, 90
Meinzer, O. E., 151
Melin, E., 69
Mendel, G., 2, 275
Menhinick, E. F., 36, 42, 43
Merriam, C. H., 173
Miller, A. A., 141
Miller, R. S., 227, 230, 231, 232, 233

Milne, A., 227
Mindermann, G., 74
Minshall, G. W., 33
Mitchell, G. F., 272
Mitchell, R., 229
Monteith, J. L., 146, 154
Moore, N. W., 93
Moreau, R. E., 291
Morgan, T. H., 278
Morrison, T. M., 69
Mourant, A. E., 316, 318
Mulla, M. S., 93
Munger, H. P., 103
Munro, P. E., 90
Murie, A., 202
Murray, D. B., 131

Newbould, P. J., 40, 67
Newman, M. T., 314, 324
Nichols, G. E., 260
Nichols, R., 131
Nicholson, A. J., 220, 221, 223,
Niklewski, B., 90
Nilsson, H., 69
Nishimura, S., 38
Nitsch, J., 133
Nobs, M. A., 288
Noyes, B., 202

Oakley, K. P., 309, 334
Oberlander, G. T., 144
Odum, E. P., 21, 22, 25, 30, 33, 34,
   39, 40, 41, 71, 99, 202, 203, 212,
   217, 234, 236
Odum, H. T., 25, 26, 32, 38
Okubo, T., 38
Olson, J. S., 40
Oosting, H. J., 131, 172, 176
Ovington, J. D., 37, 67

Park, K. J. F., 67
Parker, S. L., 202
Parsons, J. J., 338
Patrick, R., 362
Patterson, B., 309
Patton, R. T., 181
Pavlovsky, D., 288
Pearl, R., 202, 204, 205, 207
Pearl, R. Y., 170
Penfound, W. T., 36
Penman, H. L., 149, 163, 164, 166,
   170, 171
Perel'man, A. I., 53, 55, 56

Petersen, C. G. J., 197, 198
Phelps, W. H., 296
Phillipson, J., 14, 16, 25, 26, 28, 34,
   38, 44
Piper, C. S., 59
Pirie, N. W., 44
Pomeroy, L. R., 40
Porter, C. L. Jr., 36
Prothero, R. M., 354
Pugh, L. G. C. E., 326

Rabinowitch, E. I., 14
Ransom, R. M., 212
Raw, F., 68, 72, 73, 77, 81
Rawes, M., 67
Reed, L. J., 204, 207
Reid, C., 269
Rennie, P. J., 40, 87, 105
Reynoldson, T. B., 20
Rice, E. L., 90
Rice, L. A., 151
Ridley H. N., 253, 254, 255, 357
Robbins, R. G., 130
Robinson, G. W., 179
Rodin, L. E., 44, 53, 105, 106, 108,
   109
Roe, A., 226
Rosedahl, R. O., 69
Rossby, C. G., 8
Routien, J. B., 69
Rovira, D., 84
Rudd, R. L., 93
Rune, O., 183
Russell, E. J., 26
Russell, E. W., 26

Saeki, T., 37
Salisbury, E. J., 132
Salisbury, F. B., 183
Sauer, C. O., 307, 335, 336, 338, 340,
   341, 346, 347
Schmeider, P., 338
Schnell, J. H., 22
Schröder, H., 14
Schwanitz, F., 340
Sclater, P. L., 247
Semper, K., 121
Serra, L., 163
Setchell, W. A., 255
Sheard, N. M., 202
Shelford, V. E., 123, 124, 185, 198,
   210, 227
Shreve, F., 121

Sibbons, J. L. H., 163
Simons, E. L., 311
Simpson, G. C., 263, 297
Simpson, G. G., 226
Slayter, R. O., 157
Slobodkin, L. B., 29
Smith, G. W., 171
Smyth, J. D., 328, 329
Snow, D. W., 224
Sobezak, K. D., 316, 318
Solheim, W. G., 346
Solomon, M. E., 222, 227
Starkey, R. L., 72
Stebbins, G. L., 283
Steele, B., 183
Stephenson, R. E., 88
Stern, V. M., 95
Stewart, O. C., 335, 336
Strickland, J. D. M., 37
Symes, D. G., 212

Talling, J. F., 40
Tamm, C. F., 67, 88
Tansley, A. F., 5, 130, 260
Teal, J. M., 31, 32, 33
Teilhard de Chardin, P., 5
Theron, J. J., 90
Thienemann, A., 40
Thorne, R. F., 254
Thornthwaite, C. W., 160, 162, 163,
    164, 165, 166, 167, 168, 169, 179
Tinbergen, N., 226
Tobias, P. V., 309, 311, 312
Trewartha, G. T., 158, 159

Ucko, P. J., 340
Ulianova, O. M., 90
Usinger, R. L., 248

Van der Drift, J., 84
Van der Hammen, T., 338
Van Klingeren, B., 94
Van Steenis, C. G. G. J., 286
Vavilov, N. I., 341, 342
Veihmayer, F. J., 156
Verhulst, P. F., 204, 207
Vernadskii, V. I., 52, 57
Vinogradov, A. P., 55
Volterra, V., 221, 228
Vries, D. A., de, 161, 163

Waisel, V., 147
Waksman, S. A., 71
Walker, C. H., 93
Wallace, A. R., 197, 247, 256,
    313
Wangersky, P. J., 207
Ward, R. C., 149, 163
Warming, E., 121, 156
Washburn, S. L., 309
Wassink, E. C., 14
Watts, D., 357
Weaver, J. E., 37, 338
Webb, J. S., 59
Weisz, P. B., 279, 281
Went, F. W., 121
West, O., 338
West, R. G., 264, 265, 270, 271, 273,
    274
West, W., 183
Westlake, D. F., 42, 43
White, E. J., 66
White, G., 82
Whittaker, R. H., 37, 261
Whyte, R. O., 90
Wiegert, R. G., 21, 22, 39
Wiener, A. S., 317
Wiesner, B. P., 202
Wijk, W. R. van, 161, 163
Wilkinson, H. R., 86, 333
Willis, A. J., 154
Wilson, E. O., 284, 297
Witkamp, M., 79, 84
Wolfe, M., 80
Wofenbarger, D. O., 218
Wright, H. E., 296
Wright, S., 288, 289, 311, 314
Wymstra, T. A., 338
Wynne-Edwards, V. C., 221, 222,
    223, 226

Yaalon, D. H., 102
Yoshimoto, C. M., 253
Young, J. O., 20

Zimmerman, E. C., 253, 254
Zink, E., 37
Ziswiler, V., 294
Zohary, D., 343
Zukhovskij, P. M., 340

383

General Index

Italicized page numbers indicate where technical terms are defined within the text

Abies balsamifer, 146
Absorption, passive, *142*
Acer sp., 287
  A. *pseudoplatanus*, 133, 254
  A. *rubrum*, 128
  A. *saccharum*, 128, 146
Achillea millefolium, 289
Acinoyx jubatus, 295
Actinomycetes, *83*
  in nitrogen cycle, 80
  in pesticide decay, 92
  in sulphur cycle, 71
Activity-density, 22
Adaptation, 252, 255, 257, 275, 285
  to cold, 323
  to heat, 324
  preadaptation, 284
Adaptive radiation, 286
Adhesion
  in seed transportation, 255
Adret-ubac, *174*
Aedes aegyptii, 327
  A. *polynesiensis*, 332
Aerobacter sp., 79
Agriculture,
  alluvial land, 341
  and labour intensiveness, 344
  and population growth, 339, 342,
    343, 349, 357
asexual, 340–348
  conservative, 349
  destructive, 349, 350
  innovation in, 340–342, 344
  irrigated, 341, 348
  monocultural, 92, 350, 365

  origins of, 340–348
  seed planting, 341, 342, 344–346
  slash-burn, 344, 346, 349, 365
Agrostis sp., 187
  A. *tenuis*, 289
Albedo, *161*, 170
Alca impennis, 295
Alchemilla spp., 290
Alder, 43, 79, 156, 172, 273
Aldrin, 91
Alfalfa, 39, 57, 156, 293
Algae, 14, 25, 30, 31, 41, 42, 46, 80,
    83, 138, 235, 258
  nitrogen fixation of, 80
  productivity of, 41
Allee's principle, *218*
Alleles, *280*, 317, 321
  heterozygous, 280
  homozygous, 280, 286, 297
Allopolyploids, *293*
Alluvial areas, productivity in, 41
Alnus sp., 79, 156, 273
Alpine vegetation, 129, 173
Altitude,
  as limiting factor, 173, 174
Amaranthus leucocarpus, 345
Amino-acids, 54, 61, 62, 64, 68, 73,
    80
Amoeba, 83
Amphibia, 252, 253, 258
Andropogon intermedius, 278
Aneuploidy, *292*, 293
Angiosperms, 99, 289, 290, 292
Animals,
  domestication of, 333, 339–351

General Index

Animals—*continued*
  in plant dispersal, 254, 255
  mycophagous, 75
Anions, *57*
Anopheles spp., 327
Anoxic anoxia, *326*
Antibody, *20*, 330, 331, 333
Ants, 20, 21, 83, 218, 257
Apes, anthropoid, 308–310
Aphids, 20
Aquila chrysaetos, 21, 93
Arbutus unedo, 269, 271
Arctium spp., 255
Ardea cinerea, 93
Aridity index, *164*, 167, 179
Armeria maritima, 271
Arresting factors, 259
Arsenicals, 91
Artemesia tridendata, 130, 157, 174, 183,
  216, 257
Arthropods, *83*, 330
  in irradiated soils, 100
Aspect,
  as a limiting factor, 174, 175
Assimilation, net secondary, *38*
Association, plant, 5
Astralagus racemosus, 58
Asymptote, 205, 207–209, 219, 226,
  228, 237, 343
Atmometer, 125
Atta sp., 235
Auk, Great, 295
Auroch, 295
Australopithecus spp., 308, 309
  A. africanus, 309–312
  A. boisei, 309–312
  A. robustus, 309, 310, 312
Autopolyploids, *293*
Autotrophs, *14*, 32, 235
  in nitrogen cycle, 80
Auxins, *127*
Aven, mountain, 269, 271
Avena fatua, 250, 285
Avoidance, *122*, 123
Azotobacter sp., 79

Bacillus sp., 80
Bacteria, 83, 84, 89, 90, 330, 362
  aggregation of, 220
  in carbon cycle, 75
  in irradiated soils, 100
  in nitrogen cycle, 76, 78–80, 82,
    89, 90, 133

  in pesticide decay, 92
  in phosphorus cycle, 69
  in sulphur cycle, 72
  of decay, 35, 75
Baetis rhodani, 23
Balaenoptera borealis, 295
  B. physalus, 295
Banana, 134, 293, 340, 341, 356
Barley,
  cultivated, 43, 340, 342
  wall, 285
Barriers to colonization, 252, 254, 269,
  274, 275
  affecting evolution, 287
Bats, 244, 253, 258, 351
Bean, lima, 345
Bear, 295, 335
  polar, 328
Beech, 274, 350
  biomass and litter production in,
    106
  in New England, 146
  mineral elements in, 108
  productivity of, 43, 149, 259
  stem flow of, 148
Bees, 201, 217, 218, 290
Beet,
  cultivated, 340
  wild sea, 340
Beetle,
  broom, 20
  Colorado, 355
Behaviour, 199, 201, 211, 224, 226,
  283, 287
Beijerinckia sp., 79
Bellis perennis, 278
Beta maritima, 340
  B. vulgaris, 340
Betula spp., 271, 273
  B. nana, 273
Binomial system, *278*
Biochore, *4*
Biocoenosis, 5
Biocycle, *4*
Biogeochemical cycles, 52–110, 175,
  236, 242
  areal variations in, 105–110
  modification by man of, 85–105
  of carbon, 72–76
  of nitrogen, 76–82
  of phosphorus, 69–71
  of sulphur, 71, 72
  steady state of, 65

Biological clock, *120*
Biological diversity, 364–366
Biological productivity, 35–46, 364, 366
Biome, 185, *186*
Biosphere, *3–5*, 53, 55, 99, 244, 307, 326, 333, 348, 352, 359, 363, 364
Biotic area, *186*
Biotic conditions,
    as limiting factors, 184–188
Biotic potential, 206, 261
Biotin, 90
Biotope, *6*, 231, 232
Birch, 43, 184, 188
    dwarf, 172, 273
Birds, ringing of, 200
Birth rates, 201, 204–206, 211, 219–221, 223, 252
    and pesticides, 92–94
Bison bison, 294
Blackbird, 224
Blood groups and types, 315–322
Bluebell, 129
Bluebird, 355
Bog myrtle, 79
Bos primigenitus, 295
Bouteloua gracilis, 155
Bracken, 181, 186
Bramble, 186
Breckland, 187
Bronze age, 184, 347
Brown earth, 184
Buffalo, American, 294, 295, 366
Bunching, 217, 218
Burdock, 255
Buttercup, creeping, 271
Butterfly, 200
    marsh fritillary, 214, 215
    meadow brown, 216

Cacao, 332, 356
Cactospiza pallida, 310
Cactus, 139
Caddis fly, 23, 24
    net spinning, 23, 24
Caffea sp., 155, 356
Calanus finmarchicus, 57
Calluna sp., 87, 88, 181, 187, 188
    C. vulgaris, 87, 184
Cambarus tenebrosus, 34
Camel, 335, 346
Campion, red and white, 287
Canus lupus, 295

Canopy, vegetation, *129*, 130, 132
Capillary forces, *151*, 152, 178
Capsicum annuum, 345
Carbohydrates, 61, *62*, 64, 75, 325, 345
Carbon,
    cycling of, 72–76
    -14, 266
Carbon dioxide, 57, 65, 73, 75, 80, 95, 100, 102, 120, 360, 362
    at high altitudes, 326
    in leaves, 155
    recent increase in, 101, 102
Carbon monoxide, 102
Cardamine pratensis, 293
Carpinus sp., 273
    C. betulus, 275
Carrying capacity, 45, 206, *207*, 227, 343, 344
Carya spp., 267
Castanea dentata, 353
    C. sativa, 353
Casuarina sp., 79
Catena, 181
Catinella arenaria, 256
Cation, *57*
Cattle, 46, 342, 345, 346
Ceanothus spp., 289
Cell collapse, 104
Cell damage, 155, 172
Cell distortion, 93
Cell division, 61, 278, 279
Cell nucleus, 59, 61, 63, 278, 307
Cell sap concentration, 139
Cell structure, 60, 96, 138
    reorientation of, 127
Cell wall, *59*, 61, 142
Cells,
    diploid, *278*, 292
    haploid, *278*, 292
    primitive, 54
    radiation injury in, 98
    size and shape of, 59
    water reserve of, 325
Cellulose, *59*, 62, 63, 75, 83
Celtis sp., 287
Censuses, 197, 199, 200
Centipedes, 83
Cepphus grylle, 295
Cerastomella ulmi, 355
Cercopithecus sabeus, 355
Cereals, wild, 343
Cereus giganteus, 139

Cesium-137, 95, 97
Chaffinch,
    European, 292
    blue, 292
Chamaenerion angustifolium, 254
Chaparral, 289, 337
Cheetah, 295
Chemical elements,
    essential to life, 53
    exchange and transfer of, 53, 106–
        110, 259, 348, 349, 366
    in cell growth, 59–64
    in world ecosystems, 107–109
    relative abundance of, 55–59
    toxic, 90–93, 95, 360–362
Chenopodium album, 285
Chernozem, 90, 176, *178*, 179, 180,
    184
Chestnut,
    European, 353
    sweet, 353
Chestnut blight, Asiatic, 353
Chickens, 46, 346
Chickweed, 285
Chlamydomonos reinhardi, 29
Chlorinated hydrocarbons, 91, 94
Chlorophyll,
    in leaves, 37, 38
    production of, 61, 127
Chlorophyll index, *37*, 38
Chromosomes, 61, 278–281, 292, 293,
    308, 340, 341
    breakage of, 97
Ciliates, 83
Cinquefoil, 289
Circadian cycle, *120*
Citrus sp., 341, 356
Civilization,
    early, 333, 339
    hydraulic, 348
Cladonia spp., 164
Clay,
    montmorillonitic, 152
    varve, *266*
Clear Lake, California, 21, 93
Climate,
    regional, 157–159
    urban, 360
Climatic change, 10, 102, 243, 263–
    275, 342, 351
Climax, concepts of, 257–261
Clostridium sp., 79
Clover, 78, 79

Coconut, 250, 332
Cocos nucifera, 250
Coexistence, 228, 231, 233, 274
Coffee, 127, 293, 356
Colaptes auratus, 355
Cold,
    as a limiting factor, 322–323
Colloid, 151, 153
Colonization, 244, 252, 255, 257, 258,
    267, 271, 273–275, 284, 298
    barriers to, 254, 274, 275
Colony, plant, *6*
Columba palumbus, 199
Commensalism, *234*, 235
Community, *6*
    theories of development, 198
Competition, 129, 197, 223–237,
    253, 256, 270, 274, 291, 362,
    364, 365
    and succession, 235–237
    interspecific, 223, 227–234
    intraspecific, 223–227
    in tropics, 288, 289
    negative and positive, 234–235
Conservation, 366–368
Conservation areas, *367*
Conservation of energy,
    law of, *17*, 162
Continental shelf,
    productivity of, 40, 41
Continentality, *135*
Coral, 58
    productivity of reefs, 41
Cordylophora caspia, 124
Corvus frugilegus, 225
Corylus avellana, 257, 268, 271–274
Cotton, 92, 94, 150, 254, 340, 342,
    348, 350
Coypu, 355
Crabs, 235, 332
Crayfish, 34
Creosote bush, 156, 250, 288
Crinus spp., 291
Crossing over, *280*, 281, 284
Cucurbita moschata, 345
    C. pepo, 345
Cultigens, 339, 340, 352, 356
    centres of origin of, 341, 342
Culture, 307, 333, 339, 346
Curatella americana, 235
Curie, *99*
Cynodon dactylon, 150, 250
Cytoplasm, *59*, 61–63

Daisy, 278
Dandelion, 248, 254
Daphnia pulex, 29
DDT, 21, 91, 93, 94
Death rates, 201–206, 211, 219–223, 252
  and pesticides, 92–94
Decomposers, 26, 35, 38, 84, 363
Decomposition rates, 74, 75
Deer, 188, 199, 201
Deforestation, (*see* Forest removal)
Dendrochronology, *266*, 267
Denitrification, *82*
Density—dependent control, *219*–223, 227, 285
Density—independent control, *219*–223
Deschampsia flexuosa, 128
Deserts, 142, 266, 346
  biomass and litter production in, 106, 107
  climate of, 158
  competition in, 223
  dewfall in, 146
  evolution in, 284, 291
  mineral elements in, 108, 109
  nomadism in, 346
  productivity in, 40–43
Desulfovibrio aesturii, 72
  D. desulfuricans, 72
Dew, 143, 144, 146, 147
Dew point, *143*
Diarrhoea, 324, 327, 330
  Diatoms, 23 24, 31
Dicrostonyx spp., 210
Dieldrin, 21, 91, 93, 94
Diffusion pressure gradient, *143*, 155
Diptera sp., 33
Diseases, 217, 227, 315, 322, 324, 327–333, 351, 352, 355
  immunity to, 330–331
  infectious, 327–333
  intestinal 332
  nutrient-deficiency, 324
  of civilization, 326
DNA, 61
  (*see also* Nucleic acids)
Dock, curled, 285
Dodo, 296
Dog, 328, 342, 346
Dog's mercury, 257
Domestication,
  plants and animals, 333, 339–351

Dominance, *258*, 259
Dormancy, 138
Dorymyrex sp., 21
Drosphila sp., 203, 279, 287
  D. melanogaster, 202, 213
  D. pseudo-obscura, 216
Drought, physiological, *155*, 172
Dryas octopetala, 269
Duckweed, 31, 271
Dutch elm disease, 94

Eagle, golden, 21, 93
Ecesis, *257*
Ecocline, *289*, 338, 339
Ecological change, 243–275
Ecological efficiency, *29*
Ecological laws, *314*
Ecological pyramids, 24–26
Ecology, 2, 121
Ecosystems, *5*
  present status of, 362–368
  wild and tame, 364
Ecotone, *5*
Ecotonic invasion, *257*
Ectopistes migratorius, 294
Edaphic conditions, 135, 250
  as limiting factors, 175–184
Eichornia crassipes, 254
Elder, 186
Elk, 188
Elm, 273, 355
Elodea canadensis, 256
Emergents, *129*
Emigration, 211, 218–219, 313
Encelia farinosa, 184
Enchytraeids, *82*, 83
Endemics, *250*, 290, 294, 298, 331
Endemism, 183
Endogenous rhythms, *120*
  controls of, 133
Endothia parasitica, 353
Endymion non-scriptus, 128
Energetics, *7*
Energy,
  advective, 151, 154, 163, 172
  balance of, 50, 157
  chemical, 9
  exchange and transfer of, 13, 19, 30–35, 125, 157, 160, 175, 198, 230, 259, 337, 348, 364–366
    between trophic levels, 27–28, 229
    efficiency of, 19, 28–30, 31, 44, 237, 242

Energy—*continued*
    heat loss in, 19, 27
    laws of, 16–18, 35
    theory of, 27–28
  flux of, 125, 160, 171
  geothermal, 9
  heat, 9, 17, 27
  in domesticated animals, 46
  inflow, 8, 14, 15
  kinetic, 8
  law of conservation of, 17, 162
  mechanical, 8
  potential, 8–9
  radiant, 9, 10, 126
  sources of, 9–10
  states of, 8–9
Entropy, *18*
Environment,
  as selective screen, 275, 282
  concept of, 6, 120
  micro-, 124, 129
  macro-, 124
Environmental deterioration, 86–89, 184, 350, 351, 359
Environmental improvement, 89, 90
Environmental limitations,
  biotic, 184–188
  demographic, 197–237
  edaphic, 175–184
  energy/temperature-water, 157–171
  heat, 134–142
  humidity-moisture, 142–157
  light, 126–134
  on man, 323–333
  topographic, 173–175
  wind, 171–172
Environmental resistance, *206*, 207
Enzymes, *63*, 92, 100, 103, 183
  inactivation of, 138
  photooxidation of, 127
  soil, 84
Ephemeridae, 211
Ephemeroptera, 33
Epiphytes, 258
Erica spp., 87
  E. cinerea, 88
  E. mackiana, 271
  E. tetralix, 88
Eriophorum vaginatum, 182
Erodium cicutarium, 285
Erythrina spp., 295
Estuaries, productivity in, 41
Eucosma griseana, 94

Euphorbiaceae, 250
Evaporation, 89, 125, 148–153, 160–163, 167, 170, 171
Evaporative potential, 170
Evaporative power, 151, 162
Evapotranspiration, 160, 161, 165, 166, 170, 176, 178
  potential, *160*–166
  (*see also* PE)
Evapotranspirometer, 125
Evasion, *122*, 123
Evergreen forest, productivity in, 41
Evolution, 213–215, 223, 243–247, 250–253, 275–298, 306, 311, 322, 340
  and isolation, 287–291
  crop plant, 341
  in ecosystems, 275–298
  of walled towns, 342
  on oceanic islands, 253
  phyletic, *286*–294
Evolutionary traps, *297*
Evolutionary tree, 276, 277
Extinction, 206, 207, 210, 221, 222, 273, 285, 294–298, 306, 315, 364

Factor compensation, *122*
Fagus spp., 148, 273
  F. grandifolia, 146
  F. sylvatica, 259, 274
Fallout, 9, 97, 98, 100
Fallow, 344, 349, 352
Fats, 59, 61, *62*, 63, 64, 75, 325, 345
Ferns, 254, 258, 292
Fertilizer, phosphatic, 69
Festuca arundinacea, 37
  F. ovina, 22, 187
Field capacity, *152*, 166
Filariasis, bancroftian, 332
Finches,
  cactus, 310
  Galapagos island, 275, 290
Fir, 146, 173, 350
  balsam, 146
  douglas, 148, 174
  white, 148
Fire, 86, 308, 310, 313, 331–339, 348, 364
  and grasslands, 337–339
  -climax, 337–339
  -drive, 335
  man-created, 336–339

natural, 334, 336
-resistant species, 336
Flagellates, 83
Flandrian period, 263, 265, 267–274
Flea, chigger, 330
Flicker, 355
Floristic provinces, 249–251
Flounders, 220
Flux,
  energy, *125*, 160, 171
  heat, 134, 160, 170, 323
Fog, 143, 144, 160
  -drip, 144, 146, 150
  freezing, 171
Food chain, *19*–24, 52, 100, 229, 348, 364
  detritus, 33–35, 38
  -efficiency, 29
  examples of, 22–24
  grazing, 33–35, 38
  parasitic, 24, 25, 33
  predator, 33
  saprophytic, 33
  tracing of, 19–22
Food web, *19*, 23, 84, 229, 235, 236
Forest heath, productivity of, 43
Forest removal, 86, 87, 243, 273, 274, 348, 349, 360
Formation, *4*
  climax, 259, 260
Fouqueria splendens, 156
Fox, 210
Fractercula arctica, 295
Fringilla coelebs, 292
  F. teydea, 292
Frogs,
  in Australia, 291
Fulmarus glacialis, 208
Fungi, 35, 69, 83, 84, 235, 330
  in carbon cycle, 75
  in irradiated soil, 100
  in nitrogen cycle, 80, 82
  in pesticide decay, 92
  in phosphorus cycle, 69
  in sulphur cycle, 71

Galinsoga spp., 285
Gamete, *278*, 279
Gasolene compounds, 86, 104
Gecko, 253
Gene, 278–283, 313, 320
  allelic, 280
  blood group, 315–321

dominant, 280, 281, 286
haemophilic, 313
linked, 280
recessive, 280, 281, 296
Gene pool, *281*, 285–291, 296, 297, 313–315, 320, 321
Genecology, *276*
Genetic drift, *286*
Genetic reproduction, 54
Genetic variability, 213–216
Genotype, *213*, 214, 216, 279, 284, 286
Gentian, spring, 270
Gentiana verna, 270
Geological time,
  divisions of, 245
Germination, 136, 181
Germs, *330*
Gigantism, *293*, 339, 340
Glaciation, 265, 270
  features of, 265
  (*see also* Pleistocene)
Gley, *177*
Glossina spp., 332
Glucose, 9, 62, 63
Goats, 342, 345, 351
Gourd, bottle, 345
Grape, 293, 350
Grass,
  Barbados sour, 278
  bent, 187
  Bermuda, 150
  cotton 182
  Kentucky, blue, 293
  Knot, 271
  sheep's fescue, 187
Grasshopper, 21, 201
Grasslands,
  albedo of, 161
  and fire, 337–339
  Pennyroyal, 339
  productivity in, 40, 41
Gravity,
  in seed transportation, 255
Grebes, 21, 93
Growth efficiency, 46
Guano, 71
Guillemot, black, 295
Gulf stream, 135
Guttation, 154
Gymnosperms, 289
Gypsies, 315

Habitat,
    concept of, 6, 120
    -range, 198, 199
Haemotopus ostralegus, 275
Haemoglobin, 78, 326
Halophyte, *142*, 181
Hardiness, 122, 123, 293
Hard pan, 152, 177, 235
Hardwood, deciduous, 349
    evolution of, 286, 287
Hare, 94
    snowshoe, 210
Hawk, 198
Hazel, 266, 274
Heat,
    accumulated, *139*–141
    as a limiting factor, 134–142
Heat effluents, 361–362
Heat indices, *162*, 163
Heat island, *360*
Heat loss,
    between trophic levels, 19, 27
Heather, 87, 88, 184, 214, 259
    bell, 87
Hedera sp., 273
    H. helix, 139, 140
Helianthus rigidus, 184
Helichrysum spp., 290
Heliophyte, *129*, 130,
    131
Hemlock, 267
Herb paris, 275
Herbivore, grazing, *33*
Heron, 93, 209
Herpestes burmanicus, 296
Heteralocha acutirostris, 295
Heterotroph, *16*, 35, 235,
    237
Hevea brasiliensis, 255, 356
Hibernation, 133
Hickory, 267
Hippophäe rhamnoides, 79
Holly, 130, 273
Homeostatic mechanisms, *222*, 230,
    295
Homeotherm, *323*
Hominidae, *308*, 309
Homininae, *308*, 309
Homo sp., 306, 308, 309, 311
    H. erectus, 313, 334
    H. habilis, 306, 309, 311–313
    H. sapiens, 278, 313, 314
Honeysuckle, 214

Hordeum distichum, 342
    H. murinum, 285
    H. vulgare, 342
Hornbeam, 274, 275
Horse, 335, 346
Huia bird, 295
Humidity,
    as influence on PE, 162
    as a limiting factor, 142–157, 219
    index of, *164*, 167, 179
Humus, 33, 60, *75*, 82, 150, 178, 184
Hura crepitans, 255
Hyacinth, water, 254
Hybridization, 291, 293, 315, 322,
    339
Hydathode, *154*
hydrocarbon, 54, 104
Hydrological cycle, 142–157 (*147*)
Hydrolysis, *18*, 62
Hydrophyte, *156*
Hygroscopic forces, *151*, 152
Hyparrhenia dissoluta, 90
    H. filipendula, 90
Hyperthermia, *323*
Hypothermia, *323*

Ice, 143, 144
Ideas,
    diffusion of, 340, 346–348
    innovation of, 311, 334, 342–344
Ilex sp., 273
    I. aquifolium, 430
Imbibition, *142*
Impatiens parviflora, 127, 128
Indirect flow, *148*
Infiltration, 150, 151
    on coarse-textured soils, 181
Inmigration, 218, 219
Insects, 24, 83, 91, 133, 186, 204,
    209, 217–223, 253–255, 258, 284,
    287, 290, 295, 365
Interception, 144, 147–150, 351
Iodine-131, 95, 97, 100
Ion, 57
Ionic potential, *57*
Islands,
    cold climate, 287
    oceanic, 253, 255, 257, 287, 290,
    296, 297, 357, 358, 365
Isolation,
    and evolution, 287–291, 315, 316,
    320, 321
Isoptera spp., 218

Isotopic enrichment, 266
Isotopic tracing, 79
Ivy, 139, 140, 273

Jactitation, *255*
Juglans nigra, 184
Juncus sp., 156
Juniper, one-seed, 257
Juniperus monosperma, 257

Kagu, 298
Karyotype, *216*
Kenyapithecus wickeri, 311
Krakatoa, 258
Krummholz, *172*
Kudzu, 79

Lacerta sicula, 221
Lactuca serriola, 285
Lady's smock, 293
Lagenaria siceraria, 345
Lamium album, 256
Lapse rate, 173
Larrea divaricata, 250, 288
   L. tridentata, 156
Leaching,
   in nitrogen cycle, 82
   in dodsols, *176*
   on burnt forestland, 87
   on moorland, 88
Leaf,
   cuticularized layer of, 138, 157
   rosette, 174
   temperatures of, 43, 44, 48
Leaf fall, 132, 133
Leaf pore, 127
Leaf-water potential, 38
Leguminosae, 78, 250
Lemming, 210
Lemmus sp., 210
Lemna sp., 271
   L. minor, 31
Leopard, 295
Lepidoptera, 200
Leptinotarsa decemlineata, 355
Lepus europaeus, 94
Lespedeza, 36
Lettuce, prickly, 285
Leukaemia, 97
Lichen, 99, 163, 173, *235*
Life,
   -cycle, 202, 208
   -expectancy, 202

origins of, 7, 54
-table, 202
Light,
   indirect, 12
   intensity, 127–131, 154
   quality, 131, 132
Light climate,
   and stomata, 154
   and transpiration, 154
   as a limiting factor, 126–134
Lilium philippinensis, 295
Lime, 273, 274
Limiting factor, *122*
Liquidambar sp., 250
Liriodendron sp., 250
Litter,
   and food webs, 84
   and runoff, 150
   chemical elements in, 107–109
   in tropical rain forests, 73, 107–109
   production of, 42, 106–109
Lizard, wall, 221
Lobelia spp., 174
   L. deckenii, 291
Locoweed, 58
Locust, 188, 209
Locusta migratoria, 209, 210
Lonicera periclymenum, 214
Lucerne, 78
Luscinia luscinia, 291
   L. megarhynca, 291
Lusitanian flora, 269, 271
Lymphomata, malignant, 350
Lynx lynx, 210, 295
Lysimeter, 79, 125, 126

Maize, 99, 150, 340, 344
   productivity of, 42
Malaria, 327, 328, 332
Mammals, 55, 57, 82, 93, 100, 133, 186, 200, 210, 217, 221, 252, 253, 254, 258, 274, 285, 308
Man,
   and spread of organisms, 331, 332, 340, 351–359
   as a hunter, gatherer and collector, 315, 335, 339, 341–343, 346, 348, 364
   as an agent of ecosystem change, 333–362
   brain sizes of, 308–310
   Cro-magnon, 313, 334

Man—*continued*
environmental restraints on, 323–333
Java, 313
modification of element exchange
by, 85–105
origins of, 306, 309–322
Neanderthal, 313, 334
Peking, 313, 334
variants of, 308–322
Mangrove, 254
Manioc, 347
Maniola jurtina, 216
Maple, sugar, 146
Marking—recapture methods, *200*
Mayfly, 23, 24, 211
Meadowrue, alpine, 271
Measles, 329, 330, 332
Mediterranean climate, 146, 350
Megalopolis, 360
Megarhyssa spp., 234
Meiosis, *278–280*, 292, 308
Melanoplus sp., 21
Melitaea aurinia, 214
Mercurialis perennis, 257
Mesolithic period, 87, 89, 274, 331,
333, 339, 348, 359
Mesophyte, *157*
Mesquite, 156, 257, 339
Metabolism,
animal, 63
human, 323
plant, 134, 138, 139
Mice, 82, 188, 210
Miconia spp., 291
Micrococcus denitrificans, 82
Microorganisms, 4, 35, 53, 54, 65,
69, 83, 84, 95, 237, 243, 329
affected by radiation, 100
affected by pesticides, 93
as disease vectors, 327–333
in nitrogen cycle, 76–79, 82
in sulphur cycle, 71
of decomposition and decay, 64,
361
role in burnt forestlands, 88
Migration, 45, 133, 210, 218, 219,
220, 252, 285, 343, 351–359
Millipedes, 35, 83
Miocene, 245, 306, 311
Mist, 144
Mites, 35
in carbon cycle, 75
Mitosis, 61, *279*, 292

Moisture,
as a limiting factor, 142–157
-deficit, 162, *166*
-surplus, *166*
Moisture equivalent, *153*, 156
Moisture provinces, 166, 168
Molar enthalpy, *18*
Mollusc, 83, 95, 253
Monge's disease, 326
Mongoose, 296, 355
Monkey, green, 355
Moonlight,
as growth stimulator, 133
Moorland, 87, 88, 181, 182, 184, 188,
259, 367
burning of, 87, 88
fallout on, 99
use of, 87, 88, 367
Moose, 188
Mortality rates, *201*, 202
(*see also* Death rates)
Mosaic distribution, 315
Mosquitoes, 91, 208, 209, 327, 330–332
Moss, 99, 173
Moss campion, 248
Moth, larch-bud, 94
Mount Kenya, 173, 174
Mountain building,
effects of, 245
Mushrooms, 254
Mutation, 213, 281, 283, 286, 313,
326, 339
Mutualism, 234, *235*
Mycorrhiza, 69, 88
Myrica gale, 79

Natality, rates of, *201*
(*see also* Birth rates)
Natural history, 2
Natural regions, 3
Natural selection, 213, 222, 223, 234,
275, 276, 282–285, 296, 313
314, 321
Nature reserves, 367
Nauclea spp., 295
Navicula viridula, 23
Nectarina spp., 291
Nematodes, 75, 83, 328
Neolithic period, 274, 331, 333, 339
Nepenthes alata, 295
N. ventricosa, 295
Nettle, white dead, 256
Neurospora sp., 213

Newt, 291
Niche, 6, 227–234, 236, 244, 253,
     256, 284, 285, 288, 289–291, 296,
     306, 310–313, 321, 359, 362, 364
  artificial, 326
  fundamental, 231, 232
  included, 231, 234
  realized, 231
  vacant, 231
Nightingale, 291
Nitrobacter sp., 81
  N. agilis, 90
Nitrogen,
  cycling of, 76–82
  fixation of, 78–80, 133
  immobilization of, 79
  mineralization of, 76, 80
Nitrosomas sp., 81
Nomadism, 45, 346, 351, 352
Noosphere, *4*
Nucleic acids, 54, 59, 61, 68
  DNA, 61
  RNA, 61
Nutrients, 60, 82, 217
  deficiency of, 183, 324, 325
  impairment of circulation of, 86,
     87, 152
  in plants, 142
  in rain water, 66, 67
  in root exudates, 109
  leached from leaves, 109
  loss in Rough Sike catchment, 89

Oak, 43, 87, 107, 132, 161, 184, 259,
     273, 274, 348, 350, 365
  blue, 337
  cork, 336
  dwarf white, 339
  English common, 127, 283
  sessile, 130, 283
Oasis effect, *163*
Oats,
  cultivated, 293, 340
  wild, 249, 250, 285
Oceans,
  pesticides in, 95
  productivity in, 40, 41
Ocotillo, 156
Oecanthus sp., 21
Olduvai Gorge, 309, 311, 334
Olefin, 104
Oligochaets, 82
Olive, 43, 350, 360

Orchid, 254
Orconectes rusticus, 34
Organic material,
  calorific value of, 26
  production of, 65
Organisms,
  age of, 199, 211, 212
  competition between, 223–237
  cosmopolitan, 244, 247, 248, 252
  demography of, 198–223
  dispersal by man of, 351–359
  linnean grouping of, 276, 278
  natural dispersal of, 244, 248, 252–
     257, 271–275
  patterns of distribution of, 244–251
  sex of, 199, 211, 212
  spatial arrangement of, 216–218
Orographic effect, *143*
Oryctolagos cuniculus, 282
Oscillatoria aghardhii, 42
Osmosis, *142*, 325
Osmotic pressure, 142, 156, 161, 181
Osprey, 93
Ostrich, 295
Owls, 198
  snowy, 210
  tawny, 227
Oxalis acetosella, 129
  O. corniculata, 285
Oxidation, 9, 59, 104, 362
  in nitrogen cycle, 80
Oyster, 58, 95, 202, 220
Oyster catcher, 275
Ozone,
  in smog, 104
  in upper atmosphere, 10, 11, 132

Paleolithic period, 333, 334, 343
Palynology, *266*, 342
  (*See also* pollen analysis)
Pampas, 338
Panda, giant, 252
Pandanus spp., 295
Panthera panthera, 295
Paramecium aurelia, 229
  P. caudatum, 229
Paranthropus sp., 309
Parasites, 33, 83, 84, 198
  in humans, 327–333
Paris green, 91
Paris quadrifolia, 275
Parus caerulus, 221
  P. major, 226, 282, 283

Pathogen, *329*–331
PE, *160*–166, 169–171
  (*see also* Evapotranspiration,
    potential)
Peat, 88, 89, 99, 182, 184, 273
Pedalfer, *179*
Pedocal, *179*
Pepper, 345
Peptide, *62*
Periglacial conditions,
  and flora in Britain, 270, 271
Permafrost, *135*, 136, 266
Permanent wilting percentage, *155*, 156
Pesticides, 86, 90–95, 361
  decay of, 92
  fumigant, 92
  fungicide, 92
  organochlorine, 91, 93
  organophosphate, 91, 92
Pests, 90–93, 365
Petrel, fulmar, 208
pF factor, *156*
pH, 71, *79*, 182, 183
Phasianus colchicus, 208
Pheasant, 208
Phenotype, *213*–215, 279, 281–284,
    294, 313–315, 321, 323, 339, 340
Phosphorus,
  cycling of, 69–71
  deficiency of, 71
  immobilization of, 69
  seasonal imbalances of, 69
  -32, 21, 22, 200
Photoperiod,
  as limiting factor, *127*, 132–134
Photosynthesis, 9, 14, 30, 32, 33, 36–
    38, 40, 63, 73, 102, 104, 107,
    125, 127, 131, 134, 140, 155, 282,
    340, 360
Photosynthetic efficiency, *36*, 37, 44
Phragmites communis, 253
Phreatophyte, *157*
Phylogenetic tree, 277
Physiological longevity capacity, *201*
Phytodecta olivacea, 20
Phytotron, *126*
Picea abies, 150
  P. omorika, 38
  P. rubens, 146
Pig, 296, 328, 342, 345
Pigeon,
  passenger, 295, 366
  wood, 199

Pigmentation, 314
Pigweed, 285
Pilgrimages, 353, 354
Pine, 43, 107, 109, 124, 132, 161, 172,
    181, 183, 273, 336, 350
  digger, 183, 255, 337
  monterey, 285
  yellow, 183
Pinus sp., 271, 273
  P. aristata, 267
  P. flexilis, 267
  P. ponderosa, 183
  P. radiata, 285
  P. sabiniana, 183. 255, 337
Pioneer plants, 235, 256, 258, 259
Plagioclimax, *259*
Plagiosere, *259*
Plague, 327, 331
Plaice, 19
Planaria gonocephala, 233
P. montenegrina, 233
Plane, western, 133
Plankton, 19, 24, 33, 42, 95, 197, 208
Plant biomass, 106, 107
Plantago lanceolata, 128, 248, 285
  P. maritima, 256, 271
Plantain,
  common English, 248, 285
  sea, 256, 271
Plants,
  cultivated, 339–341
  domestication of, 333, 339–351
  dwarfing of, 139
  heliophytic, 129–131, 259
  hydrophytic, 156
  megathermal, 138, 139, 157
  mesophytic, 157, 259
  mesothermal, 138–140, 157
  microthermal, 138–140, 157
  non-vascular, 163
  range patterns of, 252
    arctic-alpine, 248, 269
    discontinuous, 250
    endemic, 250
    pan-tropical, 248
    temperate, 248,
  sciophytic, 129, 259
  xerophilous, 156, 157
Plasmodium spp., 327, 328, 330
Platanus occidentalis, 133
Pleistocene, 135, 242, 245, 250, 256,
    262–267, 274, 276, 291, 306–312,
    333, 335

Pliocene, 245, 267, 306, 311, 312
Pondiceps cristatus, 93
Podsol, *176*, 177, 179, 180, 184, 273
Pollen analysis, 271, 346
  (*see also* palymology)
Pollution, 86, 101–105, 127, 359–362,
  364, 367
  and carcinogenic elements, 102–104
  industrial, 360, 361
  pulp-mill, 220
  water, 361–362
Polygonum aviculare agg., 271
Polyploidy, *292*, 293, 307, 339
Pongidae, 308
Population,
  age structure of, 211
  and genetics, 213–216
  decline, 204, 207, 210, 219
  definition of, *198*
  density—dependent control of, 219–
    223
  density—independent control of,
    219–223
  equilibrium of, 207–210, 219, 220,
    223
  -explosions, 256, 285, 308, 359
  fluctuations of, 207–210, 213, 215,
    220, 226, 256, 365
  founder, *288*
  growth of,
    and agricultural dev't., 342, 344,
      349, 350
    city, 260
    patterns of, 204–210
    rates of, 201–204, 206
    world, 339, 359
  oscillations of, 207–210, 220, 221
  sampling of, 200
  stability and instability of, 211, 219,
    230
  surveys of, *197*
Populations, 6, *198*, 199
  allopatric, *233*, 234, 282
  contiguously allopatric, *233*, 234
  hybrid, 322
  parapatric, *291*
  single, and multi-species, 365
  sympatric, *233*, 234, 282
Portulaca oleracea, 285
Potato,
  Irish, 43, 91, 293, 355
  sweet, 347, 348
Potentilla glandulosa, 289

Prairie, 4, 36, 37, 43
  productivity of, 43
  -soils, 180
Preadaptation, *284*
Precipitation, atmospheric, 143–147,
  165, 166, 176, 178,
  219, 360
Primates, 306
Prisere, *258*
Proales decipiens, 202
Production,
  in root growth, 36, 37
  of world ecosystems, 40–44
  primary, *14*, 16, 36, 38–44
Productivity,
  potential, 37
  secondary, *38*
  of world ecosystems, 44–46
Prosopis sp., 156, 257
Protein, 44, 45, 54, 59, 61–64, 68,75,
  80, 127, 325, 336, 343
  precipitation of, 139
  synthesis of, 61
Proteolysis, *80*
Proteus vulgaris, 80
Protocooperation, *234*, 235
Protozoa, 83, 229, 330
  aggregating activity of, 220
  in carbon cycle, 75
Prunus sp., 287, 288
Pseudomonas sp., 80
  P. denitrificans, 82
Pseudotsuga taxifolia, 174
Psychrometer, *125*
Pyschrometric constant, *170*
Pteridium aquilinum, 186
Pueraria lobata, 79
Puffin, 295
Pumpkin, 345
Purslane, 285
Pyramids, ecological, 24–26
  age-sex, *211*, 212
  of biomass, *25*, 26
  of energy, *26*
  of habitat, *198*, 199
  of numbers, *24*, 25
Pyrheliometer, 125

Quercus sp., 273, 287
  Q. douglassi, 337
  Q. petraea, 130, 283
  Q. robur, 127, 132, 259, 283
  Q. rubra, 128

Rabbit, 22, 45, 82, 186–188, 199, 201, 282, 355
Radiation,
  cosmic, 9, 11, 96, 281, 326
  gamma, 100
  infra-red, 10, 12, 160
  ionizing, 9
    from atomic tests, 97
  natural, 96, 97
  ultraviolet, 10, 131, 132, 326
Radiation balance, 160, 162, 163, 167
Radiative cooling, 144
Radioactivity,
  in rocks, 9, 96, 281
  in soil, 200
  in other substances, 95–101
Radiocarbon,
  dating, *266*
  tracers, 63, 200, 266
Radiosensitive body organs, 97
Radish, wild, 285
Ragwort, 244,
  Oxford, 256, 284
Rain, 143–145, 147–150, 157, 158, 171, 209, 264, 265
Ramapithecus punjabicus, 311
Randomization, *252*
Range,
  ecological, *252*
  tolerance, *252*
  (*see also* Plants, range patterns of)
Ranunculus lappaceus, 289
  R. repens 271
  R. scleratus, 271
Raphanus sativus, 285
Rats, 202, 244, 253, 258, 296, 327, 328, 331, 332, 351, 355
Rattus exulans, 332
Redwood, California, 24, 142, 144, 250, 267, 336
Reed,
  common, 253
  -swamps, productivity in, 42
Reflection, 14
  (*See also* Albedo)
Refuges, 250, 269, 275, 296, 365
Regions,
  climatic, 158
  natural, 3
  thermal, 163, 164
Regulatory factors, *122*
Relict species, 265, *294*

Reproduction, differential, 282
Reptiles, 221, 253, 254, 258
Resistance,
  towards injurious factors, 123
Respiration, 32, 36–39, 73, 76, 131, 134, 140, 183, 237, 282, 329
Rhaphus cucullatus, 296
Rhizobium leguminosarum, 78
  R. radicicola, 78
Rhizophora mangle, 254
Rhizopods, 83
Rhizosphere, 4, *84*, 155
Rhyacophila sp., 23
Rhynochetus jubatus, 298
Rice, 80, 293, 344
Rime, 144
RNA, 61
  (*See also* Nucleic acids)
Robins, 94
Rontgen, *96*
Rook, 225
Root constant, *164*
Roots,
  permeability of, 138
Rosa sp., 287
Rotifers, 202, 292
Rough Sike catchment, 88, 89
Rubber, para, 255, 356
Rubus fruticosus, 186
Ruderals, 271
Rumex acetosa, 128
  R. crispus, 285
Runoff, 87, 144, 150–153, 162, 166, 181, 243, 351

Sagebrush, 130, 157, 174, 183, 216, 257, 339
Salix sp., 157, 273, 287
  S. herbacea, 248, 271
Salmon,
  pesticides in, 94
Salt spray, 71, 172
Salts,
  mercuric, 92
  mineral, 57, 59, 61, 64, 68, 69, 181
Salvinia natans, 134
Sambucus nigra, 186
Sandbox tree, 235, 255
Saprophyte, 33, *83*, 84
Saturation,
  of air, 143, 155
  of plant communities, 150–152
  of soil, 148

Savannas, 4, 311
    animals in, 45
    as fire climax, 337, 338
    biomass and litter production, 106,
        107
    in Guyana, 235
    mineral elements circulating in, 108,
        109
    productivity of, 43
Saxifraga nivalis, 270
    S. oppositifolia, 271
Saxifrage,
    alpine, 269
    purple, 271
Scabiosa pratensis, 214
Sciophyte, *129*
Sclerophylly, *157*
Scolytus multistratus, 355
Seablite, 271
Sea buckthorn, 79
Sea pink, 271
Seaweed, 58
Sedimentation processes, 65, 76
Seeds, transportation of,
    endozoic, *254*
    epizoic, *254*, 255
Selection pressure, 286
Selective chemical enrichment, *58*
Self-fertilization, 285, 289
Senecio sp., 174, 244, 290
    S. squalidus, 256
Sensors, 125
Sequoia sp., 24, 267
    S. sempervirens, 142, 250
Sere, *258*, 259
Sericea lespedeza, 36
Serpentine, 250
    soils on, 183
Settlement, high altitude, 326
Sewall Wright principle, *288*, 311,
        314, 315
Shade,
    adaptation to, 128–131
    tolerance of, 128–131
Sheep, 21, 87, 88, 99, 202, 209, 342,
        345, 351, 367
Sheep's fescue, 187
Shivering, 323
Shrews, 82, 198
Sialis sialis, 355
Silene acaulis, 248
    S. alba, 287
    S. dioica, 287

Skinks, 253
Skuas, 92
Sleeping sickness, 330, 332
Slope,
    as a limiting factor, 175
Smallpox, 329–332
Smog, 101
    photochemical, *104*
Snakes, 253, 275, 355
Snow, 143, 144, 148, 150, 157
Snowberry, 214
Society, *6*
Soil,
    acidic, 79, 82, 83, 99, 176, 182, 184
    alkaline, 81
    and infiltration, 151
    bleached, *177*
    clay, 181
    conditions of,
        aerobic, *71*, 81, 152
        anaerobic , *71*, 72, 79, 152
    erosion of, 187, 243, 307, 349, 350,
        360
    gley, 177
    granitic, 182
    horizons in, *176–177*
    intrazonal, *181–183*
    limestone, 182
    organisms in,
        and biogeochemical cycle, 80–85
        and irradiation, 100
        food-web role of, 84
    peaty, 182
    profile of, *176*
    salty, 181
    sandy, 90, 181
    serpentine, 183
    structure of, *182*
    texture of, 152, 181, *182*
    water in, 151, 152
    waxy, 182
    zonal, *176*, 179, 180
Solar constant, *10*, 363
Solar radiation,
    absorption by atmosphere, 10–12
    conversion by plants, 14
Sorrel, wood, 129
    yellow, 285
Spartina spp., 33, 43, 356
    S. alterniflora, 257, 356
    S. maritima, 356
    S. townsendii, 257, 356
Speciation, *286–294*

Species,
concept of, 282
fugitive, *284*
variation in, 275, 276
Spectrum,
electromagnetic, 10
Sphagnum sp., 273
biomass and litter production in, 106
mineral elements circulating in, 108
productivity of, 43
Spiders, 21, 22, 83
Spinach, 340
Sponges, 58
Springs, hot, 135, 138
Springtail, 75
Spruce, 146, 149, 173, 266
Norway, 150
red, 146
Squash, 345
Standing crop, 27, 32, 34, 235
Starling, 221, 355
Stefan's law, *170*
Stellaria media, 285
Stem flow, 148–150
Steppes, 4, 346
and chernozem, 176, 178, 184
as fire climax, 337, 338
biomass and litter production in, 106, 107
climate of, 158
Khuzistan, 344
mineral elements in, 108, 109
productivity of, 43
Stomata, 102, *127*, 155, 156
water, 154
Storksbill, 285
Stratification of vegetation, 129, 130, 237
and interception, 150
and temperature, 138
Strawberry tree, 269
Strix aluco, 227
Strontium-90, 95–100
Struthio camelus, 295
Strurnus vulgaris, 355
Suaeda maritima, 271
Subboreal forests, 337
Subtropical deciduous forest,
biomass and litter production in, 106
mineral elements in, 108, 109

Subtropical evergreen forest, 287
Succession, 235–237, 257–261
Sugar beet, 14, 146
Sugar cane, 293, 341, 350, 356
productivity of, 42, 43
Sulphur,
cycling of, 69, 71, 72
-35, 200
Sulphur dioxide, 101, 102, 104
Sunflower, 184
Survival curves, *202*, 203
Sweet gum, 250
Swidden, *344*
Sycamore, 133, 254
Symphoricarpus rivularis, 214

Taiga spruce forests,
biomass and litter production in, 106, 107
mineral elements in, 108
productivity of, 42, 43
Taraxacum agg., 248, 254
Technology,
acquisition of, 333–334
Temperature,
accummulated, 139–141
and transpiration, 154
as a limiting factor, 134–142, 219, 322–324
extremes of air, 135
hot spring, 135
leaf, 138, 139, 153
mean air, 135–139
megathermal, 134, 135
mesothermal, 134, 135
microthermal, 134, 135
plant, 138,
soil, 135, 137, 138
Termites, 83, 218, 235
Territoriality,
concept of, 223–227
Testudo elephantopus, 295
T. gigantea, 295
Thalictrum alpinum, 271
Thermal efficiency index, 179
Thermodynamics, laws of, 16–18, 35, 38
Thiobacillus thiooxidans, 71
Thorium-230, 266
Thorn bush,
animal communities in, 45
Throughfall, 148, 150
Thrush, song, 224

Tilia sp., 273
  T. cordata, 274
Tit,
  blue, 221
  great, 226, 227, 282, 283
Tobacco, 92, 350
Tolerance,
  concept of, *122*–124
  -range, 233, 235, *252*
Tools, 306, 308, 310, 311, 313, 334, 335
Topography,
  as a limiting factor, 173–175
Tortoises, giant, 295
Trace elements,
  in human diet, 325
Tracers, radiocative, 20, 21
Trading,
  and disease, 331–333, 351
  and dispersal of ideas, 346–348
  and dispersal of organisms, 353–359
Transhumance, 352
Transpiration, 123, 125, 138, 142, 144, 152–157, 160, 164, 171, 172, 325
  cuticular, *154*
  direct, *154*
  potential *170*
  stomatal, *154*
Trapa natans, 267, 268
Trichinella spiralis, 328, 329
Trichinosis, 329
Triclads, 20, 233
Triticum aestivum, 342
  T. dicoccum, 342
Triturus cristatus, 291
  T. marmoratus, 291
Trophic levels, 19, 24–27, 29, 30, 32, 198, 199, 229, 230
Tropical rain forest, 107, 158, 250, 258, 289, 296, 314, 337, 365
  biomass of, 106, 108
  climate of, 158
  competition in, 223
  litter production in, 73, 106, 108
  mineral elements in, 108, 109
  productivity in, 42, 45
  rates of root decay in, 73
  stratification in, 129, 130
Trout,
  and heat effluents, 362
  and pesticides, 94
Tsetse fly, 200

Tsuga sp., 267
  T. canadensis, 128
Tulip tree, 250
Tundra, 173, 180, 266, 271, 273
  biomass and litter production in, 106
  climate of, 158, 164, 258
  mineral elements in, 108, 109
  productivity in, 40, 42, 43
Tunga spp., 330
Turbulence, atmospheric,
  effect on PE, 167, 170
Turdus merula, 224
  T. philomelos, 224
Turnover time,
  of elements, 67
  of organisms, *26*, 296, 298
Turtle, 253

Ulmus sp., 273
  U. americana, 355
Uranium-235, 266
  U-238, 266
Urbanization, 359–360
Ursus arctos, 295
Urtica dioica, 128

Vacuole, *61*
Van Allen belts, 11, 12
Vapour pressure gradient,
  leaf to atmosphere, 38, 143, 160, 167, 170
Vector, *327*–333, 351, 355
Vegeculture, *341*, 346, 349
Viola spp., 129
Violet, 129
Virus, *84*, 329, 330
Vitamins, 64, 324, 325, 345
  Vitamin-D, 314, 325
Vole, 188, 210, 212

Wallace's realms, 246, *247*
Walnut, 184
Wasp, ichneumon, 234
Waste disposal,
  problems of, 86, 359, 360
Water,
  in plant dispersal, 254
Water balance, 157, *166*
  of plants, 61
Water need,
  in man, 325

Water table, 151, 152, 157, 243
  perched, 152
Water vapour, 143, 147, 153
Waterweed, Canadian, 256
Weapons, 310, 335
Weeds, 248, 271, 285, 293, 351, 356
Weichselian period, 267, 269, 271,
  273, 274, 293
Whales, 19, 295, 328
Wheat, 43, 133, 146, 293, 340, 360
  bread, 342
  emmer, 342
  wild, 343, 344
Willow, 157, 248, 273
  dwarf, 172, 271
Willowherb, rosebay, 254
Wilting, 93, 138, 155, 172
Wind,
  and evapotranspiration, 154, 161
  and salt spray, 172
  as a limiting factor, 171, 172
  as affecting PE, 162, 167
  -bevelling, *172*
  -dispersal of seeds, 254
  -chill, *172*
  -erosion, 187

-kill, *172*
-throw, 171
Winter chilling, 140
Wolves, 202, 295
Woodlands, elfin, 172
Worms, 35, 75, 82, 83, 94
Wucheraria bancroftii, 333

X-rays, 12, 96, 281
Xeromorphs, *156*, 172
  on serpentine, 183
Xerophytes, *156*, 175, 259

Yam, 341
Yarrow, 289
Yellow fever, 327, 330, 332

Zea mays, 345
Zebra, 45
Zinjanthropus sp., 309
Zonation,
  altitudinal, 173, 175
  biotic, 186, 187
Zone of intermittent saturation, *152–*
  154, 157
Zygote, *278*

PRINTED AND BOUND IN ENGLAND BY
HAZELL WATSON AND VINEY LTD
AYLESBURY, BUCKS